Introduction to Wood and Natural Fiber Composites

Wiley Series in Renewable Resources

Series Editor

Christian V. Stevens – Faculty of Bioscience Engineering, Ghent University, Ghent, Belgium

Titles in the Series

Wood Modification – Chemical, Thermal and Other Processes
Callum A. S. Hill

Renewables – Based Technology – Sustainability Assessment
Jo Dewulf & Herman Van Langenhove

Introduction to Chemicals from Biomass
James H. Clark & Fabien E.I. Deswarte

Biofuels
Wim Soetaert & Erick Vandamme

Handbook of Natural Colorants
Thomas Bechtold & Rita Mussak

Surfactants from Renewable Resources
Mikael Kjellin & Ingegärd Johansson

Industrial Application of Natural Fibres – Structure, Properties and Technical Applications
Jörg Müssig

Thermochemical Processing of Biomass – Conversion into Fuels, Chemicals and Power
Robert C. Brown

Biorefinery Co-Products: Phytochemicals, Primary Metabolites and Value-Added Biomass Processing
Chantal Bergeron, Danielle Julie Carrier & Shri Ramaswamy

Aqueous Pretreatment of Plant Biomass for Biological and Chemical Conversion to Fuels and Chemicals
Charles E. Wyman

Forthcoming Titles

Bio-Based Plastics: Materials and Applications
Stephan Kabasci

Cellulosic Energy Cropping Systems
Doug Karlen

Cellulose Nanocrystals: Properties, Production and Applications
Wadood Y. Hamad

Lignin and Lignans as Renewable Raw Materials: Chemistry, Technology and Applications
Francisco García Calvo-Flores, José A. Dobado, Joaquín Isac García & Francisco J. Martin-Martinez

Introduction to Chemicals from Biomass, 2nd edition
James Clark & Fabien Deswarte

Introduction to Wood and Natural Fiber Composites

DOUGLAS D. STOKKE
Department of Natural Resource Ecology and Management
Iowa State University, Ames, USA

QINGLIN WU
School of Renewable Natural Resources
Louisiana State University AgCenter, Baton Rouge, USA

GUANGPING HAN
College of Material Science and Engineering
Northeast Forestry University, Harbin, China

WILEY

This edition first published 2014
© 2014 John Wiley & Sons, Ltd

Registered office
John Wiley & Sons Ltd, The Atrium, Southern Gate, Chichester, West Sussex, PO19 8SQ, United Kingdom

For details of our global editorial offices, for customer services and for information about how to apply for permission to reuse the copyright material in this book please see our website at www.wiley.com.

Library of Congress Cataloging-in-Publication Data

Stokke, Douglas D.
    Introduction to wood and natural fiber composites / Douglas D. Stokke, Qinglin Wu, Guangping Han.
        pages cm. – (Wiley series in renewable resource)
    Includes bibliographical references and index.
    ISBN 978-0-470-71091-3 (hardback)
1. Engineered wood.   2. Composite materials.   3. Natural products.   I. Wu, Qinglin.   II. Han, Guangping.
III. Title.
TS875.S76 2014
674′.8–dc23

                                                                                                2013025429

A catalogue record for this book is available from the British Library.

ISBN: 9780470710913

Set in 10/12pt Times by Aptara Inc., New Delhi, India

1   2014

Contents

Series Preface

Renewable resources, their use and modification are involved in a multitude of important processes with a major influence on our everyday lives. Applications can be found in the energy sector, chemistry, pharmacy, the textile industry, paints and coatings, to name but a few.

The area interconnects several scientific disciplines (agriculture, biochemistry, chemistry, technology, environmental sciences, forestry, ...), which makes it very difficult to have an expert view on the complicated interaction. Therefore, the idea to create a series of scientific books, focusing on specific topics concerning renewable resources, has been very opportune and can help to clarify some of the underlying connections in this area.

In a very fast changing world, trends are not only characteristic for fashion and political standpoints, also science is not free from hypes and buzzwords. The use of renewable resources is again more important nowadays; however, it is not part of a hype or a fashion. As the lively discussions among scientists continue about how many years we will still be able to use fossil fuels, opinions ranging from 50 to 500 years, they do agree that the reserve is limited and that it is essential not only to search for new energy carriers but also for new material sources.

In this respect, renewable resources are a crucial area in the search for alternatives for fossil-based raw material and energy. In the field of the energy supply, biomass and renewable-based resources will be part of the solution alongside other alternatives such as solar energy, wind energy, hydraulic power, hydrogen technology, and nuclear energy.

In the field of material sciences, the impact of renewable resources will probably be even bigger. Integral utilization of crops and the use of waste streams in certain industries will grow in importance leading to a more sustainable way of producing materials.

Although our society was much more (almost exclusively) based on renewable resources centuries ago, this disappeared in the Western world in the nineteenth century. Now it is time to focus again on this field of research. However, it should not mean a "retour à la nature," but it should be a multidisciplinary effort on a highly technological level to perform research toward new opportunities, to develop new crops and products from renewable resources. This will be essential to guarantee a level of comfort for a growing number of people living on our planet. It is "the" challenge for the coming generations of scientists to develop more sustainable ways to create prosperity and to fight poverty and hunger in the world. A global approach is certainly favored.

This challenge can only be dealt with if scientists are attracted to this area and are recognized for their efforts in this interdisciplinary field. It is therefore essential that consumers recognize the fate of renewable resources in a number of products.

Furthermore, scientists do need to communicate and discuss the relevance of their work. The use and modification of renewable resources may not follow the path of the genetic engineering concept in view of consumer acceptance in Europe. Related to this aspect, the series will certainly help to increase the visibility of the importance of renewable resources.

Being convinced of the value of the renewables approach for the industrial world, as well as for developing countries, I was myself delighted to collaborate on this series of books focusing on different aspects of renewable resources. I hope that readers become aware of the complexity, the interaction and interconnections, and the challenges of this field and that they will help to communicate on the importance of renewable resources.

I certainly want to thank the people of Wiley from the Chichester office, especially David Hughes, Jenny Cossham, and Lyn Roberts, in seeing the need for such a series of books on renewable resources, for initiating and supporting it and for helping to carry the project to the end.

Last but not least, I want to thank my family, especially my wife Hilde and children Paulien and Pieter-Jan for their patience and for giving me the time to work on the series when other activities seemed to be more inviting.

Christian V. Stevens
Faculty of Bioscience Engineering
Ghent University, Belgium
Series Editor Renewable Resources
June 2005

Preface

The objective of this book is to draw together widely scattered information concerning fundamental concepts essential to the manufacture of wood- and natural-fiber composites. Information from the scientific literature was synthesized into a form that is intended to provide a context and content presentation that will be understandable by upper-level undergraduate students, graduate students, and practicing professionals. The topics addressed include basic information on the chemical and physical composition of wood and other lignocellulosic materials, the behavior of said materials under thermocompression processes, fundamentals of adhesion, specific adhesive systems used to manufacture composite materials, the industrial technology of said material manufacturing, and a chapter devoted to fiber/plastic composites. To my knowledge, there is no contemporary book that attempts to bring both fundamental and applied science together to form a textlike presentation of these subjects.

The vision for undertaking this project originated in my experience of teaching several iterations of an upper-division undergraduate course on wood-based composites. During the iterative development of my course, I found no adequate textbook to address the key topics that I think are essential for a foundational understanding of lignocellulosic composites. Among these are a basic overview of lignocellulose composition, behavior of such materials under thermocompressive processing (the major technological umbrella under which most extant manufacturing processes may be categorized), and fundamentals of adhesion, combined with an overview of manufacturing technology. In order to address these topics, it was necessary to draw readings from the scientific literature. My experience was that although I found many excellent research papers that I deemed applicable and instructive, most undergraduate students were unable to adequately digest the information. It is my belief that even though these students should be able to comprehend and apply the scientific literature, they need a degree of interpretation and what I call "contextualization" to enable the learning process. This is particularly true for the technology-oriented (rather than engineering oriented) students with whom I work. These students are representative of those traditionally enrolled in wood science/forest products curricula, industrial technology, or manufacturing. Students who gravitate to these majors tend to be visual and active learners with a high propensity for hands-on learning experiences. They tend not to learn as readily when faced with an "inordinate" degree of theoretical concepts. However, it is crucial that these students master important fundamental, theory- and experimentally-based concepts as it is often these very students who will be tasked with operating, monitoring, troubleshooting, and improving industrial processes when they enter the professional world. Likewise, those already in employment may be required to function in operational areas of wood- and natural-fiber processing, with little background on either material properties or basic manufacturing processes. It is thus my hope that this book will be valuable to both students and professionals in analogous situations.

As outlined, this book could be viewed as consisting of four essential components or modules, with the first two largely fundamental or foundational in character, and the remaining two of a more applied nature. Following an introductory chapter, the first major topic addressed is an overview of lignocellulosic material (its chemical and physical nature) and the response of this material when subjected to heat and pressure. In this section, relatively recent research on the application of polymer theory to lignocellulose is significant. This module is contained in chapters two, three, and four. The second fundamental component or module is a discussion of adhesion/adhesion theory. This is contained in Chapter 5. The third component, of an applied nature, is a discussion of the specific types of adhesives used to bond lignocellulosic composites. This is presented in Chapter 6. The final module, also of an applied nature, is an overview of the manufacturing technology of major product types (Chapter 7) and a consideration of the growing field of fiber/plastic composites (Chapter 8).

It was our earnest goal to compose a text with the major aims of readability and comprehension for technology-oriented students. That is, our main audience is not students or practitioners of engineering analysis, but it is for those studying the field of forest products and wood science who need to both understand and manage the materials and technology at an operations management level.

Douglas D. Stokke

Acknowledgments

The preparation of this book involved the contributions of many individuals and organizations in sundry capacities. First, my thanks to Sarah Hall, our Commissioning Editor at John Wiley & Sons, for much encouragement and guidance through the initial proposal and review process, during which the comments of several anonymous reviewers provided further reassurance that this endeavor was worthwhile. Sarah Tilley, Senior Project Editor in the Chemistry, Physics, and Material Sciences division of Wiley, was eminently patient when the reality of the task extended the project well beyond the initial deadline. I am also grateful to the office of the Vice President for Research and Economic Development at Iowa State University for approving the subvention publication grant that provided funding for my invaluable student assistants, Gabriele Frerichs, Alexa Dostart, and Crystal Krapfl. Gabriele (Gabbi) proved a most able Editorial Assistant, reading and thoughtfully commenting on draft manuscripts. Gabbi also prepared most of the graphs and computer graphics in this text, including the chemical structures, all the while injecting a healthy "can do" attitude. I honestly think this book never would have been completed without her assistance. Alexa Dostart created the exquisite original art that adorns the book. Her pencil drawings showcase her marvelous talent as an artist, and bring a fresh visual appeal to the project. Thank you, Alexa! Crystal provided valuable assistance with reference citations and tables.

I am indebted to my coauthors, Dr. Qinglin Wu and Dr. Guangping Han, for allowing me to draw them into a project that entailed much more time and effort than we had dreamed. I am grateful for their expertise, contributions, and friendship.

To the many scientists and engineers whose splendid work we have sought to capture and package for our intended audience, thank you for sharing your results and insights through your publications and presentations. I am also obliged to my students on whom I have tried various means of communicating scientific and technical information cognate to our topic.

Finally, I owe my utmost thanks and gratitude to my wife, Jodie, and our daughter, Melody, for their unfailing love and patience. Thank you for granting me the freedom to expend seemingly countless hours on the research and preparation of this work, and for allowing me to pile so many books and papers in our home office while I monopolized our computer! Jodie, your gentle encouragement and understanding have always underpinned our marriage, and your patience allowed me to complete this project even when I did not think I could. I thank God for you and for granting me life and health sufficient for the task.

Douglas D. Stokke
Story City, Iowa
April 2013

1

Wood and Natural Fiber Composites: An Overview

1.1 Introduction

Societal needs for materials and infrastructure demand transformational change in processes and materials, driven by requirements for lighter-weight, energy-efficient, environment-friendly, carbon-sequestering, renewable, and sustainable solutions. This includes the methods and resources used in industry, construction, and consumer products, including those made from wood and other natural fiber. Although solid-sawn lumber and wood products remain vital from an economic and utilitarian standpoint, it is increasingly essential that the technology to create composite materials from renewable resources such as wood be understood, applied, and improved. Changes in the forest resource worldwide [1, 2], along with growing demands for products due to economic development and increased human population, dictate that we produced and use materials having solid metrics of environmental performance [3].

In order to meet these challenges, wood is routinely combined with adhesives, polymers, and other ingredients to produce composites, thereby improving material properties while making more efficient use of the wood resource. Furthermore, although wood is still the primary plant-derived substrate for manufacturing structural building materials, cabinets, furniture, and a myriad of other products, many other botanical fiber sources are sought and utilized as supplements, substitutes, or alternatives to wood, often in some type of composite material. It is thus important for students of wood science and industrial technology to understand both the similarities and differences in the vast array of natural, plant-based raw material from which useful products can be manufactured.

1.2 What Is Wood?

If you were to ask the question "what is wood?" you might be greeted with incredulous looks. After all, is not the answer obviously simple? In some sense, the answer is yes; there is a

Introduction to Wood and Natural Fiber Composites, First Edition. Douglas D. Stokke, Qinglin Wu and Guangping Han.

certain simplicity with which the question may be approached, given the general familiarity with wood and its ubiquitous exploitation in many common applications. But, as is often true for many such "simple" topics, there can be much more to the answer than that which is immediately apparent. As an example, although it would be generally recognized that the trunk of an oak tree contains wood, what might be said of the material obtained from the stem of a coconut palm? While it may in some respects resemble the wood of trees, it is generally agreed that monocots such as coconut palm stems do not contain wood *per se*, but are said to be composed of "woody material."

In light of questions such as that illustrated by the coconut palm, the following definition of wood serves the purposes of this book: "Wood is the hard, fibrous tissue that comprises the major part of stems, branches, and roots of trees (and shrubs) belonging to the plant groups known as the gymnosperms and the dicotyledonous angiosperms. Its function in living trees is to transport liquids, provide mechanical support, store food, and produce secretions. Virtually all of the wood of economic significance is derived from trees and, as timber, it becomes a versatile material with a multitude of uses. In addition, there are woody materials found in the stems of tree forms in another plant group, the monocotyledonous angiosperms, including the bamboos, rattans, and coconut palms. These woody materials are similar to wood in being lignocellulosic in composition, but they differ substantially in their anatomic structure" [4]. We thus distinguish between "wood" and "woody materials" primarily on the basis of anatomy.

1.3   Natural Fibers

1.3.1   Fibers

Generally, a "fiber" is an object that is elongated, with a length-to-diameter (L/D) ratio of somewhat greater than one. Although the term is used variously, one may think of natural fibers as any fibrous material that is extracted from the environment, be it a fiber derived from plant, animal, or mineral sources (Table 1.1). Natural fibers may be processed by hand, simple tools, or sophisticated industrial processes to render them useful for some purpose, but they are clearly distinguished from manufactured fibers that are synthesized. Rayon fibers, for example, are reconstituted chemically from the cellulose obtained from plants, but they are not generally viewed as "natural fibers" because of the degree of chemical processing required to form them. Similarly, fibrous materials generated from petrochemicals and carbon are obviously made from materials found in the environment, but since the raw material are not fibrous in their native state and are not extracted directly from living organisms, they are not considered as natural fibers.

This text focuses on principles applicable to plant materials ("vegetable" fibers) appearing on the left side of Table 1.1. Thus, the term "natural fiber" as used in this book will be understood to apply only to a subset of all possible natural fibers, that is, to fibers obtained from botanical sources. These sources may include either woody or nonwoody plants. Furthermore, the term "fiber" will be used in the broadest sense possible, in that it will be used to refer generically to plant substance in just about any geometric form one could imagine. As an example, it is common practice to refer to whole logs supplied to an industrial processing site as the fiber supply, even if the wood is intended for use in relatively whole form, such as lumber. This generic usage is in contrast to the more specific definition of a fiber as an object having a degree of slenderness defined by a specified ratio of length to diameter.

Table 1.1 A classification of natural fibers.

Natural fibers								
Vegetable (cellulose or lignocellulose)						Animal (protein)		Mineral
Seed	Fruit	Bast (or stem)	Leaf (or hard)	Wood	Stalk	Cane, grass and reed	Wool/hair	Silk
Cotton	Coir (coconut hull fiber)	Flax	Pineapple (PALF)	Softwood	Wheat	Bamboo	Lamb's wool	Tussah silk	Asbestos
Kapok		Hemp	Abaca (Manilla hemp)	Hardwood	Maize	Bagasse	Goat hair	Mulberry silk	Wollastonite
Milkweed		Jute	Henequen		Barley	Esparto	Angora wool		Fibrous brucite
		Ramie	Sisal		Rye	Sabei	Cashmere		
		Kenaf			Oat	Phragmites	Yak		
					Rice	Communis	Horsehair etc.		

Source: Reproduced from [5] with permission of Taylor and Francis Group LLC-Books © 2005.

These basic terms or usages present an example of the perplexing nature of scientific and technical terminology. Many terms, such as fiber, may in fact have a specific scientific or technical definition, but within the industry or general field of practice, the term may be used in a broad nontechnical sense, and even in a manner contradictory to a technical definition. Indeed, the main focus of this text is on the concepts basic to the manufacture of composites via hot-pressing technology of "fiber" feedstocks that may be geometrically isodiametric, fibrous, laminar, or prismatic, and which may range in size from submicroscopic to large entities such as timber laminates.

1.3.2   Lignocellulosic Materials

Regardless of species-specific anatomic variations, vegetable or plant-based fibers share a similarity, in that their fundamental organic structural component is cellulose. Most vegetable fibers also contain, in varying chemical composition and content, the organic structural polymers hemicellulose(s) and lignin(s). Given these three major organic constituents, plant fibers are thus collectively termed lignocellulosic fiber(s), lignocellulosic material(s), or simply lignocellulose. Although it has become almost commonplace in some quarters to use the term lignocellulose as a noun, researchers caution against such usage, primarily because this implies that "lignocellulose" is sufficiently uniform as a material to be regarded as a singular substance. This notion is to be avoided, as one of the great challenges in the industrial use of lignocellulosic material is to understand the widely varying composition and attributes resulting from species differences, as well as the often considerable variation between individuals with a species and even within individual plants. In an effort to avoid oversimplification, this book will typically employ longer but preferred terms such as "lignocellulosic fibers" or "lignocellulosic materials" as opposed to the somewhat problematic term, "lignocellulose."

There are many similarities in the chemical and physical characteristics of lignocellulosic fibers obtained from the myriad of botanic sources, but at the same time, it should be recognized that each plant source has its own unique chemical composition, anatomic structure, and resultant chemical, physical, and mechanical properties. However, in lieu of exploring in detail the characteristics of all natural fibers of potential industrial interest, wood will be used in this text as the primary example or exemplar of the properties and behavior of natural fiber. We do this for two reasons: (1) the availability of a considerable volume of research literature on wood chemistry, structure, properties, and utilization; and (2) the significance of wood as the leading plant material used by human societies worldwide.

1.3.3   Worldwide Lignocellulosic Fiber Resources

Table 1.2 presents estimates of plant fiber resources in the world. Some examination and explanation of these estimates are in order. First, notice that sum or total fiber estimate (bottom of table) is substantially different for the two columns. The primary reason for this difference is that the left column of Table 1.2 does not contain data for crop residues (e.g., straw and crop stalks). If crop residues are omitted from the right column, each estimate total is within approximately 7.5%.

Estimates of the kind we are considering here are often based on data from a variety of sources, including extrapolations based on cropland area, harvest data, and utilization surveys [9]. Thus, such estimates are not to be viewed precise, but as relative indicators of fiber

Table 1.2 World production of natural fibers useful for composite materials.

Fiber source	Species	Plant origin	World production dry metric tons $\times$ 10^3	
			Sources: [6–8]	Rowell, 2008 [9]
Wood	>10,000 species	Stem	1,750,000	1,750,000
Straw	Wheat, rice, oat, barley, rye, flax, grass	Stem	–	1,145,000
Stalks	Corn, sorghum, cotton	Stem	–	970,000
Sugarcane bagasse	*Saccharum* sp.	Stem	–	75,000
Reeds	–	Stem	–	30,000
Bamboo	>1,250 spp.	Stem	10,000	30,000
Cotton staple	*Gossypium* sp.	Fruit	–	15,000
Cotton linters	*Gossypium* sp.	Fruit	18,450	1,000
Core	Jute, kenaf, hemp	Stem	–	8,000
Papyrus	*Cyperus papyrus*	Stem	–	5,000
Bast	Jute, kenaf, hemp	Stem	–	2,900
Jute	*Corchorus* sp.	Stem	2,300	–
Kenaf	*Hibiscus cannabinus*	Stem	970	–
Flax	*Linum usitatissimum*		830	–
Esparto grass	*Stipa tenacissima; Lygeum spartum*	Leaf	–	500
Sisal	*Agave sisilana*	Leaf	378	–
Leaf	Sisal, abaca, henequen	Leaf	–	480
Roselle	*Hibiscus sabdariffa*	Stem	250	–
Hemp	*Cannabis sativa*	Stem	214	–
Sabai grass	*Eulaliopsis binata*	Leaf	–	200
Coir	*Cocos nucifera*	Fruit	100	–
Ramie	*Boehmeria nivea*	Stem	100	–
Abaca	*Musa textilcs*	Leaf	70	–
Total	–	–	1,783,662	4,033,080

supplies. Even in this light, it is apparent that in relative terms, the supply of wood fiber dwarfs that of any other single type of natural plant fiber. Observe that wood fiber comprises more than 98% of the total, omitting crop residues (Table 1.2, first data column). With straw and stalks included (second data column), wood comprises about 43% of the total. In either case, the significance of wood fiber is observed, making wood a worthy teaching exemplar on this basis alone. It is perhaps due to the abundance and resulting widespread use of wood that the preponderance of research information on wood as a material provides the major reason why wood is the primary teaching example we will use to highlight the characteristics of lignocellulosic materials in general.

1.3.4   Wood as a Teaching Example

The reader should keep in mind that when we discuss the various attributes of wood, many of the basic concepts may be applied to other lignocellulosic materials obtained from nonwood plant sources. As an example, because of its natural chemical composition, wood is a hygroscopic (water-attracting) material. Other lignocellulosic materials are likewise hygroscopic, thus the fundamental concepts of hygroscopicity illustrated by wood's behavior are applicable to other natural fibers. The degree of hygroscopicity, however, is likely to vary between fiber sources due to differences in chemical composition. Many of the other principles illustrated with wood as a composite substrate (e.g., regarding thermal consolidation, adhesion) are also applicable or adaptable to other lignocellulosic fibers, with the caveat that specific material characteristics and behavior will likely vary with material source.

1.4   Composite Concept

1.4.1   Composites Are Important Materials

Composites are materials manufactured via human ingenuity. In addition, materials found in nature, such as wood, may be considered as natural polymer composites in their own right [10]. Composite materials, both natural and manufactured, are found in a multitude of applications in today's world. From building materials, automobile components, and medical devices to aircraft bodies and spacecraft, composites encompass an expanding universe of engineered materials and products. In order to achieve the material performance required by a specific application, some composites are a combination of synthetic fabrications with naturally occurring, lignocellulosic composite material (Figure 1.1).

Figure 1.1   *Advanced synthetics meet wood, nature's cellular polymeric composite: Massive electrical-generation wind turbines blades, up to 50 m long, are made of glass fiber-epoxy matrix composite blades containing a central, thin layer of end-grain balsa wood for improved flexural properties. Photo by D.D. Stokke.*

Impressive structures such as a wind turbine highlight the fact that many of our most important composites are fabricated from a wide variety of polymer resins, organic or inorganic fibers and fillers, adhesives, and other sophisticated constituents. In discussion of this broad class of materials, those derived from natural, plant-based fiber are sometimes overlooked. Nevertheless, lignocellulosic composite materials are important particularly in structural applications as building components, and in nonstructural applications for windows, doors, cabinetry, and furniture, and in packaging and shipping containers. For example, in the United States, the mass of wood used annually is greater than that of all metals and plastics combined [11]. On a global basis, the mass of wood utilized by human societies equals that of steel, and on a volume basis, wood usage exceed that of steel by one order of magnitude [11, 12]. Our consideration of the data in Table 1.2 also underscores the significance of other plant fiber worldwide, some of which is used in composite fabrications. We might therefore conclude that the dedicated study of wood- and natural-fiber composites is of global significance.

1.4.2   What Is a Composite?

A composite is a combination of at least two materials, each of which maintains its identity in the combination. A mixture of clay and rocks could therefore be considered a composite, but ordinarily, our minds turn to more exotic systems. Combinations of synthetic polymers with advanced engineering fibers, or plant fibers as amalgamations of the natural polymers, cellulose, hemicelluloses, and lignin, provide examples of sophisticated composites.

Classification of composites may be approached in a variety of ways. One of the simplest classifications is division into "natural" or "synthetic" composites. Materials such as wood are viewed as a polymer composite of nature as opposed to polymer composites manufactured via chemical syntheses. Another two-way division is to consider "traditional" versus "synthetic." In this case, the traditional category includes things like wood, but also entails simple composites such as Portland cement mixed with sand or gravel aggregate, whereas the synthetic category would include glass-epoxy, carbon fiber-reinforced polymers, and other sophisticated materials.

Many synthetic composites consist of two phases, one of which is dispersed within the other. Commonly, the dispersed phase is represented by a material that is either particulate or fibrous. A continuous matrix phase encapsulates the dispersed phase. We would thus call such materials matrix composites. In this case, one definition of a composite is given as follows: "Composites are artificially produced multiphase materials having a desirable combination of the best properties of the constituent phases. Usually, one phase (the matrix) is continuous and completely surrounds the other (the dispersed phase)" [13]. Based on this understanding of a composite, one approach to the classification of composite materials is shown in Figure 1.2. In a taxonomy such as this, the general interpretation is that those materials under the headings of particle-reinforced and fiber reinforced are matrix composites, whereas those under the structural heading are composites of a different type.

1.4.3   Taxonomy of Matrix Composites

Groover [14] describes perhaps the most common taxonomy or classification scheme for composite materials as that based on the characteristics of the matrix phase. This approach results in the following categories: (1) metal matrix composites, (2) ceramic matrix composites,

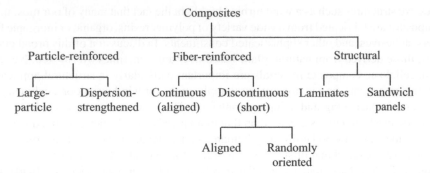

Figure 1.2 *A general classification scheme for composite materials. Reproduced from [13] with permission of John Wiley & Sons, Inc © 2001.*

and (3) polymer matrix composites. In this system, the matrix is considered the primary phase, and the reinforcement (or dispersed phase), the secondary. The primary phase is generally responsible for providing the bulk form of the material as a whole. The primary phase encloses or surrounds the secondary phase, and shares or transfers imposed mechanical load to and from the reinforcing phase. The reinforcing or secondary phase material may be in the form of discontinuous flakes, particles, fibers, whiskers, or nanoparticles. Alternatively, the reinforcing phase may be of a continuous nature, that is, as long fibers or woven mats of fiber. In either case, directional orientation of the reinforcing phase has a profound effect on composite material properties (Figure 1.3). Wood-plastic composites (Chapter 8) and inorganic-bonded materials (Section 7.8) are examples of matrix composites that employ natural fibers as the reinforcing phase.

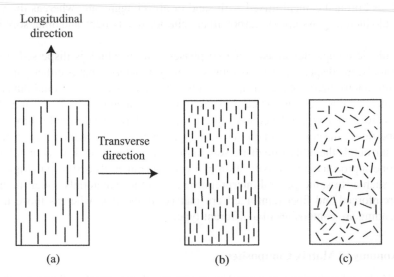

Figure 1.3 *(a) Continuous and aligned, (b) discontinuous and aligned, and (c) discontinuous and randomly oriented fiber-reinforced composites. Alignment of the fiber phase improves composite material performance. Reproduced from [13] with permission of John Wiley & Sons, Inc © 2001.*

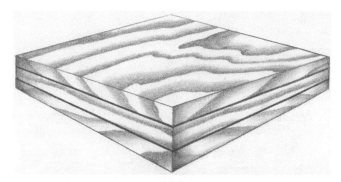

Figure 1.4   *Macrostructure of three-ply, crossbanded plywood, representative of a laminate composite consisting of wood veneer bonded by continuous adhesive bondlines, here represented by the two parallel, dark horizontal lines. The grain direction, that is, the primary orientation of the wood fibers comprising the wood, is perpendicular in adjacent layers. Artist's rendering by Alexa Dostart.*

1.4.4   Laminar Composites

Laminar composites, or laminates, occupy a different classification niche than the matrix composites (refer to the Structural composites at the right side of Figure 1.2). "A laminar composite is composed of two-dimensional sheets or panels that have a preferred high-strength direction, such as is found in wood and continuous and aligned fiber-reinforced plastics" [13]. Plywood is a classic example of a laminar composite and is also an adhesive-bonded wood composite (Figure 1.4). Alignment of the wood grain of the surface plies of veneer in the long direction of the plywood panel results in optimal bending properties. This amounts to alignment of the fundamental anatomic fibers constituting the wood itself, and that of the natural cellulose polymers comprising the wood fibers. Adjacent wood veneers are oriented with their grain direction perpendicular to the adjoining layers, providing improved strength in the transverse direction and just as important, contributing to within-plane dimensional stability of the panel.

A laminar composite is often a material that is crossbanded, as in plywood. Oriented strand board (OSB), composed not of complete lamina, but of many thin, individual wood flakes or strands, is a variation on the concept. OSB may be considered a laminar composite, wherein multiple, small laminates are aggregated to form the panel material (Figure 1.5). OSB also incorporates the advantages of crossbanding, as the face layer flakes are oriented parallel to the long dimension of the panel, with the core layer flakes oriented perpendicular to the faces.

Some laminates are made with all of the layers oriented in the same direction. An example of a unidirectional laminar material within the world of wood-based composites is that of glued laminated timber, consisting of face glued lumber with the grain of all lumber pieces parallel to one another. Structural composite lumber (SCL), such as laminated veneer lumber (LVL), parallel strand lumber (PSL), and laminated strand lumber (LSL), also represent variants of adhesive-bonded laminar composites.

Engineers and scientists utilize the taxonomic descriptions of composites as a basis for developing models of material properties. Sophisticated mathematical modeling is used to optimize material properties and manufacturing processes, and to develop new products. Among the approaches used, micromechanical modeling and classical laminate theory represent two

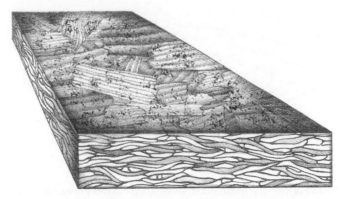

Figure 1.5 *Structure of oriented strand board (OSB), a composite composed of thin wood strands bonded by droplets of adhesive, the latter represented here by tiny dots on the wood surfaces. Note that the strands on the surface are primarily oriented left to right, parallel to the original long axis of the OSB panel. Artist's rendering by Alexa Dostart.*

widely employed methodologies, which may be used to estimate composite performance at differing scales. In some instances, composites may be modeled as matrix materials at a nano- or microscale, with the output from this level as input to modeling at a macroscale. An active community of researchers worldwide continue to expand the scope of modeling research and application for wood- and fiber-based composites.

1.4.5   Taxonomy of Wood and Natural Fiber Composites

Table 1.3 presents a wood-and-fiber-composite classification based on the form (size and geometry) of the lignocellulosic substrate. The taxonomy used here is based on a progression of wood substrate size, from largest (lumber) to smallest (nanocellulose), and on the form of the final product (lumber-like or panel). The wood elements are understood as defined in ASTM terminology [15–18].

As you think through, the material forms represented in Table 1.3 recognize the uniqueness of wood relative to almost all other plant fiber sources, in that wood may be utilized in large— even massive—net-shape objects, manufactured with relatively simple technologies. Wood may also be reduced to smaller elements and then reconstituted into larger forms, but the option of using wood largely intact remains. Most other plant materials must, of necessity, be reduced and reconstituted simply because they do not have either the size or intrinsic anatomic properties that permit them to be manufactured into lumber. The stems of palms may be sawn into lumber-like materials, but as monocotyledons, palm "wood" is fundamentally different in its inherent structure as compared to the secondary xylem (wood) of trees, thus limiting it as a direct substitute for large wooden members. Bamboo may be used as a structural material in whole form, but because its stem is hollow, it cannot be sawn into lumber. It must be separated into narrow strips and glued together to form lumber-like products. Accordingly, wood is, in many ways, a wholly unique material, owing to its intrinsic size and anatomic structure. These features make it difficult to substitute other materials for wood in a number of its technical applications.

Table 1.3 A taxonomy of wood and fiber composites[a].

Lignocellulosic raw material/substrate/furnish	Lumber-like products	Panel products
Solid-sawn lumber	Glue-laminated timbers	Cross-laminated timber (CLT)
Veneer	Laminated veneer lumber (LVL)	Plywood
Strands	Parallel strand lumber (PSL)	Oriented strand board (OSB)
	Laminated strand lumber (LSL)	
Wafers	NA	Waferboard
Particles	Com-ply[b]	Particleboard
Wood wool (excelsior)	NA	Acoustical insulation board or tiles
Fibers	NA	Medium density fiberboard (MDF)
		Hardboard
		Insulation board
Wood flour	Extruded fiber/plastic composite lumber[c]	Compression-molded products
Nanocellulose	Nanocomposites, conceivably formable into any material geometry desired	Nanocomposites

NA, not applicable.
[a]The lists of both lumber-like products and panel products are intended as representative only; they are not exhaustive lists.
[b]"Com-ply" refers to lumber substitute materials consisting of a particleboard core with veneer/plywood faces.
[c]"Fiber-plastic composites" are true matrix composites, but are not adhesively bonded in the sense of the other materials represented in this table. However, principles and mechanisms of adhesion do apply with respect to the interaction of the plastic matrix with the filler. Although many of these materials are typically manufactured with wood flour, they are nonetheless routinely referred to as fiber/plastic composites.

Despite the relative uniqueness of wood, its similarities with other lignocellulosic materials are more evident as it is reduced in size. Comminution of wood and other plant materials into flakes, particles, flours, or fibers tends to introduce geometric similarities to disparate species sources. Chemical composition variations generally remain and may be significant. Such compositional variations may be removed to some extent by chemical treatment or processing of plant cell wall substance. Isolation of nanoscale crystalline cellulose likely introduces the greatest homogeneity possible for natural fiber-starting materials and yields the smallest elements represented in Table 1.3. Nanotechnology, the science and practice of producing, measuring, classifying, and utilizing materials at a nanoscale, that is, particles with dimensions at the nanometer level (1 nm = 10^{-9} m), represents the newest frontier of materials science, perhaps opening the door to greater levels of substitution of nonwood, plant-based materials for wood than previously possible.

Our focus of study within this text will be on lignocellulosic materials that are combined, most typically, with thermosetting adhesive polymers. In the strictest sense, one would speak of "fiber composites" as those materials in which at least one component is incorporated in fibrous form, that is, having an L/D ratio greater than one and typically equal to or greater than 10. In many composites, a fiber L/D ratio in the range of 20–150 is required to effect the desired reinforcement effect [13]. Interestingly, individual wood fibers (distinct anatomic elements or ultimates) typically have an L/D of approximately 100.

In practice, the term fiber is often applied to materials in which the reinforcing phase actually has a form that is more particulate (geometrically, isodiametric) than elongated, as in "fiber-plastic composites." Composites made from lignocellulosic material, particularly those from wood, are often made from elements that are more accurately described as laminates, but generically, may be referred to as a wood fiber-based material. An example would be glue-laminated lumber, composed of large, solid-sawn lumber glued together on the wide faces to form a massive structural beam or column. Other examples would include plywood, flake- or strand-board, and other panel products that are lignocellulosic, but not necessarily fibrous *per se*, except at most fundamental anatomic level of structure. The point here is that there are both specific and generic uses of the term fiber. Both of these usages are employed in this text.

1.4.6   Composite Scale

We have observed that a composite is a combination of at least two materials, each of which maintains its identity in the combination. Dietz [19] observed that "a precise definition of composites is difficult to formulate," in that the structural scale at which one observes the identity of individual components may be arbitrarily selected. Is a molecule composed of two types of atoms a composite? What about materials that are blends of two types of polymers? Neither of these examples would ordinarily be considered composites, but if one specified an atomic or molecular level of structure as that scale at which the identity of materials in the combination is to be observed, then atoms or polymer blends could be considered as composites in the strictest sense. This is not to say that these two examples are necessarily reasonable, but it is important to note that we need to have some structural scale in mind to make the definition sensible.

Despite the difficulties in formulating a precise definition, Dietz [19] defined composites as "combinations of materials in which the constituents retain their identities in the composite on a macroscale, that is, they do not dissolve or otherwise merge into each other completely, but they do act in concert." This limitation to macroscale structure is useful in further defining a composite. It is easy to envision materials such as fiberglass, consisting of a woven mat of glass fiber infused and surrounded by a matrix of polymeric epoxy resin, as a representative example of a composite. Though distinctly different in assembly and appearance, adhesive-bonded wood products such as plywood, OSB, and engineered structural lumber also meet this definitional requirement of a composite because the wood component and the adhesive component are discernible upon inspection of their macrostructure.

Plywood and OSB are thus examples of wood composites that are easily observed as materials that retain their identities on a macroscale. Returning, however, to our discussion of the implications of scale in our definition of composite, we find that limitation to the macroscale excludes nanocomposites, in which nanoparticles, -spheres, -whiskers, or -fibers are dispersed in a matrix to yield extreme synergy in material properties. While much of the contemporary research and development in the field of nanocomposites is directed toward materials such as nanoclay particles dispersed in polymers, there is also a burgeoning field of work on materials based on nanocellulose [20]. An arbitrary limitation of scale to the macrolevel also obviates from our consideration the notion of natural materials such as wood in its unaltered state as a polymer composite structure at the molecular level. As we will further observe in subsequent chapters, wood can and should be viewed as a natural composite at the anatomical and ultrastructural levels.

The point of this discussion of the definition of the word, composite, is not to become wrapped up in semantics, but to simply acknowledge that our notion of a composite may be somewhat more complicated than just the idea of combining two dissimilar materials, and that placing strict limitations on the structural scale at which dissimilar materials may be discerned in the combination is not always conceptually advantageous.

1.5   Cellular Solids

1.5.1   Natural and Synthetic Cellular Solids

Wood and natural fiber are inherently of a composite composition, and when combined with other materials, may themselves be used to manufacture semisynthetic composites. In either case, these materials exhibit the characteristics of a cellular solid. Gibson and Ashby [21] defined a cellular solid as a substance "made up of an interconnected network of solid struts or plates which form the edges and faces of cells." Cellular solids may be closed- or open-celled, depending on whether adjacent spaces are sealed off from one another.

Cellular solids may be natural, for example, wood, or synthetic, for example, polystyrene foam. Lignocellulosic materials, whether solid (e.g., lumber) or reconstituted (e.g., as part of a composite), are, by nature, cellular materials. Fundamentally, natural fiber is the product of living organisms. In the course of their development, plant cells divide, grow, and generate cell walls that are constructed outside of the living cytoplasm. One may in fact think of the cell wall as the extracellular product of the living cell, a product rightly considered as a natural composite material. Once the cell wall has been formed, plant cells eventually die. The space once occupied by the living cell is called a lumen, which may contain water, solutes, water vapor, and other gasses. Cells, or more correctly to the biologist, cell walls, are cemented to one another by lignin and form an interconnected network of hollow structures, ranging in geometric form from nearly spherical to tubular. To the engineer or materials scientist, these interconnected, hollow structures are the cells comprising materials of surprising mechanical efficiency and versatility. Wood and other woody material may be thus described and modeled as inherently cellular solid material [22]. An appreciation for the cellular nature of wood may be gained by viewing its cross-sectional surface (Figure 1.6).

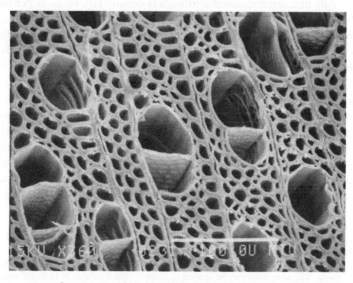

Figure 1.6   *Cross-section of aspen (Populus tremuloides) wood, a natural cellular solid. Scanning electron micrograph by D.D. Stokke.*

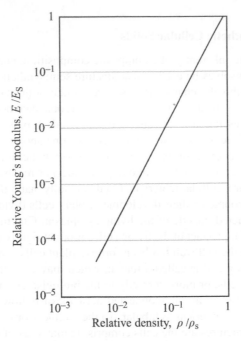

Figure 1.7   *Regression of literature data for relative Young's modulus, E/E_s, versus relative density, ρ/ρ_s, for a variety of synthetic polymer and ceramic foams. Adapted from [23] with permission of Elsevier © 1989.*

1.5.2   Relative Density

One of the most easily measured material attributes is density (ρ), or mass per unit volume.[1] The density of both closed- and open-celled cellular solids is dependent upon ρ_s, the density of the solid comprising the solid struts or plates, that is, the cell walls, and the thickness and geometric configuration of the cells, that is, the size and shape of the cells that define the amount of lumen space in the cellular solid. The ratio of bulk material density to wall solid density, ρ/ρ_s, is the relative density. Relative density has been called "the single-most important feature of a cellular solid" [21, 23]. "Models describing the properties of cellular materials in terms of relative density, cell wall properties and cell geometry can be used to select the most appropriate material for a particular engineering application" [23].

Relative density is of key significance, in that this ratio is strongly and directly correlated to other physical properties of a cellular solid. As one example, consider the plot of relative Young's modulus[2] versus relative density for a variety of synthetic polymer and ceramic foam materials (Figure 1.7). As demonstrated by this figure, relative density is a strong indicator of

---

[1] Note that material density is represented by the Greek letter rho (ρ), a convention that will be used throughout this text. Other authors have sometimes used ρ^* to represent density, particularly of a cellular solid.

[2] Young's modulus is a material property that measures resistance to deformation under tensile strain; it is a measure related to material stiffness. The term may also be applied to compression modulus in isotropic materials. Relative modulus is the ratio of the Young's modulus of the material, E, to the Young's modulus of the solid cell wall substance, E_s.

this important physical property, seen here as a linear relationship between the two parameters. In similar fashion, relative density is a reliable predictor of many other physical and mechanical properties of cellular solids.

As you consider Figure 1.7, note that the materials represented here are cellular solids manufactured via human ingenuity. These materials have wide-ranging applications, categorically enumerated as thermal insulation, packaging, structural, and buoyancy applications [21]. These practical applications of cellular solids are largely dictated by the properties resulting from the relationship of material structure to relative density. Wood has been used historically and contemporarily in similar manners. Given these observations, it is reasonable and instructive to conceptualize and model natural biological materials such as wood, cork, coral, and trabecular bone as cellular solids [12].

In summary, composites are the combination of two or more materials that retain their identities in the composite. The scale at which materials retain their identity is often deemed the macroscale, but this need not be the case, vis-à-vis nanocomposites. Natural materials, such as wood and other plant-based fiber, are composites in and of themselves, and are furthermore cellular solids. The composites manufactured from natural fibers are also cellular solids. The concept is useful, given the relationship of the relative density of cellular solids to material properties.

1.6   Objectives and Organization of This Book

The organization of this book is, we hope, straightforward. In this chapter, we have briefly introduced wood, natural fibers, and composites. We have also attempted to make the case that adhesive-bonded wood materials are composites in their own right though often overlooked in many discussions of composites in general. Chapter 2 describes the basic nature of lignocellulosic materials, providing an overview of wood and selected other plant fibers. In Chapter 3, we seek to establish wood as the primary teaching example or exemplar for the remainder of the text. We consider that if one understands the essential attributes of wood, this knowledge may be applied to the use of other natural plant fiber in composites.

Chapters 4–6 present fundamental information that is crucial to a sound understanding of the technology used to manufacture lignocellulosic composites. We might go so far as to say that these three chapters form the core of this text, and in fact, the desire to communicate this information in a manner and context relevant to students of wood science provided the overall impetus for this writing project. It is our hope that these chapters will show some progress toward this goal. First, fundamentals concerning the behavior of lignocellulosic materials during consolidation under heat and pressure are discussed (Chapter 4), followed by essential concepts of adhesion (Chapter 5), and the chemical nature of specific types of adhesives (Chapter 6). This presentation is intended to give the reader a background that will provide clues as to "why" these systems and materials work, in addition to the "how" these things are made and used. It is hoped that this will provide a somewhat unique packaging of information, drawing together and interpreting important scientific literature, proving useful to instructors in their presentation and explanation of fundamental concepts specific to lignocellulosic composites.

The fundamental material found in the first six chapters is followed in Chapter 7 with an overview of the industrial technologies used to manufacture major product categories.

Chapter 8 is devoted to the burgeoning field of natural fiber-plastic materials. A useful index concludes the book.

We have intentionally aimed much of our discussion in a manner that limits the use of mathematics. Our experience with most students of wood science suggests that this is the approach that will communicate most clearly to individuals drawn to this field of study. Instructors or independent readers of this text who desire a more rigorous quantitative explanation of concepts presented herein should be able to find further information along these lines in the literature references included for each chapter.

Our efforts to provide a readable text necessarily limited the scope of examples used. For better or worse, this text largely presents a North American bias in terms of the examples, procedures, and standards referenced. Readers in other areas may substitute data or standards applicable to their region. Numeric data is presented with commas separating units of one thousand, thus, $1,000 = 1 \times 10^3$. A period is used to indicate decimals. Though not standard practice in the US industry, the SI system is used throughout this book, with exceptions noted in the text.

References

1. Food and Agriculture Organization of the United Nations. *Global Forest Resources Assessment 2005: Progress Towards Sustainable Forest Management*. FAO Forestry Paper 147. Rome, Italy: FAO; 2006.
2. Food and Agriculture Organization of the United Nations. *FAO Yearbook: Forest Products 2010*. Rome, Italy: FAO; 2012.
3. Bowyer JL. The green movement and the forest products industry. *Forest Products Journal*, 2008;58(7.8):6–13.
4. Schniewind AP. An introduction to wood and wood-based materials. In: Schniewind AP, editor. *Concise Encyclopedia of Wood and Wood-Based Materials*, 1st ed. Cambridge, MA: Pergamon Press; 1989. pp xvii–xx.
5. Bismarck A, Mishra S, Lampke T. Plant fibers as reinforcement for green composites. In: Mohanty AK, Misra M, Drzal LT, editors. *Natural Fibers, Biopolymers, and Biocomposites*. Boca Raton, FL: Taylor & Francis/CRC Press; 2005. pp 37–108.
6. Suddell BC, Evans WJ. Natural fiber composites in automotive applications. In: Mohanty AK, Misra M, Drzal LT, editors. *Natural Fibers, Biopolymers, and Biocomposites*. Boca Raton, FL: Taylor & Francis/CRC Press; 2005. pp 231–259.
7. Bolton AJ. Natural fibers for plastic reinforcement. *Materials Technology*, 1994;9(1):12–20.
8. Bolton J. The potential of plant fibres as crops for industrial use. *Outlook on Agriculture*, 1995;24: 85–89.
9. Rowell RM. Natural fibres: types and properties. In: Pickering KL, editor. *Properties and Performance of Natural-Fibre Composites*. Cambridge, UK: Woodhead Publishing; 2008. pp 3–66.
10. Dinwoodie JM. *Wood: Nature's Cellular, Polymeric Fibre-Composite*. London, Brookfield, VT: Institute of Metals; 1989.
11. Shmulsky R, Jones PD. *Forest Products and Wood Science, An Introduction*, 6th ed. Wiley-Blackwell; 2011.
12. Gibson LJ, Ashby MF, Harley BA. *Cellular Materials in Nature and Medicine*. Cambridge, UK: Cambridge University Press; 2010.
13. Callister WD Jr. *Fundamentals of Materials Science and Engineering: An Interactive eText*, 5th ed. New York: John Wiley & Sons, Inc.; 2001.
14. Groover MP. *Fundamentals of Modern Manufacturing: Materials, Processes, and Systems*, 4th ed. John Wiley & Sons, Inc.; 2010.

15. ASTM International. D1554-01(Reapproved 2005): Standard terminology relating to wood-base fiber and particle panel materials. In: *Annual Book of ASTM Standards*. West Conshohocken, PA: ASTM International; 2010.

16. ASTM International. D1038-83 (Reapproved 2005): Standard terminology relating to veneer and plywood. In: *Annual Book of ASTM Standards*. West Conshohocken, PA: ASTM International; 2010.

17. ASTM International. D9-09a: Standard terminology relating to wood and wood-based products. In: *Annual Book of ASTM Standards*. West Conshohocken, PA: ASTM International; 2010.

18. ASTM International. E2456-06: Standard terminology relating to nanotechnology. In: *Annual Book of ASTM Standards*. West Conshohocken, PA: ASTM International; 2006.

19. Dietz, AG. Composite materials: a general overview. In: Jayne BA, editor. *Theory and Design of Wood and Fiber Composite Materials*. Syracuse, NY: Syracuse University Press; 1972. p 418.

20. Newman RH, Staiger MP. Cellulose nanocomposites. In: Pickering KL, editor. *Properties and Performance of Natural-Fibre Composites*. Cambridge, UK: Woodhead Publishing, Maney Publishing; 2008. pp 209–217.

21. Gibson LJ, Ashby MF. *Cellular Solids: Structure and Properties*, 2nd ed. Cambridge, NY: Cambridge University Press; 2001.

22. Easterling KE, Harrysson R, Gibson LJ, Ashby MF. On the mechanics of balsa and other woods. *Proceedings of the Royal Society of London Series A, Mathematical and Physical Sciences*, 1982;383(1784):31–41.

23. Gibson LJ. Modelling the mechanical behavior of cellular materials. *Materials Science and Engineering: A*, 1989;110:1–36.

2

Lignocellulosic Materials

2.1 Introduction

Contemporary focus on the development of biorenewable energy sources and sustainable materials have made plant-based feedstocks, or lignocellulosic materials, common topics of discussion. As a ubiquitous generic classification, lignocellulosic materials broadly encompass biomass derived from both nonwoody and woody plants. Our goal in this chapter will be to build a basic understanding of the chemical composition and anatomical structure of these materials. This understanding is required to fully conceptualize the behavior of lignocellulosic cellular solids when subjected to heat and pressure during manufacturing processes, their interaction with polymeric adhesives, and as final products in specific uses.

2.2 Chemical Composition of Lignocellulosic Materials

2.2.1 Polymers: Structure and Properties

A molecule is a chemical entity or unit that is composed of atoms, such that the entity exhibits unique or predictable properties. Molecules are made of at least two atoms that are bonded together by covalent bonds. An example of such a molecule is oxygen, O_2. More complex molecules may consist of many types of elements (atoms) bonded together to form a fundamental, yet minute, whole. Indeed, the word "molecule" means "extremely minute particle." A molecule represents or exhibits the fundamental characteristics of a compound.

An organic molecule is simply a molecule containing carbon or having a "carbon backbone." Thus, the simplest organic molecule is methane, CH_4. Polymers are complex molecules that are made up of many repeating chemical structural units. The word "polymer" is derived from *poly*, meaning many, and *meros* (shortened to *mer*), meaning part or (repeating) unit. Thus, an organic polymer is a carbon-backboned chemical structure having many repeating parts or units.

Introduction to Wood and Natural Fiber Composites, First Edition. Douglas D. Stokke, Qinglin Wu and Guangping Han.
© 2014 John Wiley & Sons, Ltd. Published 2014 by John Wiley & Sons, Ltd.

An analogy may be helpful in understanding polymers and one useful descriptor of their size. Think about how a chain is made. It consists of many individual links, connected together to form a complete chain, which may vary in length. The length of the chain is defined by the size of the individual links and the number of links comprising the chain. If we count the number of "mers" in our chain (i.e., number of links), we could then say that the total number of links represents the "degree of polymerization" or d.p. of the chain. A short chain might have only 100 links, with a corresponding d.p. of 100, whereas a very long chain may have a d.p. of 10,000. Keep this concept in mind as you study cellulose, which is an excellent example of a long-chain, natural polymer.

2.2.2   Lignocellulose

Lignocellulose refers to the structural substance produced by growing plants that forms the basis for plant cell walls. Herbaceous (nonwoody) and woody plants (e.g., shrubs and trees) all produce lignocellulosic substance from which the plant roots, stems, branches, and leaves are constructed. In general, all plants contain cellulose, hemicellulose, and lignin, the three of which constitute the three major organic constituents of plant cell walls. Collectively, these natural structural polymers are known as lignocellulose.

A dictionary definition is: "Lignocellulose, n. Any of several combinations of lignin and hemicellulose, forming the essential part of woody tissue – lignocellulosic, adj." [1]. This definition is a good starting point for our understanding of lignocellulosic materials, but it is obviously limited. Notice that cellulose per se is omitted from the definition, yet it is widely accepted that lignocellulosic substances contain some combination of lignin, hemicellulose, and cellulose. Some plant materials, such as cotton bolls, contain little, if any, lignin but are nonetheless often classified broadly as a lignocellulosic material. It is apparent that common usage in scientific and technological communities goes beyond that found in a dictionary.

Although the word "lignocellulose" is defined in the dictionary as a noun, its general use as such in scientific circles is often discouraged, as careless usage tends to dull the fact that there are differences in the chemical and structural composition of materials from various plant families, genera, and species. Variation of lignocellulose is recognized in the definition under consideration by the phrase, "Any of several combinations . . ." but the fact remains that the assumption of intrinsic variability is often lost in communication. It is therefore more precise to speak of the lignocellulose associated with a specific type of plant, a crop harvest, or a crop or wood source processed by a specific technology, rather than to broadly cast all plant materials into a singular categorization as lignocellulose. "Lignocellulosic materials" is thus preferred as a means of broad reference to plant materials in general, with lignocellulose qualified by reference to a specific plant material source (Figure 2.1).

Although we may rightly state that plants are composed of lignocellulosic material, we must recognize that substantial differences may be observed between diverse types of plants. Table 2.1 shows a summary of the gross chemical composition of the above-ground portion of nonwoody and woody plants used as inputs into industrial processes such as pulping or manufacture of composite materials. The variation in chemical composition has an impact on how different plant material may be used in manufacturing processes. As a result, an understanding of and appreciation for the diversity in the basic chemical and anatomical composition of lignocellulosic materials is important. We will spend the remainder of this chapter in discussion of these topics.

Figure 2.1 *Lignocellulosic materials may be derived from a wide range of plants. Depicted here, left to right, are representative nonwoody plants: flax, wheat, switchgrass, miscanthus, kenaf, corn. Four representative woody plants, bamboo, palm, and softwood and hardwood trees, are shown at the right. Though not producing wood per se, bamboo and palm are considered woody plants, as discussed in Section 1.2. Artist's rendering by Alexa Dostart.*

2.2.3   Cellulose

Cellulose is the most abundant biopolymer on earth, accounting for upward of 40% of all fixed carbon in the biosphere [3]. One estimate suggests that on a global basis, 10^{11} tons of cellulose are synthesized and destroyed annually [2]. Anselme Payen is credited with coining the term "cellulose" in 1838, when he described the "resistant fibrous solid that remains behind after treatment of various plant tissues with acid and ammonia" [4]. Although cellulose is best known as a fundamental product of land plants, it is also made by algae, phytoplankton, the bacterium *Gluconacetobacter xylinus* (formerly, *Acetobacter xylinum*) and by strange sea animals known as Tunicates.

It is difficult to imagine human society without cellulose. It is found in structural building materials, packaging, paper, textiles, insulation, and many other industrial and consumer products. Cellulose is used as a food additive and as a chemical starting material for synthetic textiles, adhesive films, photographic films, explosives, and a myriad of other products. Cellulose has been and continues to be the intriguing subject of a vast amount of research.

Table 2.1 *Gross chemical composition of woody and nonwoody industrial feedstocks.*

Component	Nonwoody feedstock	Woody feedstock
	Percent of dry weight	
Carbohydrates	50–80	65–80
Cellulose	30–45	40–45
Hemicelluloses	20–35	25–35
Lignin	10–25	20–30
Extractives	5–15	2–5
Proteins	5–10	<0.5
Inorganics	0.5–10	0.1–1

Source: Reproduced from [2] with permission of Paperi-Insinöörit/Paper Engineers' Association.

Our intention in the following few pages is to introduce some salient facts concerning this fascinating and most useful material.

2.2.3.1   Molecular Structure

Cellulose is remarkably similar in its basic chemical structure between different species of land plants, owing to the biosynthetic mechanism whereby cellulose is manufactured. Plants produce the simple sugar, glucose, by the process of photosynthesis. Living plant cells then link up to 15,000 glucose molecules together to form a linear polymer known as cellulose. During the assembly process, one molecule of water is lost each time a glucose molecule is added to the cellulose molecule or glucan chain under assembly, characteristic of a condensation reaction. Thus, each building block or subunit within cellulose is rightly called a glucose anhydride unit.

The glucose anhydride residues are linked together by covalent bonds in a manner known as a beta bond (β) configuration. This bonding arrangement results in the linear structure of each cellulose molecule. Any two glucose anhydride residues so linked are rotated 180° relative to one another. The two-unit, β-bonded glucose anhydride residues form a repeating structure known as cellobiose, with an axial dimension of 1.03 nm.

Although the two-unit structure known as cellobiose is a regular, repeating feature of cellulose, the basic repeat unit is nevertheless considered as a single glucose anhydride. It is these glucan residues that scientists count to determine the degree of polymerization (d.p.) of a cellulose molecule. In wood cellulose, there are approximately 10,000 glucose anhydride units comprising one cellulose molecule, thus the d.p. is 10,000. Though the basic structure of cotton cellulose is similar to wood, its d.p. is approximately 15,000 [2, 5]. Given this range of d.p. for wood and cotton cellulose, the molecular masses are on the order of 1.6 to 2.4×10^6 Daltons, with molecular lengths of 5.2 and 7.7 µm, respectively [2]. Cellulose may thus be described as a long-chain polymer that is composed of glucose anhydride monomers linked end-to-end by β-1,4 glycosidic bonds. Its fundamental structure may be represented graphically in a variety of ways, as illustrated in Figure 2.2.

2.2.3.2   Supramolecular Structure: Cellulose I

As plant cells produce cellulose, the individual cellulose molecules are organized within the cell wall in close association with one another, resulting in a "supramolecular" level of structure. Cellulose molecules are bonded laterally within the plane of the glucose residue ring structures by intermolecular hydrogen bonds. Intramolecular hydrogen bonding also occurs within cellulose molecules in an axial direction, stabilizing the linearity of the structure (Figure 2.3).

The hydrogen bonding between and within cellulose polymers has two important outcomes. First, the intermolecular hydrogen bonds serve to stabilize the structure laterally, thus reinforcing the bundle of cellulose from side to side, while the intramolecular bonds add linear stability. Moreover, the fact that the OH groups (hydroxyl sorption sites) are mutually satisfied means that these sites are largely unavailable for sorption of moisture (water, water vapor) or other polar molecules. This renders the crystalline regions of cellulose much less hygroscopic (i.e., attractive to water) than might be expected simply based on inspection of the chemical composition of the molecule.

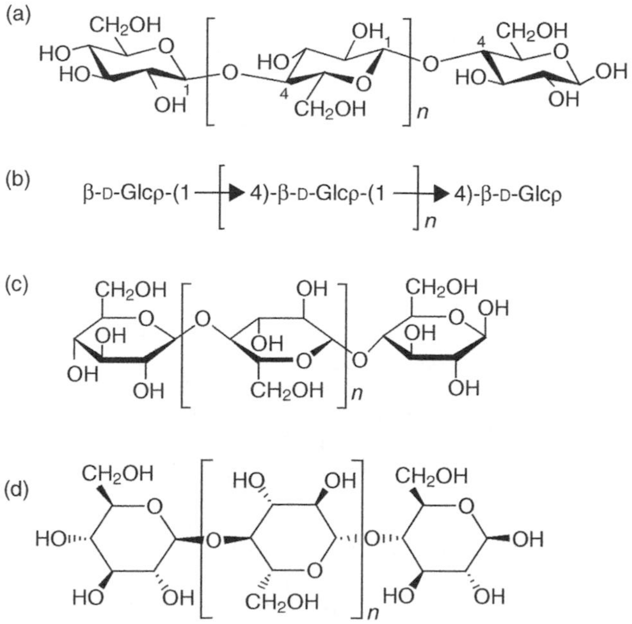

Figure 2.2   *Four typical representations of the fundamental structure of cellulose. (a) Sterochemical; (b) abbreviated; (c) Haworth perspective; (d) Mills'. Glcp = Glucopyranose (i.e., glucose in the form of a six-membered ring containing five carbon atoms and one oxygen atom in the ring). Literature suggests n = 10,000 for wood cellulose.*

One researcher has described the intermolecular and intramolecular interactions within the supramolecular architecture of cellulose as hydrophilic ("water loving") and hydrophobic ("water repelling") sites or zones parallel and perpendicular to the plane of the glucosyl rings, respectively [8]. The net result is the widespread formation of intermolecular hydrogen bonding between adjacent cellulose molecules sharing the same plane, intramolecular hydrogen bonding along the chain axis, and hydrogen bonds plus van der Waals interactions between cellulose molecules occupying adjacent planes, producing an extremely stable and chemically resistant supramolecular structure, consisting of sheets of cellulose (Figure 2.4).

Native cellulose, or cellulose as found *in situ* within plant cell walls, is known as Cellulose I. Cellulose I is said to have a parallel configuration, meaning that all of the cellulose chains are parallel to one another and "pointed" in the same direction. In chemists' terminology, a parallel configuration exists when all reducing ends of the glucan chains are at the same end of an aggregation of cellulose molecules. Individual cellulose molecules have one terminus with a closed ring structure and another terminus with an aldehyde or ketone group that is not fixed into the ring structure. The closed end of the glucan chain is called the nonreducing end, and the open end is the reducing end.

Interestingly, if cellulose is dissolved by chemical means and then subsequently chemically regenerated in a "dissolving pulp" process (e.g., as in rayon manufacturing), the synthetic cellulose chains so-produced assume an antiparallel configuration known as Cellulose II. Other modified forms of cellulose, called polymorphs, are also known. At least seven cellulose polymorphs are recognized, designated as cellulose Iα, Iβ, II, III$_1$, III$_2$, IV$_1$, and IV$_2$ [10].

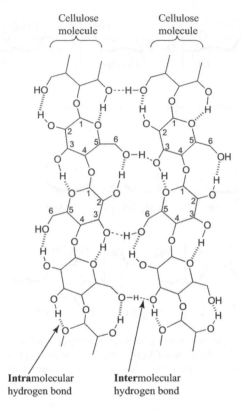

Figure 2.3 *Intermolecular and intramolecular hydrogen bonding between and within two adjacent cellulose molecules. Carbon atoms are numbered 1–6 within representative glucose anhydrides. This illustration also demonstrates parallel chain configuration of cellulose. Reproduced from [6] with permission of Elsevier © 1990.*

Within the plant cell wall, individual cellulose chains are organized together into definite spatial order, forming unit cells. In woody plants, these unit cells are approximately 0.8 by 1 nm in cross section, with one polymer chain at each of the four corners. The most stable crystalline lattice conformation is cellulose Iβ. It is this form that is considered most representative of native plant cellulose, although native cellulose consists of a blend of Iα and Iβ which varies according to plant source [10].

The integrity of the unit cell is maintained by the intermolecular hydrogen bonds and van der Waals forces [5, 8, 10]. Although hydrogen bonds and van der Waals forces have low strength relative to covalent bonds, the large number of hydrogen bonds and van der Waals interactions that develop between adjacent cellulose molecules results in a structure that is extremely durable and resistant to common solvents. Alkali cannot dissolve native cellulose. Native Cellulose I is so resistant to dissolution that only very strong acids, for example, 72% sulfuric, are able to solubilize it at room temperature. Cellulose may be hydrolyzed by dilute (0.4–0.6%) sulfuric acid at temperatures in excess of 130°C. Other cellulose solvents are also known, primarily certain quaternary ammonium compounds. The overall chemical stability of cellulose is a direct result of its fundamental molecular and supramolecular structure.

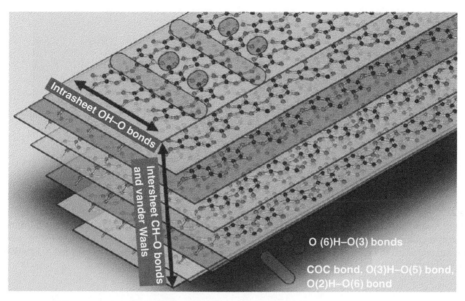

Figure 2.4 *Cellulose I is aggregated into sheets bonded together by hydrogen bonding and van der Waals forces. Republished from [9] with permission of the American Chemical Society © 2011.*

2.2.3.3 Supramolecular Structure: Microfibrils

Cellulose, in unit cell form, is further aggregated into larger linear structures known as microfibrils. At this level of supramolecular structure, a high degree of spatial order is maintained between the linear microfibrils. The microfibril diameter or cross-sectional dimensions vary with region of the cell wall from which the cellulose is derived (e.g., primary or secondary wall) and plant species. Researchers also use variations in terminology that produce differing interpretations of the level of structure constituting a microfibril. Within the literature, one will encounter terms such as protofibril, elementary fibril, subelementary fibril, microfibril, and macrofibril, all with somewhat differing interpretation regarding the scale defined by each term. Authors may use some of these terms synonymously. Thus, it is important to note the actual dimensions of structures described by the use of these terms. In addition, the dimensions observed vary with the measurement technique used.

As a first approximation, cellulose microfibrils consisting of bundles of 30–50 individual cellulose chains are about 3 nm in diameter [7]. Zugenmaier [10] summarized data for lateral dimensions of Cellulose I microfibrils from algae, bacteria, cotton, ramie, hemp, flax, and wood dissolving pulp as measured by two methods, "x-ray line-width method of reflection d(110)" and by electron microscopy. In general, the x-ray method yields microfibril diameters in the 3–5 nm range for seed plants, 5 nm for bacteria, and up to 10 nm for algae, whereas measurements via electron microscopy show diameters ranging from 25 nm to 100 nm across this same range of cellulose sources [10]. Contemporary literature on wood cellulose indicates a fundamental microfibril cross-sectional diameter in the range of 10–20 [5] to 10–30 nm [3, 10–12].

Although the crystallite regions are generally estimated to be in the range of 30–60 nm in length [2, 11], most researchers are reluctant to specify a determinant length of a microfibril

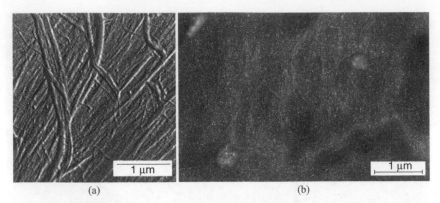

Figure 2.5   *(a) Carbon replica of water-extracted* Pinus resinosa *(red pine) longitudinal tracheid cell wall, S_2 layer. Microfibril bundles and lamellae of various sizes are apparent. Transmission electron micrograph by D.D. Stokke. (b) Atomic force micrograph of the primary wall of macerated* Pinus taeda *(loblolly pine) tracheids. Microfibril bundles ranging in width from ~25 to 65 nm are seen here. Reproduced from [13] with permission of John Wiley & Sons, Inc © 2006.*

in its entirety. Nevertheless, the fibrous nature of the cell wall at the most fundamental level of structure is undeniable (Figure 2.5).

2.2.3.4   Degree of Crystallinity

The crystalline regions of cellulose, also known as micelles or crystallites, are defined by linear aggregations of cellulose polymers. A high degree of spatial order is maintained with the crystallites, such that x-ray diffraction patterns for cellulose indicates a regular, crystalline order not unlike materials such as gemstones. The crystalline packing of cellulose is an important determinant of its solvent resistance and mechanical strength.

The degree to which the spatial order is maintained along the length of a given aggregation of cellulose polymers determines the degree of crystallinity of the cellulose. If all cellulose polymers within a unit cell were perfectly aligned throughout their entire length, the degree of crystallinity would be 100%. However, there are within native cellulose, regions of relative disorder punctuating the orderly unit-cell structural arrangement, thus accounting for crystallinity values of less than 100%. These are known as semicrystalline and amorphous regions, as shown in Figure 2.6.

Various figures have been reported concerning the linear dispersion of the crystalline features within cellulose. Some measurements on wood pulp have shown the amorphous regions occurring at longitudinal intervals of approximately 10–80 nm [14]. Other data indicate crystallites ranging from 100 to 250 nm in length [5]. The crystalline regions of cellulose may contain between 5,000 and 300,000 glucose residues [15]. Thus, it is in this characteristic, that is, degree of crystallinity, where most variation in the structure of plant cellulose is seen.

The degree of crystallinity in cellulose varies not only between plant species but also by location within plant, age of plant, moisture content, and physical or chemical treatment applied to the material. Crystallinity is typically measured by x-ray diffraction, but other techniques, such as Raman spectroscopy, are also employed to study this structural feature.

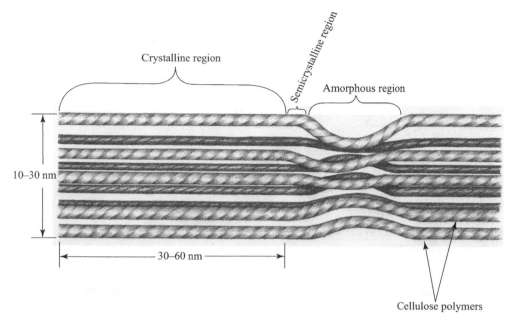

Figure 2.6 *Stylized longitudinal view of a short section of a cellulose microfibril, showing crystalline, semi-crystalline and amorphous regions. Each microfibril is an aggregation of numerous cellulose molecules. The crystalline regions are approximately 30–60 nm in length. Microfibrils within the woody cell wall have cross-sectional dimensions of 10–30 nm. Artist's rendering by Alexa Dostart.*

Cotton fiber, derived from the flowering part of the plant *Gossypium* spp., has a crystallinity index as measured by x-ray diffraction ranging from 0.70 to 0.86 (i.e., 70–86% crystalline) for various commercial cotton textile fibers [16], and may be up to 91% crystalline. Coir, a highly lignified seed fiber from the hull of the coconut (*Cocos nucifera*), has a native crystallinity of 25.72% [17]. The crystallinity indices of kenaf (*Hibiscus cannabinus*) bast fibers was shown to decrease with plant age from a high of 84.25% at 56 days after planting to 72.34% at 196 days after planting [18]. Wood cellulose, derived from the secondary xylem of trees, is generally accepted to be approximately 70% crystalline, although measurements on wood pulps range from 45% to 92%, depending on measurement technique [5]. Recent research suggests that *in situ* cellulose may in fact be less crystalline than currently accepted: "linear parallel segments" (of nanoscale cellulose elementary fibrils) "which are artifacts of isolation processes are easily mistaken for naturally occurring crystalline domains" [19]. Regardless of whether the variation in cellulose crystallinity is either native or induced, this difference is one important factor determining the relative properties of diverse types of lignocellulosic material.

Cellulose consists of linear, unbranched molecules aggregated in predictable form at the supramolecular level. Since the long axis of individual cellulose molecules is parallel to that of the microfibril of which they are a part, the microfibrils as a whole exhibit properties consistent with those characteristic of cellulose, such as outstanding tensile strength and resistance to degradation by most common organic solvents. These properties, in turn, are imparted to the cell wall, a structure built up by aggregations of cellulosic microfibrils, hemicellulose,

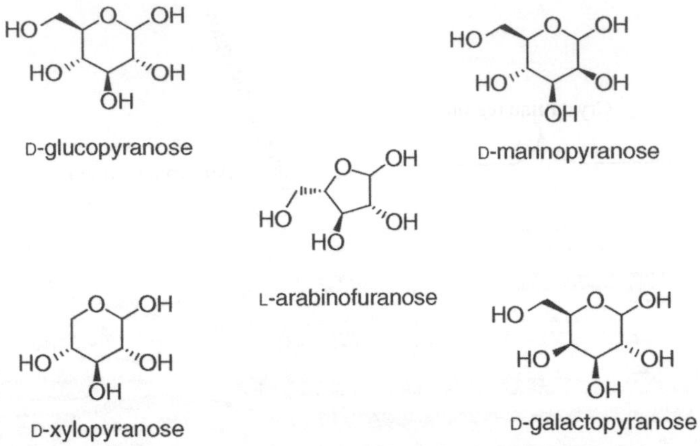

D-glucopyranose

D-mannopyranose

L-arabinofuranose

D-xylopyranose

D-galactopyranose

Figure 2.7   *Some monosaccharides comprising the hemicelluloses of higher plants. The suffix pyranose or fura-nose denotes a ring structure.*

and lignin. Given their regular, linear structure, cellulosic microfibrils thus provide the basic framework for lignocellulosic cell walls throughout the plant kingdom.

2.2.4   Hemicelluloses

2.2.4.1   Monomers Comprising Hemicelluloses

It is commonly stated that the plant cell wall is composed of cellulose, hemicellulose, and lignin (note singular form of these key words). It would be more precise, however, to refer to hemicelluloses, as the following discussion will show. Like cellulose, hemicelluloses are natural polymers that are composed of carbohydrate monomers. But unlike the homopolysaccharide cellulose, hemicelluloses are constituted from a variety of carbohydrate monomers including five-membered furanose and six-membered pyranose ring structures. Thirteen different monosaccharides have been identified in the primary cell walls of plants [7]. Representative of these sugars comprising hemicelluloses are glucose (Gluc), mannose (Man), arabinose (Ara), xylose (Xyl), and galactose (Gal), as illustrated in Figure 2.7.

2.2.4.2   Hemicelluloses in Grasses

Grasses are higher plants classified in the family *Poaceae*. This family of plants is also known as the *Graminae*. Grasses are important sources of biomass for energy as represented by switchgrass (*Panicum virgatum*) and miscanthus (*Miscanthus* spp.). Cereal grain crops such as wheat (*Triticum* spp.) maize or corn (*Zea mays*) and rice (*Oryza* spp.) are grasses. People are sometimes surprised to learn that "woody" plants such as the bamboos, with approximately 75 genera and over 1,250 species worldwide [20], are also grasses. Palms, like bamboos, are monocotyledonous plants, but are not classified in the grass family. Palms are in their own unique family known as the *Arecaceae*.

Like wood-producing plants (trees and shrubs), grasses contain cellulose. The cellulose found in woody versus nonwoody plants differs not so much in primary structure as in

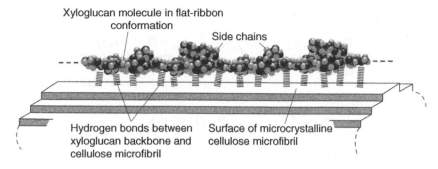

Xyloglucan molecule in flat-ribbon
conformation

Side chains

Hydrogen bonds between   Surface of microcrystalline
xyloglucan backbone and   cellulose microfibril
cellulose microfibril

Figure 2.8 *Molecular model of the conformation of xyloglucan (xylan) in association with a cellulose microfibril. Note the open, branched, and untwisted structure of xylan. Hydrogen bonds provide the mechanism of xylan attachment to the cellulose. Reproduced from [7] by permission of Garland Science/Taylor & Francis LLC © 2010.*

secondary factors such as degree of crystallinity. Considerably greater variation is found in the hemicelluloses and lignins comprising grasses as compared to that in woody plant tissue.

The major hemicellulose of the monocotyledonous grasses, accounting for up to 40% of the dry weight of cell walls, is arabinoxylan. This hemicellulose consists of a chain of xylopyranose residues with side chains of xylosyl, arabinosyl glucosyluronic acid, and 4-*O*-methyl glucosyluronic acid residues [7]. Woody plants also contain significant amounts of xylan-backboned hemicelluloses, but they are characterized by different side-branch groups than those found in the grasses. The linear backbone and side branching of a xyloglucan in association with cellulose is shown in Figure 2.8. When isolated in solution, xylan assumes a more twisted conformation than exhibited *in situ* within the cell wall.

2.2.4.3   Hemicelluloses in Wood

In the woody tissue of trees, carbohydrate monomers are assembled in varying proportions into relatively short and generally branched polymers to form hemicelluloses. Wood hemicelluloses have a d.p. ranging from 100 to 200, low indeed compared to cellulose with a d.p. of 10,000. Wood is generally composed of 25–35% hemicelluloses by dry weight. Because hemicelluloses have a more open structure than cellulose, they are more hygroscopic (i.e., they attract water molecules more readily) and are considerably more soluble. Their structure also makes them more heat-labile, that is, more prone to thermal degradation, as compared to cellulose. Hemicelluloses are soluble in dilute alkali (e.g., 1% sodium hydroxide) and are also hydrolyzed by weak acids. Some may even be solubilized by hot water.

The hemicellulose from the wood of coniferous trees consists primarily of galactoglucomannan (about 15–20% of dry weight) and arabinoglucuronoxylan. As shown in Figure 2.9a, these glucomannan consists of a linear backbone of glucopyranose and mannopyranose, which are partially acetylated (i.e., contain a –CH₃CO group), with about one acetyl group per 3–4 hexose units. The glucomannans vary in galactose:glucose:mannose ratio, with the galactose-poor fraction having a ratio of 0.1–0.2:1:3–4, and the galactose-rich portion with a ratio of 1:1:3 [2]. The backbone of arabinoglucuronoxylan (Figure 2.9b) is comprised of unacetylated xylopyranose. The wood of larches (*Larix* spp.) has a large amount, up to 35%, of water-soluble arabinogalactan hemicellulose, which is found only in low amounts in other types of wood [21].

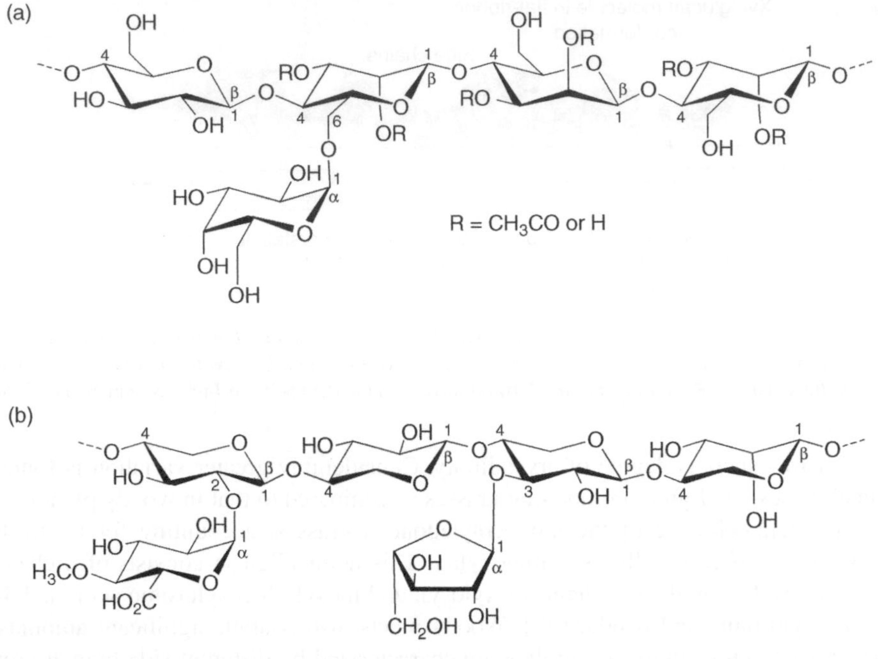

Figure 2.9   *Structures of major hemicelluloses found in the wood of gymnosperm (coniferous or softwood) trees. (a) Galactoglucomannan (or simply, glucomannan) has a backbone (main chain) composed of glucopyranose and mannopyranose, with side-branch, 1→6 linked galactopyranose residues. (b) Arabinoglucuronoxylan (xylan) has a xylopyranose backbone, with glucuronic acid (1→2) and arabinofuranose (1→3) side branches. Adapted from [22] with permission of Oy Keskuslaboratorio, KCL, Espoo, Finland © 2005.*

Hardwood hemicelluloses consist mainly of glucuronoxylan, about 20–30% of wood dry weight, and glucomannan (<5% of wood dry weight). The xylan backbone is similar to that in softwoods, but has considerably fewer and more evenly distributed uronic acid side groups (Figure 2.10). The xylan backbone contains about 3.5–7 acetyl groups per 10 xylose units [2]. Hardwood glucomannan is unsubstituted and is not acetylated.

2.2.5   Pectins

Grasses typically have a high proportion of hemicelluloses, in the range of 40–50% of the dry weight of biomass. Pectins, which are carbohydrate oligomers and polymers akin to hemicelluloses, are important components especially of primary (first-formed) cell walls. As a proportion of the dry cell wall, pectins are present in low amounts (<10%) in wood, are intermediate in grasses at 10–15%, and may range upward to 30% in some plants. Hemicelluloses and pectins are components of dietary fiber, and are used as food additives, thickeners, emulsifiers, and as substrates for fermentation processes. In the context of composite material manufacturing, hemicelluloses, and possibly pectin (depending on plant material) are important in a negative sense as they are both substantially hygroscopic and thermally labile, that is, they attract moisture and are readily degraded by heat.

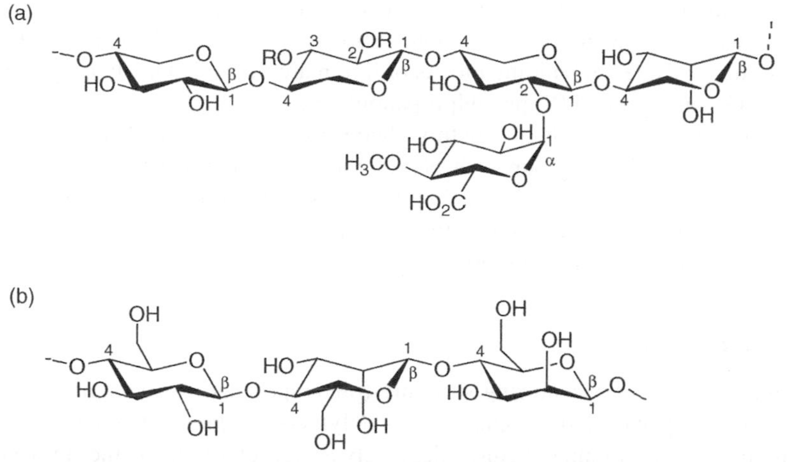

(a)

(b)

Figure 2.10 *Structures of major hemicelluloses found in the wood of angiosperm (hardwood) trees. (a) Glu-curonoxylan (xylan) is composed of a xylopyranose backbone with glucuronic acid side branches. (b) Glucomannan is a straight-chain hemicellulose, made of glucopyranose and mannopyranose residues in a ratio of 1:1–2. Adapted from [22] with permission of Oy Keskuslaboratorio, KCL, Espoo, Finland © 2005.*

2.2.6   Lignin

The third major structural organic polymer of lignocellulosic material is lignin. Lignin is highly significant, accounting for approximately 30% of organic carbon in the biosphere [23]. Unlike the carbohydrate-based cellulose and hemicelluloses, lignin is an amorphous molecule containing aromatic structures in combination with aliphatic chains. The biosynthesis of lignin involves free radical polymerization of phenylpropane building blocks, resulting in an extremely variable lignin structure, even within the same plant [24]. Lignin imparts rigidity to the cell wall while serving as the "glue" that holds individual cells together.

2.2.6.1   Lignin Precursors

Though highly complex as a complete molecule, lignin is constructed from three relatively simple starting compounds, or precursors (Figure 2.11). Notice that each of the three lignin

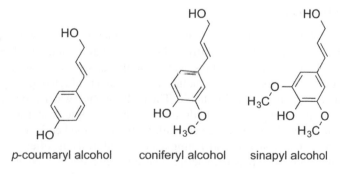

p-coumaryl alcohol   coniferyl alcohol   sinapyl alcohol

Figure 2.11 *Lignin precursors or monolignols.*

precursors, p-coumaryl, coniferyl, and sinapyl alcohol, consist of an aromatic, benzene ring (phenyl unit) with a three-carbon (propanoid) aliphatic chain attached. From these three monolignols, the fundamental building blocks of lignin are derived, namely p-hydroxyphenyl (H), guaiacyl (G), and syringyl (S) phenylpropanoid units.

Coniferyl and sinapyl alcohols are methoxylated, as are their G and S phenylpropanoid derivatives, with one methoxyl group ($-OCH_3$) on the former and two on the latter. These are important sites for cross-linking as the lignin polymer is built during the biosynthetic process. As we shall see during the study of synthetic adhesives, methoxylated benzene rings are also an important component of many thermosetting adhesive resins.

2.2.6.2   *Lignin Structure*

The basic aliphatic chain-benzene ring structure (phenylpropane) is maintained, as the precursors are built by plants into complex lignin polymers. Even so, it is nearly a practical impossibility to represent lignin structure adequately as a result of the manner in which plants assemble these subunits into lignin macromolecules. However, although the supramolecular structure of lignins remains elusive, much is known about the types and frequency of bonds between phenylpropanoid building blocks. It is on this basis that representative models of lignin structure may be proposed, as shown in Figure 2.12.

Despite the uniqueness among lignin structures, it is essentially standard practice to refer to lignin, singular. However, as with hemicelluloses, it is probably more accurate to speak of lignins, plural. One reason for this stems from the astonishing complexity of lignin, resulting from the free radical polymerization mechanism by which lignin is assembled by growing plants [23]. A single lignin molecule may have a molecular weight of 20,000 to 70,000. Given that the basic phenylpropanoid units have molecular weights under 200, it is apparent that there may be hundreds of such units comprising a single lignin molecule. Studies of model lignin compounds consisting of only 20 monomeric building block units (Figure 2.13) suggest 2^{38} optical isomers and half that many physically distinct isomers [24]. Think, therefore, of the staggering number of possibilities of lignin structure, and this not from the combination of hundreds of phenylpropane building blocks, but from a "mere" 20! Such considerations have led researchers to posit that for lignin, "the 'randomness' of linkage generation . . . and the astronomical number of possible isomers of even a simple polymer structure, suggests a low probability of two lignin macromolecules being identical" [24].

2.2.6.3   *Lignin Descriptors*

Lignin is often described as an amorphous, aromatic, and thermoplastic natural polymer. It is amorphous (without definite form) owing from the mind-boggling number of possible structures. It is aromatic due to the presence of resonance-stabilized, benzene-ring structures, that is, the phenyl substituents. Because of the aromatic structures and the high degree of crosslinking within a lignin molecule, lignin is hydrophobic. The hydrophobic (water-repelling) characteristic of lignin contrasts sharply with the hydrophilic or hygroscopic (water-attracting) nature of the carbohydrate-based fraction of the cell wall, particularly the hemicelluloses. Although lignin serves as a natural adhesive and stiffening agent in cell walls, it softens at elevated temperatures, behaving much like a plastic material. This characteristic may be used to advantage in the manufacture of composite materials.

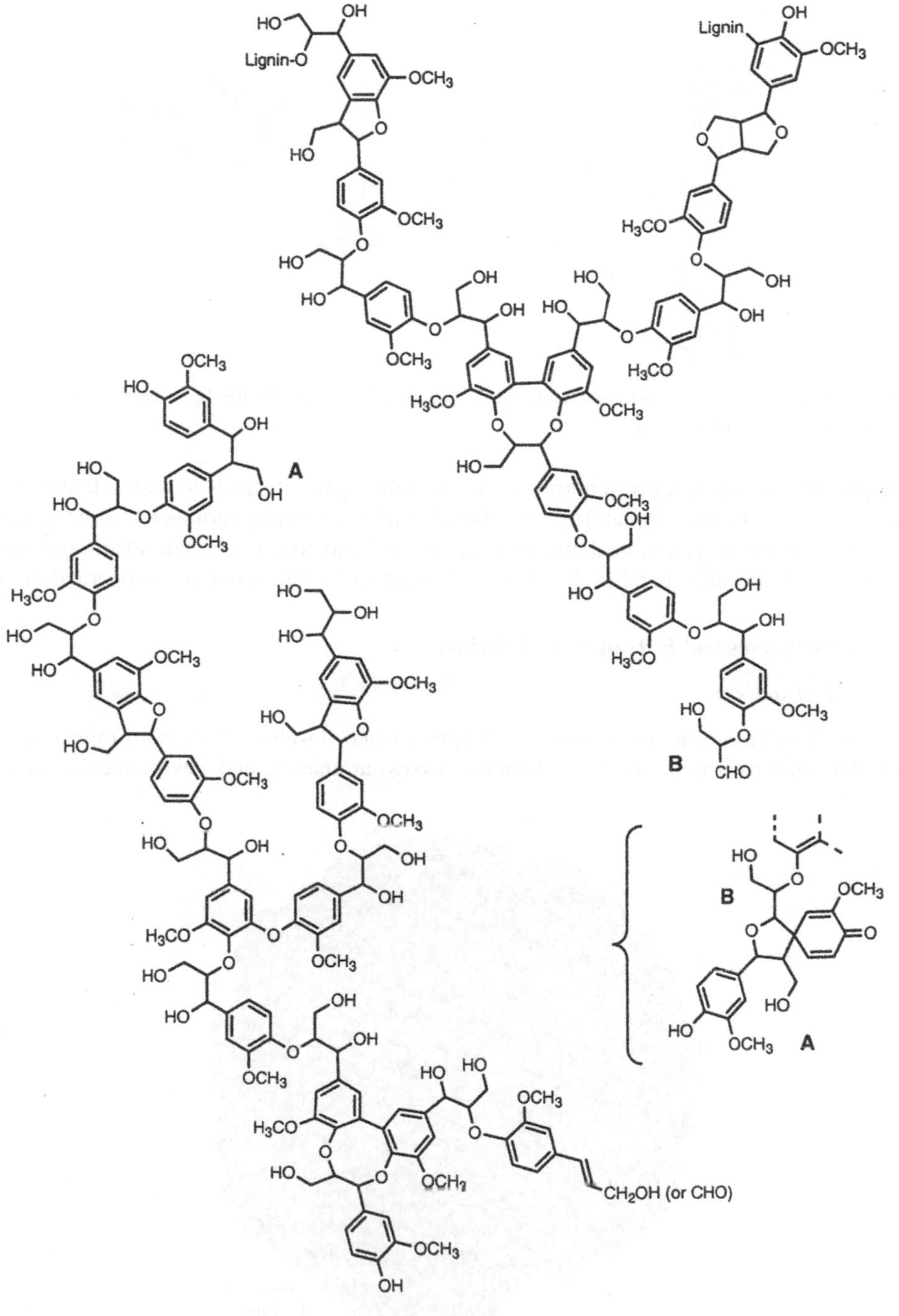

Figure 2.12 *Proposed model of softwood lignin structure, showing two side branches resulting from cleavage of a lignin molecule. The bracketed structure at the lower right is proposed as a possible link between side branches. Reproduced from [25] with permission of Wiley-VCH Verlag GmbH & Co. KGaA © 2001.*

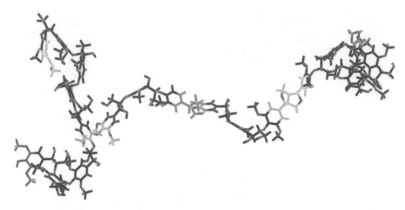

Figure 2.13   *Computer-generated framework model section of wild type Poplar lignin consisting of twenty monomeric units. From [26].*

Despite the use of the term "amorphous" to describe lignin, it stands to reason that it must have some shape. Isolated lignin fragments have been visualized by various microscopy techniques. In addition, computer modeling has been extended to postulate structural arrangements such as those shown in Figure 2.14. In this model, note that the lignin assumes a spherical shape.

2.2.7   Extractives and Extraneous Materials

2.2.7.1   *Extractives*

In addition to cellulose, hemicelluloses and lignins, plant cell walls contain many other organic molecules such as pectins, proteins, tannins, waxes, aromatics, and low-molecular-weight

1 nm

Figure 2.14   *Computer-generated spherical model of lignin consisting of 1,147 monolignol subunits, molecular weight 180,000. The sphere represented here is approximately 8 nm in diameter. Reproduced from [27] with permission of The Pulp and Paper Technical Association of Canada © 1995.*

carbohydrates. These materials are considered nonstructural, that is, though present within the cell wall, they do not form the fundamental structure of the wall as do cellulose, hemicelluloses, and lignins. Nevertheless, they do have a profound effect on the chemical nature of lignocellulosic materials. These natural chemicals are typically soluble in water and/or organic solvents, and are thus generically classified as extractives, meaning they may be removed or extracted from plant material by relatively mild chemical treatment. With respect to composite manufacturing, extractives are of key importance in dictating the surface chemical interactions that affect adhesive bonding.

2.2.7.2   *Extraneous Materials*

Plant cell walls also contain inorganics such as calcium and silica. These materials may present additional difficulties in processing. For example, plant materials with high silica content are abrasive, leading to rapid tool wear during machining (e.g., rapid dulling of saw blades). Some wheat straw particleboards, for example, have mineral content that literally results in "flying sparks" as the board is cut with a circular saw.

2.3   The Woody Cell Wall as a Multicomponent Polymer System

2.3.1   Skeletal Framework Polymers

The skeletal framework of plant cell walls is composed of the natural polymers cellulose, hemicelluloses, and lignins. The composition of these framework polymers and their organization relative to one another varies, sometimes significantly, within and between species. Lignocellulose, while an encompassing term for plant materials in general, also conveys marvelous diversity reflective of the variety found throughout the plant kingdom.

The study of plant cell walls is interesting in and of itself. Our purpose here, however, is to understand that the cell wall structure determines the physical and mechanical properties of significance in the manufacture of lignocellulosic composites. Despite the presence of extractives and inorganic substances within plant cell walls, the three major framework polymers dominate the physical properties of plant biomass as an industrial material. Table 2.2 shows a representative weight percent composition of the secondary xylem (wood) of trees, demonstrating that woody plants may be considered as multicomponent polymer systems. These systems not only have particular chemical structure and ultrastructural arrangement, they also

Table 2.2   *Basic chemical composition of the secondary xylem of softwoods and hardwoods.*

	Chemical component				
	Cellulose	Glucomannan	Xylan	Other polysaccharides[a]	Lignin
Type of tree	Percent of dry weight of extractive-free wood				
Softwood	33–42	14–20	5–11	3–9	27–32
Hardwood	38–51	1–4	14–30	2–4	21–31

Source: Adapted from References 5, 28, 29, and 30.
[a]"Other polysaccharides" generally consists of pectin(s), which are low-molecular-weight carbohydrates typically found in the primary wall of plants.

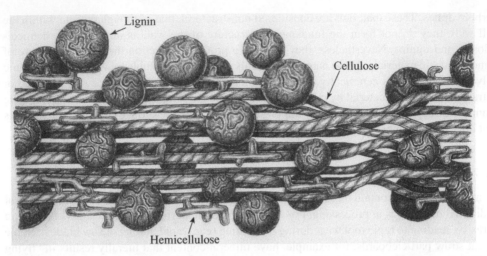

Figure 2.15 *Conceptual rendering of the reinforced matrix theory of woody plant cell wall structure. This figure combines concepts of cellulose, hemicellulose, and lignin structures as represented in Figures 2.6, 2.8, and 2.14, respectively. In order to reveal the underlying cellulose microfibril framework, lignin and hemicellulose are represented with lower frequency than would be seen in the cell wall. Artist's rendering by Alexa Dostart.*

function as anisotropic, viscoelastic materials. These characteristics, in addition to the cellular nature of natural fiber, confer properties to the gross material and to composites manufactured from them. These concepts should be kept in mind as we continue to consider the structural attributes of lignocellulosic materials.

2.3.2 Reinforced Matrix Theory

A useful concept for understanding cell wall structure is a summary referred to as the reinforced matrix theory, which may be simply stated as follows: "The cell wall of a (woody) plant consists of a high tensile-strength material (cellulose), surrounded hygroscopic material (hemicellulose), and embedded in a thermoplastic matrix (lignin)" (Figure 2.15). A loose analogy is to liken the fundamental structure of plant cell walls to reinforced concrete. In this analogy, cellulose is represented by steel reinforcing rods embedded in the concrete (lignin) matrix. In the cell wall, hemicelluloses serve as a bridge between cellulose and lignin. In the concrete analogy, a sizing or bonding agent intended to enhance the interaction or bonding of steel to concrete would represent hemicellulose. With this concept in mind, let us look more closely at the elements of this theory to build our understanding of lignocellulose ultrastructure. We will begin with the arrangement of the central skeletal component, cellulose microfibrils.

2.3.3 Cell Wall Ultrastructure

The "skeletal" structure of the cell wall of woody plants is one in which microfibrils are arranged in various, but predictable, orientation with respect to the long axis of the cell itself. Moreover, the microfibril orientation is seen to vary in discernible layers of the cell wall.

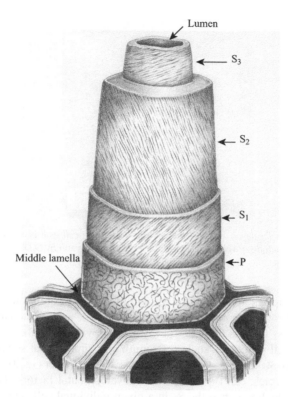

Figure 2.16   *Layered cell wall structure of a wood fiber, modeled after a softwood longitudinal tracheid. The tracheid is an elongated, hollow tube with a layered wall structure. This cutaway drawing shows the relative orientation of cellulose microfibrils within the primary (P) and secondary (S₁, S₂, and S₃) wall layers The lignin rich middle lamella cements the fiber to other adjacent fibers, with portions of three such cells shown in the foreground, each with its own layered cell wall enclosing a central space (lumen). Artist's rendering by Alexa Dostart.*

Consider the layered representation of the cell wall structure of a "typical" fiber in mature secondary xylem (wood) of a tree (Figure 2.16). In this diagram, a cell wall of a softwood longitudinal tracheid (an ultimate fiber, or simply "fiber" in industrial terminology) is cutaway to show the expected layers comprising the wall, starting with the middle lamella (ML), a highly lignified layer that bonds adjacent, separate cell walls together. Next, beneath the shared ML layer is the primary cell wall, a thin (approximately 0.1 μm in thickness) loose network of cellulosic microfibrils. Observe that each individual line represents a microfibril, or an aggregation of individual cellulose polymers.

Three underlying layers comprise what is generally known as the secondary cell wall. The secondary wall consists of layers designated S_1, S_2, and S_3. In some literature, the S_3 is alternatively called "T" for "tertiary" layer. In order to describe the general orientation of microfibrils within any given layer, a microfibril angle (MFA) is defined as the angle of microfibril orientation with respect to the long axis of a fibrous cell. Thus defined, a low microfibril angle corresponds to microfibrils largely aligned with the cell's long axis, whereas a high MFA describes an orientation tending toward perpendicular to the cell's longitudinal axis.

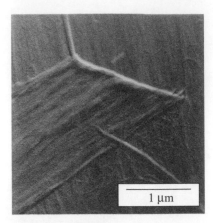

Figure 2.17   *Cellulosic lamella (layer, sheet of microfibrils) within the cell wall of red pine (Pinus resinosa). Carbon replica of water-extracted sapwood. Transmission electron micrograph by D.D. Stokke.*

The following descriptors apply to normal, mature wood of coniferous trees: In the S_1, a layer approximately 0.1–0.3 μm in thickness, the microfibrils are oriented at a relatively high MFA of 50–70°. The S_2 layer is the thickest layer, occupying the bulk of cell wall volume, and as such, dictates cell properties such as strength and dimensional stability. The S_2 has a microfibril angle of 10–30° and the layer may be a few micrometers to 20 μm or more thick, depending on wood species, cell type, and other developmental factors. The S_2 is typically 6–25 times greater in thickness than the S_1 in a given individual cell. The S_3, or T (tertiary) layer, like the S_1, is also thin, about 0.1 μm thick with a typical MFA ranging from 60° to 90°.

The direction of winding of microfibrils can alternate between S or Z helices, sometimes within a wall layer. In some softwood (coniferous) species, there may be a bumpy surface lining the cell lumen (open space in the center). If present, this bumpy surface is called the "warty layer" (W).

Microfibrils are also seen to aggregate into sheets or lamellae within the cell wall layers (Figure 2.17). With this additional structural complexity, it is observed that the S_1 consists of 3–4 lamellae, the S_2 of 30–40 or even up to 150 lamellae, and the S_3 consisting of several lamellae [5]. This layering of cell wall substance is reminiscent of the nanoscale layering observed in the supramolecular structure of cellulose that we considered in Figure 2.4.

2.3.4   Cell Wall Structure Dictates Physical Properties

The organization of microfibrils with definite microfibril angle and lamellar structure provides the ultrastructural basis for the physical properties of the cell wall. The importance of microfibril angle to fiber and wood properties is demonstrated in Figure 2.18. Here, it is clearly seen that S_2 MFA has a significant and predictable affect on the elastic modulus of both wood pulp fibers and solid wood. The microfibrillar skeleton of the cell largely determines material behavior at structural scale from the anatomic level (individual fibers and their natural aggregations within the plant) to the macroscopic bulk material (e.g., strands, veneers, solid wood in lumber form, etc.).

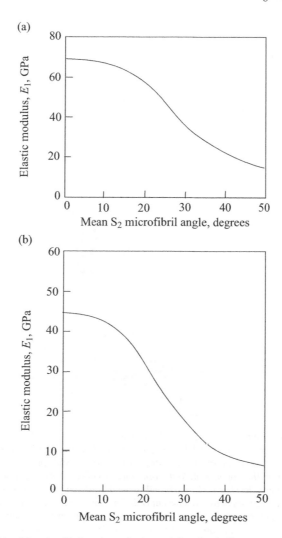

Figure 2.18 *Relationship of S_2 microfibril angle to elastic modulus of pulp fibers (a) and solid wood (b). Adapted from [14] with permission of Taylor and Francis Group LLC-Books © 1986.*

2.3.5 Cell Wall Structure from Molecular to Anatomic Level

Putting all of the pieces together, a picture emerges of the polymeric multicomponent system that is the cell wall. Cellulose, as the primary structural skeleton, is closely associated with the second major carbohydrate-based wall polymer, hemicellulose, which is in turn covalently bonded to and encased in the lignin which serves to cement the entire composite together. A simplified view of this entire system is presented in Figure 2.19. While generally reflective of what is known of cell wall structure, it should be noted that it is likely that a greater degree of intermixing of the wall polymers exists in the native structure than is suggested by this rendering. Nevertheless, it is apparent that the cell wall, as the extracellular product of

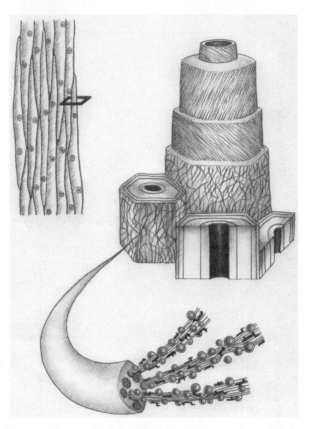

Figure 2.19 Cell wall structure from molecular to ultrastructural to anatomic scale. Three cellulose microfibrils with associated lignin polymers and hemicellulose oligomers are shown in the lower foreground. Bundles of microfibrils are aggregated together to form the ultrastructure of the cell walls of four different fibers, shown in various cutaway illustrations in the upper right. In the upper left, a collection of several secondary xylem (wood) cells are shown at the anatomic level of scale, each with several bordered pits (concentric circles) that provide passages for flow of liquids and gasses between adjacent cells. The horizontal square suggests an area represented by the molecular to ultrastructural drawings. Artist's rendering by Alexa Dostart, based on concepts shown in Figures 2.15 and 2.16 of this text, and Figure 4.1 of Esau [31] and Figure 5.2 of Niklas [32].

the living cell, is a sophisticated structure well suited for the functional purposes it serves in the plant body. Our knowledge of cell wall structure will also serve us well as we seek to understand the behavior of lignocellulosic materials under thermo-compression processes typical of many composite manufacturing operations.

2.4   Anatomical Structure of Representative Plants

2.4.1   Plant Cell Walls Are Not Solitary Entities

Although our consideration of plant cell wall structure has conceptually involved a discussion of one cell, these cells obviously do not exist in living organisms as solitary entities. Rather, cells are aggregated together in patterns inherent in any given species of plant, that is, the cells are arranged in specific anatomic composition. In some cases, industrial processes may seek to reduce plant material to essentially solitary entities as in the case of pulp for papermaking or fiberboard manufacturing. Even in these instances, it is difficult to isolate all cells individually; aggregations of fibers (fiber bundles or shives) will inevitably remain. Aside from complete chemical dissolution of plant material into chemical separations or purified fiber, it is obvious that the plant raw material that is used as input into composite manufacturing will retain some degree of its original anatomical structure. Often, this structure is altered substantially by manufacturing's thermo-compression processes, but in other cases, the anatomical structure remains intact (e.g., in plywood and glue-laminated lumber). It is thus important to have at least some familiarity with the range of variation in cell types and anatomical arrangement that might be expected in plant stems.

Plant stems contain a variety of cell types arranged into functional tissues. Parenchyma cells are generally small, ranging from nearly spherical to cubical to somewhat elongated, with either square or tapered ends. Parenchyma are generally thin-walled and are important cells for food storage and plant metabolism. Parenchyma often comprise a significant volume of the stem in nonwoody plants, contributing to the overall low density and relatively low strength of herbaceous plant stems. Fibers are generally thick-walled cells with tapered and closed ends. They are elongated, with a length-to-diameter ratio generally between 10 and 100. The primary function of fibers is to provide mechanical strength to the plant stem. Fibers are present in relatively low volume in herbaceous stems, but may comprise over 90% of the volume of the wood of trees. Thick-walled cells, including fibers, are sometimes described by the term sclerenchyma. Botanists sometimes recognize a class of cells with thickened walls intermediate between parenchyma and sclerenchyma. They call this intermediate class of cells collenchyma. Vessel elements are specialized cells that are stacked vertically in the stem to form the vessels that conduct water and nutrients. Vessels are the primary anatomic feature that distinguishes the wood of hardwood trees, which have vessels, from the wood of softwood trees, which do not have vessels. Phloem, or inner bark cells, conduct photosynthate.

Despite differences in chemical composition of cell walls, the density of plant cell wall substance is remarkably similar throughout the plant kingdom, at about 1.5 g/cm^3. Given this fact, and the observation that plant stem density varies between species from less than 0.1 to more than 1 g/cm^3, it is apparent that difference in gross anatomy is the key determinant of density. This is significant in that density is positively correlated with the physical properties that govern material behavior during the manufacturing and subsequent use of lignocellulosic composites.

In the same way that it is impossible, in a few short pages, to explore all of the intricacies of plant cell wall chemical composition and structure, it will be outside of our ability here to introduce all of the complexities of plant anatomy. However, it is our goal to acquaint the reader to the simple fact that though land plants show many similarities, there are at least as many variations in the specific structural arrangement of cells within plant tissues and between plant species. Such variation needs appreciation in that anatomical structure has an influence on processing characteristics of lignocellulosic raw material and ultimately on the properties of products manufactured from these materials. In addition, students of this subject will hopefully find that the observation and study of plant anatomy is simply marvelous! We find great beauty in plant structure and the joyful experience of viewing these marvels through the microscope. We trust that you will have opportunity to experience this as well in the course of your studies.

The approach we have chosen here is to provide, in the following sections, two examples from each of five categories of plant sources of lignocellulosic materials: Grain crop stems, biomass crops, bast fiber crops, woody monocotyledons, and the wood of trees. Our selections are based on the representative characteristics of these categories of plants and their relative importance as current or projected sources of lignocellulosic biomass for industrial purposes. We think this strategy will provide an interesting overview of important plant fiber sources, but admittedly, some categories had to be omitted. We have deliberately chosen to omit discussion of leaf fibers, seed fibers, fruit fibers (coir, from coconut husks), and the fiber of canes and reeds. Many excellent works describe these classes of plant fibers in the context of composite materials, among them Mohanty et al. [33] and Pickering [34].

2.4.2   Structure of Grain Crop Stems

2.4.2.1   *Morphological Features of Grasses*

Grasses are taxonomically classified in the family *Poaceae* (also known as the *Graminae*). Worldwide, the *Poaceae* is an important source of food, forage, and fiber. Included in the grass family are grain crops such as rice, wheat, and corn. These crops are important not only for the value of the grain but also as sources of fiber, which we know as lignocellulosic material.

Grasses share similarities in the morphology (form) of their stems. Some of the morphological features of grass stems are illustrated in Figure 2.20. The stem of a grass is known as a culm, from the Latin word for stalk, *culmus*. When utilized as fiber, the culms of grasses are the primary components of interest. Culm structure of the extending internode provides insight into the anatomy and types of fiber found in these plants.

2.4.2.2   *Wheat*

Wheat (*Triticum* spp.) is a staple grain crop throughout much of the world. Wheat culms, collected as straw, are used as raw material for fiberboard and particleboard manufacturing. The waxy cuticle of the culm presents a significant challenge to adhesive bonding when wheat straw is used to make particleboard (Figure 2.21). Worldwide, the major use of wheat straw is for pulp. In Asia, wheat straw accounts for 46% of fiber used for the production of paper [35].

2.4.2.3   *Maize*

Maize (*Zea mays*), or corn as it is better known in some parts of the world, is a highly significant grain crop of the family *Poaceae*. Although the corn stover (stems and leaves) is often left in fields following grain harvest, large amounts of stover are collected for use primarily as animal bedding (Figure 2.22). However, research and development are increasing concerning the use of corn stover as a raw material for materials manufacturing and biofuels production (lignocellulose to ethanol). In the United States alone, median estimates of sustainable and profitable corn stover supplies range from approximately 94×10^6 dry tons to 164×10^6 dry tons per year (85 to 149×10^6 metric tons)[1] over the next 20 years [36].

---

[1] One ton = 2,000 pounds. One metric ton = 1,000 kg.

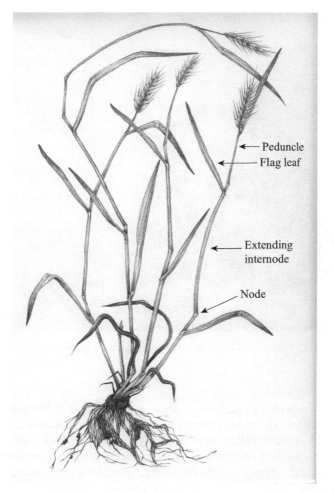

Figure 2.20 *Culm (stem) morphology of the grass family (Poaceae) as exemplified by wheat (Triticum spp.). Artist's rendering by Alexa Dostart.*

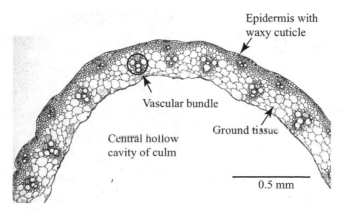

Figure 2.21 *Cross-section of the extending internode of a wheat (Triticum sp.) culm. Approximately half of the circular culm is shown here. The waxy cuticle is resistant to bonding by adhesive resins used in making wheat straw particleboard. Light micrograph by D.D. Stokke.*

Figure 2.22   *Large 500 kg bale of corn stover awaiting collection in a Central Iowa, US field. Photograph by D.D. Stokke.*

Young corn stems provide an excellent example of the anatomical structure of grass culms in general (Figure 2.23a). The stem consists of a mass of parenchymatous ground tissue that surrounds scattered vascular bundles. Approximately two rows of thick-walled sclerenchyma (fibers) underlie the outermost epidermal cells. The vascular bundles contain phloem cells for conducting photosynthate throughout the plant, and xylem cells that conduct water from the roots to the leaves (Figure 2.23b). Sclerenchyma accompany the conductive cells within each

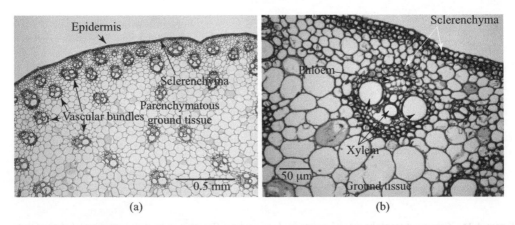

Figure 2.23   *(a) Cross-section of young Zea mays (corn or maize) stem. (b) Magnified view of a vascular bundle in a corn stem. Light micrographs by D.D. Stokke.*

vascular bundle. Since the fibrous sclerenchyma are confined to the vascular bundles and the outer periphery of the stem, the fiber content of corn stems is very low.

Mature corn culms have a low-density pith, consisting of parenchyma cells with a high cellulose content. The pith is surrounded by a relatively narrow zone of ground tissue, in which the vascular bundles are embedded. Thus, unlike wheat and many other grasses, the corn stem is not generally hollow, although plants under growing stress may develop hollow culms. Like wheat, the epidermis of the corn stalk has a waxy cuticle that is difficult to bond with adhesives.

2.4.3   Structure of Herbaceous Biomass Crop Stems

2.4.3.1   *Dedicated Biomass Crops*

Biomass is any organic, carbon-based material produced by living organisms. Therefore, all living things produce some type of biomass. Plant biomass is an important subset of biomass in general. It may seem somewhat odd, therefore, to speak of biomass crops, unless one realizes that a dedicated biomass crop is one that is selected and cultivated expressly for the purpose of optimal biomass production, as opposed to a primary purpose of grain production. In contemporary thought, biomass production is often equated with generation of feedstocks for fuels and chemicals. However, biomass could just as easily be conceived and purposed to produce fiber for use in lignocellulosic composites.

Many species of plants may be considered as candidates for biomass production. Key considerations in the selection of biomass crops include productive capacity, pathogen resistance, availability of suitable land, knowledge of cultural practices, and economic incentives. In the United States, a number of herbaceous and woody species have been proposed for biomass energy production, including switchgrass and other perennial grasses, giant miscanthus, sugarcane, sorghum, poplar, willow, *Eucalyptus* sp. and Southern pines [36]. Several of these currently are used or are under consideration for use in materials manufacturing as well.

2.4.3.2   *Switchgrass*

Switchgrass (*Panicum virgatum*) is a perennial, warm season grass native to the tall-grass prairies of North America. Its high biomass productive capacity and ability to grow on soils unsuitable for grain crops have made it a subject of ongoing research and development. Though generally regarded as a biomass energy crop for direct combustion, it has also been evaluated as a feedstock for production of synthesis gas (syn gas, an analog of natural gas made from gasification/pyrolysis of biomass), liquid fuels such as ethanol, or as a filler in fiber-plastic composites.

When viewed in cross-section, a switchgrass culm is hollow (Figure 2.24a). Vascular bundles, containing xylem, phloem, and thick-walled, fibrous sclerenchyma are embedded in a parenchymatous ground tissue in much the same manner as seen in a young wheat stem. In one research study, significant variation in the amount, cell wall thickness, and degree of lignification of sclerenchyma cells was observed in high- and low-lignin content genotypes of switchgrass culms sampled in the second internode beneath the peduncle [37]. In the same study, variation in the amount of parenchyma, xylem, and phloem cells in vascular bundles was observed between genotypes. These findings suggest that there may be important differences

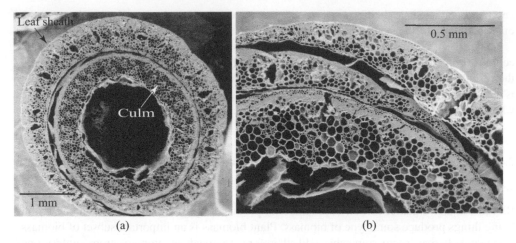

Figure 2.24   *Cross-section of switchgrass culm and leaf sheath. (a) Hollow culm surrounded by leaf sheath. (b) Portion of culm (lower part of micrograph) and overlapping leaf sheath. Scanning electron micrographs courtesy of M.I. Kuo.*

between genotypes with respect to suitability for use of switchgrass as an industrial fiber supply. As with other grasses, relatively high inorganic ash content in the culm is a concern.

2.4.3.3   Miscanthus

Miscanthus (*Miscanthus* spp.) is another dedicated biomass crop with high productive capacity. Giant miscanthus, *Miscanthus x giganteus* is hybrid of *M. sinensis* and *M. sacchariflorus*, native to Japan. The "Giganteous" hybrid has shown promise as a herbaceous biomass crop in Europe and North America (Figure 2.25). A review of published studies of biomass yield from mature (>3 years old) plantings showed that miscanthus outperformed switchgrass in biomass productivity by a factor of more than two (22 Mg/ha vs. 10 Mg/ha) [38]. Given its productive capacity, Miscanthus may be scrutinized more closely in the future as a possible source of lignocellulosic feedstock for composite materials. As a member of the *Poaceae* family, one might expect that the stem anatomy of Miscanthus resembles that of switchgrass, which it does [39].

2.4.4   Structure of Bast Fiber Stems

Bast fibers provide a classic example of technical fibers, which are aggregations of a multitude of ultimate fibers (individual cells or anatomic elements). Technical fibers are extracted from suitable plants as long, fibrous strands, usable for both woven and nonwoven composite materials. Bast fibers are derived from the phloem, located in the outer periphery of the stem. Human societies have used bast fibers from antiquity for cordage (ropes), matting, and textiles.

Although most bast fibers are obtained from herbaceous plants, their name is derived from the basswood tree, *Tilia* spp. The phloem or bast fibers of *Tilia* are arranged in dense layers within the inner bark (secondary phloem) of the stem (Figure 2.26). Long strips of the bast technical fibers may be obtained from a felled basswood by removing the bark and then

Figure 2.25 *Stand of Giant miscanthus (Miscanthus x giganteus), which may reach heights of 2.44–3.66 m (8–12 feet). Artist's rendering by Alexa Dostart.*

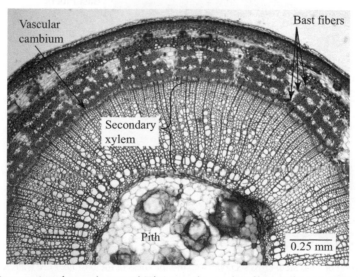

Figure 2.26 *Cross-section of young basswood (Tilia) stem showing bast fibers. Light micrograph by D.D. Stokke.*

pounding on the stem to crush the cortical and phloem parenchyma surrounding the bast fiber bundles, followed by stripping of the loosened technical fibers.

2.4.4.1   *Flax*

Flax (*Linum usitatissimum*) is an annual plant of the family *Linaceae*, native to the Mediterranean region and eastward to India. It is not a grass, but is an angiosperm that, like trees, is capable of producing xylem through cell division in a cambial zone. It is currently cultivated in many countries, including Canada, China, India, the United States, Ethiopia, Bangladesh, Russia, Ukraine, France, and Argentina. Flax is the source of the fiber used to make linen textiles and has been known and used by human societies for many thousands of years. Flax produces seeds from which linseed oil used in furniture finishes is derived. Oilseed flax is also cultivated for production of oils for human consumption, as the oil is rich in omega-3 fatty acids. Flax technical fibers are used in composite materials, and to impart strength and moisture resistance to tea bags, currency, and Bible paper. Flax fibers are perhaps the strongest of all natural fibers derived from plants.

Flax bast fibers, like bast fibers in general, are located near the periphery of the plant stem (Figure 2.27).

Flax bast fibers are obtained by a postharvest process of retting, breaking or scutching, hacking, and other fiber refining processes. Retting involves the initial softening of the plant stem and loosening of the bast fibers from the surrounding tissue by means of natural biodegradation. In pond retting, stems are submerged in water for days, weeks, or months. Field retting, whereby cut stems are left lying in the field for a period of time, is generally favored today due to water quality problems associated with pond retting. Breaking or scutching is a mechanical

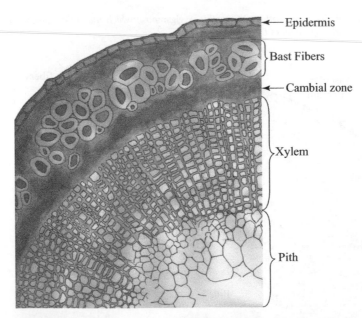

Figure 2.27   *Portion of a flax stem cross section. Note the thickness of the bast fiber cell walls relative to other cell types. Artist's rendering by Melissa Hymen.*

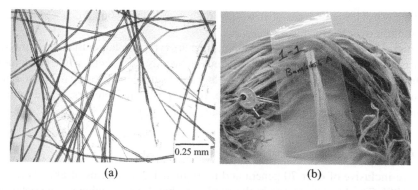

(a)   (b)

Figure 2.28   (a) Kenaf ultimates obtained by the Kraft chemical pulping process. Light micrograph by D.D. Stokke. (b) High-quality kenaf bast technical fibers from Bangladesh. Photograph by D.D. Stokke.

process intended to free the bast fiber bundles from the remainder of the plant tissue. This process is sometimes referred to as decortication. The nonbast component of the decorticated or fractionated stem is called the hurd. Hacking is a subsequent mechanical process designed to reduce the fiber bundles to a diameter typically in the range of 50–100 μm, thereby yielding usable technical fibers. Depending on intended use, other refining or fiber cleaning steps may be employed. Approximately 27–30% of flax technical fibers produced by mechanical means fall in the range of 20–200 mm in length, with an average fiber length of 80 mm specified for many industrial applications [40]. Fibers obtained from oilseed flax are typically shorter than those from plants bred for linen production. Furthermore, overall fiber yield from oilseed varieties is less than that of linen varieties.

2.4.4.2   Kenaf

Kenaf (*Hibiscus cannabinus*) is an annual to rarely perennial plant of the family *Malvaceae*, with stems growing to approximately 2 cm in diameter and up to 4 m in height. It is thought to be native to Asia, and is cultivated throughout Asia, Africa, and parts of the United States and Europe for fiber, oilseeds, and paper pulp (Figure 2.28a). In some parts of the world, particularly in Bangladesh and India, kenaf is field or pond retted and then processed largely by hand, yielding very long and high quality bast technical fibers that are used in composite material applications ranging from office furniture systems to automotive parts (Figure 2.28b). Up to 40% of the stem volume is composed of bast fibers. The remainder, mostly the parenchymatous pith, is very low in density but high in cellulose content.

2.4.5   Structure of Woody Monocotyledons

2.4.5.1   The Woody Monocots Include the Bamboos and Palms

Monocotyledons are flowering land plants whose seeds have a single primordial leaf, or cotyledon. Grasses are monocotyledons (monocots), as are palms. Most grasses (e.g., wheat, corn, switchgrass, miscanthus) are herbaceous, that is, nonwoody, but even these have some lignification that provides sufficient stiffness to grow to the size characteristic of each. Bamboo is considered a woody monocotyledon and is also classified as a grass (i.e., member of the

Poaceae family). The palms are also woody monocotyledons, but are not grasses. They classified in their own family, the *Arecaceae*.

Woody monocotyledons exhibit growth habits that are typically larger than most grasses and in some ways resemble those of trees. The stems of woody monocots, while fundamentally different from trees in anatomical structure, are large and sufficiently strong to be used in some of the same applications as wood, for example, poles, posts, and timbers. Woody monocots are thus an important group of plants having many practical uses.

2.4.5.2   *Bamboo*

Bamboos are inclusive of over 70 genera and more than 1,250 species of grasses (*Poaceae*) worldwide [20]. The largest grasses in the world are bamboos, reaching a diameter of 15–20 cm and heights in excess of 30 m. Most bamboos of industrial interest are on the order of 3–12 m in height. Bamboo is used as food (young stems), in Chinese medicine (extracts), for musical instruments, paper pulp, textiles, furniture, construction (poles for scaffolding or as structural timbers), flooring, and other glued-laminated composites.

Although bamboo may be used in some applications similar to wood, its fundamental structure is quite different. The stems (culms) of bamboo, like other grasses, are hollow within the internode, with a solid diaphragm at each node (Figure 2.29a). Because of the hollow stem, bamboo cannot be sawn into lumber in the same manner as trees, but lumber-like products are made by glue-laminating thin strips of bamboo together. The woody wall of the culm is composed of a dense ground tissue, in which vascular bundles containing xylem and phloem cells are scattered and may have associated with them a substantial fiber sheath (Figure 2.29b). A close view of the vascular bundle of bamboo is noteworthy of itself (Figure 2.29c). For further information on the interesting and beautiful anatomic structure of bamboo, please refer to the work of Liese [20].

2.4.5.3   *Palms*

Palms are tropical woody monocotyledons of the family *Arecaceae*, consisting of over 200 genera and greater than 2,500 species. Palms have been used by humans perhaps further into antiquity than flax for things such as food (dates, coconuts), oils, Carnauba wax, rattan (furniture, baskets), charcoal, fiber, and many other local uses. Oil is extracted from the fruit pulp of the African oil palm (*Elaeis guineensis*). These plants grow up to 20 m high, but the stems are considered of little use. The fruit is the object of cultivation, as this is the source of the oil. Although significant amounts of oil palm fiber is generated by oil processing, relatively few economic uses of the fiber have been put into practice. With respect to use in composites, "the academic community is searching for applications of oil palm fibers as a reinforcement for polymers" [30]. In contrast, the fruit fiber of the coconut palm, known as coir, is important from a utilitarian perspective. However, its use will not be considered here as we have elected to confine the present discussion to plant stem-derived fiber.

Coconut palms (*Cocos nucifera*) are cultivated throughout the Pacific region, East Africa, the Caribbean, Mexico, and South America, in about 80 countries worldwide. They typically reach a diameter of 30–40 cm with a height of 20–25 m. Because of their growth habit and the size attained, palm timber is obtained from coconut palms, and has a density ranging from <0.4 to >0.6 g/cm^3 from stem core to periphery [41].

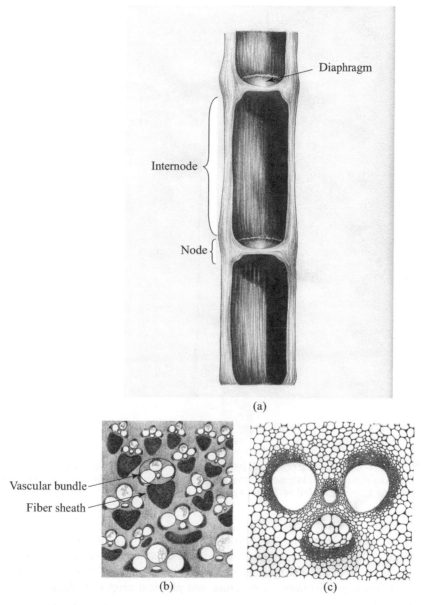

(a)

Vascular bundle
Fiber sheath

(b)   (c)

Figure 2.29   *(a) A bamboo culm cut in longitudinal section. (b) Vascular bundles are scattered throughout the ground tissue in this cross-section of the culm of the bamboo, Phyllostachys edulis. Bundles are somewhat smaller and more closely spaced near the outer periphery of the culm (top). (c) Closeup view of a vascular bundle of Phyllostachys edulis. Artist's renderings by Alexa Dostart.*

Figure 2.30   *Partial cutaway drawing of coconut palm stem. Stippling represents approximate density variation of palm wood in both transverse and longitudinal directions in stems up to 25 cm radius and 20 m high, according to data presented by Killmann and Fink [41]. Artist's rendering by Alexa Dostart.*

In contrast to the bamboo culm, the stem of the coconut palm is not hollow (Figure 2.30). It is a solid mass of parenchyma ground tissue, with vascular bundles scattered mainly toward the periphery. Because of this structure, palm wood is more "stringy" than the wood of trees, but it also exhibits less anisotropy in structure and physical properties than the secondary xylem of trees do.

2.4.6   Wood

2.4.6.1   *Wood is Principally the Secondary Xylem of Trees*

In the strictest sense, "wood" is the secondary xylem of trees. Secondary xylem is also produced by shrubs and woody lianas (vines). Unlike the grasses and palms, trees and other woody plants exhibit secondary or thickening (diameter growth). The diameter of grass culms or palm stems is established at the base of the culm or stem; no further growth in diameter occurs in these types

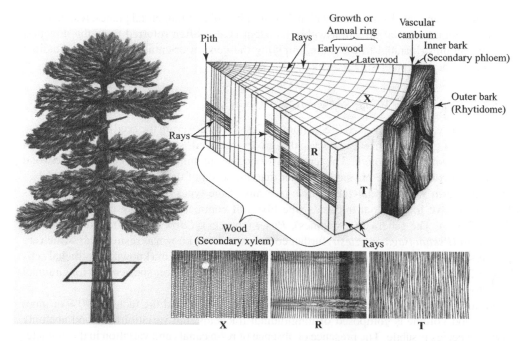

Figure 2.31   *Three principal planes of wood structure: cross-sectional or transverse (X), radial (R) and tangential (T). A pie-shaped wedge of wood, extracted from the main stem (bole) of a Ponderosa pine (Pinus ponderosa) is shown at the upper right, with important features noted. At lower right are low-magnification light micrographs of the wood structure. Artist's rendering by Alexa Dostart and micrographs by D.D. Stokke.*

of plants. Trees, however, have a vascular cambium that produces secondary xylem (wood) and secondary phloem (inner bark) cells, enabling tree species to reach impressive diameters.

Because of their secondary growth, trees have anatomic structures that differ substantially from grasses and palms. Anyone who has ever observed a tree stump or the end of a log or piece of lumber knows that the wood is divided into concentric circles representing annual (or, in the case of tropical trees, seasonal) growth (Figure 2.31). The annual rings may be further characterized as having zones of earlywood and latewood, representing wood formed in the early, that is, spring–summer, or late, that is, summer–fall, portion of any given growing season, respectively. Also in contrast to grasses, wood contains rays, which are ribbon-like composite structures composed primary of parenchyma, oriented radially like spokes on a wheel from the center of the tree stem. These rays provide for radial transport and storage of photosynthate and are sites of active metabolism in the living tree.

Apparent in Figure 2.31 is the structural anisotropy of wood. The word "anisotropic" means that a material property varies with the direction of measurement. Wood is an anisotropic material, owing to its anisotropic anatomic structure. It is clear from the illustration and micrographs that there are observable (and predictable) anatomic differences in three principal planes, namely cross-sectional, tangential, and radial. Note that the cross-sectional plane is also referred to as the transverse plane. Be careful not to confuse "transverse" with "tangential." Since the radial and tangential planes are oriented parallel to the main tree stem axis, these may be referred to as axial or longitudinal planes. We therefore may speak of the "radial-longitudinal plane" or the "tangential-longitudinal plane" or simply radial or tangential planes, respectively. The longitudinal or axial direction within the stem is also often referred to as the direction along the grain or parallel to the grain, implying the general orientation of the longitudinal tracheids ("fibers") comprising the wood.

2.4.6.2   Gymnosperm Wood: Softwoods

Trees that carry their seeds naked (i.e., without a fleshy fruit covering the seed) in a woody cone are conifers or gymnosperms. The trees in this group are called "softwoods" and the wood from these trees is called "softwood." The term, softwood, is something of a misnomer, as the wood of some softwoods may be physically harder than the wood of their counterpart hardwoods.

Gymnosperm wood is relatively simple in terms of the types of cells present. All species of conifers have longitudinal tracheids (the fibers of commerce) and ray parenchyma cells (Figure 2.32a). The genera *Pinus* (pines), *Picea* (spruces), *Larix* (larches), and the species Douglas fir (*Pseudotsuga menziesii*) further contain structures known as resin canals, which are openings in the wood structure surrounded by specialized parenchyma known as epithelial cells (Figure 2.32b). Other cell types that may be present, depending on species, are longitudinal parenchyma and ray tracheids.

Due to the limited number of cell types found in softwood, and the fact that 90% or more of softwood volume is composed of longitudinal tracheids, the variation in wood anatomy between species is subtle. The presence or absence of resin canals and variation in the diameter and wall thickness of the longitudinal tracheids are the main variables that define gross wood anatomy.

2.4.6.3   Angiosperm Wood: Hardwoods

The hardwoods are trees distinguished by botanical features of generally broad leaves and bearing their seeds in fleshy fruit, fleshy seeds, or in nuts. In the temperate zone, hardwoods are also deciduous. In terms of wood anatomy, the primary distinguishing feature that sets hardwoods apart from softwoods is the presence of specialized water-conducting cells called vessel elements. In softwoods, the longitudinal tracheids function both as conduits for transport

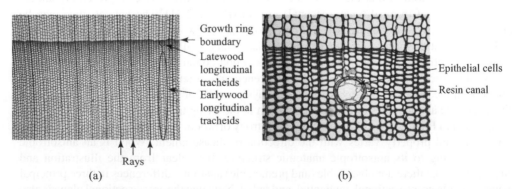

Figure 2.32   *(a) Longitudinal tracheids and ray parenchyma are the two cell types found in all coniferous wood, as illustrated here in balsam fir (Abies balsamea). (b) Resin canals, structures found in select coniferous genera, are structures delineated by epithelial cells, which produce the sticky resin contained within the canal, shown here in Eastern white pine (Pinus strobus). Light micrographs by D.D. Stokke.*

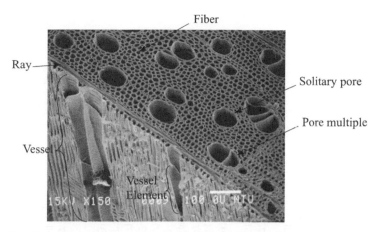

Figure 2.33 *Scanning electron micrograph of the secondary xylem (wood) of red maple (Acer rubrum). The transverse (cross-sectional) and radial surfaces are shown, upper right and lower left, respectively. Scanning electron micrograph by D.D. Stokke.*

of water and as the main source of the wood's mechanical strength. In hardwoods, fibers provide the mechanical strength, and most of the water is conducted through the vessel elements. The individual vessel elements, often less than 1 mm in length, are stacked vertically in the tree stem to form vessels which may be several meters long. The ends of vessel elements are open. As a result, the vessels formed from them are like long, narrow tubes for water flow. Vessels, when viewed in cross-section, are called pores. Thus, hardwoods are called "porous woods" due to the presence of this type of cell (Figure 2.33).

Hardwoods also contain ray parenchyma, with much variation in the ray structure, height, and width observed in different species of trees. Hardwoods may have longitudinal parenchyma or other cell types, depending on species. These factors tend to make the wood anatomy of hardwoods more variable than that observed in softwoods.

The growth rings in hardwoods may be divided into earlywood and latewood zones. In some species, the earlywood pores are noticeably larger in diameter than pores found in the latewood. If this is the case, the wood is described as "ring porous" (Figure 2.34a). If the pores

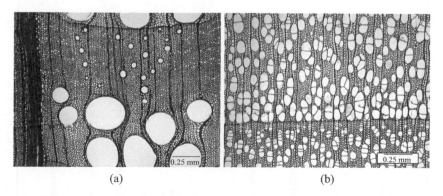

(a)   (b)

Figure 2.34 *(a) Cross-section of Northern red oak (Quercus rubra), a ring-porous hardwood. (b) Cross-section of bigtooth aspen (Populus grandidentata), a diffuse-porous hardwood showing portions of two growth rings. Light micrographs by D.D. Stokke.*

Table 2.3 Chemical composition and ultimate fiber dimensions for representative lignocellulosic plant stems described in Sections 2.4.2–2.4.6 and illustrated in Figure 2.1.

Common name	Scientific name	Percent weight composition of stem				Fiber length, mm	Fiber diameter, µm
		Cellulose	Hemicellulose	Lignin	Ash		
Wheat	Triticum sp.	29–39.9 [42]	16–28.2 [pentosans] [42]	15.6–21.5 [42]	–	1.5 [43]	15 [43]
Corn/maize	Zea mays	32–34.2 [42]	20–27 [pentosans] [42]	5–34; 5–14 typical [42]	1.2–7 [42]	1–1.5 [43]	20 [43]
Switchgrass	Panicum virgatum	32.4 [42]	26.3–29.3 [pentosans] [44]	20.1–23 [44]	4.8–6.1 [44]	0.78 [45]	–
Giant miscanthus	Miscanthus x giganteus	50.3–52.1 [45]	24.8–25.8 [45]	12–12.6 [45]	2.7 [45]	0.97 ± 0.08 [46]	14.2 ± 2.5 [46]
Flax	Linum usitatissimum	71 [30]	18.6–20.6 [30]	2.2 [30]	–	4–77, ave 33 [ultimates] 20–140 cm [technical] [17]	5–76 [17]
Kenaf	Hibiscus cannabinus	51–59 [47, 48]	21.5 [47, 48]	15–19 [47, 48]	5 [47, 48]	1.5–11, ave 3.4 [17]	12–36, ave 24 [17]
Bamboo	Phyllostachys sp. and others	42.3–49.1 [47, 48]	24.1–27.7 [47, 48]	23.8–26 [47, 48]	1.3–2 [47, 48]	2.7–4 [43]	15 [43]
Coconut palm	Cocos nucifera	62.3 [42]	9.6 [pentosans] [42]	30.9 [42]	2.7 [42]	–	<18–46 [49]
Softwoods	–	40–45 [50]	7–14 [pentosans] [50]	26–34 [50]	<1 [50]	2.7–4.6 [Southern yellow pine, Pinus sp.] [50]	32–43 [Southern yellow pine, Pinus sp.] [50]
Hardwoods	–	38–49 [50]	19–26 [pentosans] [50]	23–30 [50]	<1 [50]	0.7–1.6 [aspen, Populus sp.] [50]	10–20 [aspen, Populus sp.] [50]

Literature citation number indicating source of data appears in brackets.

are approximately equal in diameter and are evenly distributed throughout the growth ring, the wood is called "diffuse porous" (Figure 2.34b). Regardless, the pore diameter and frequency has a decided effect on wood density.

2.5   Comparison of Representative Plant Stems

The purpose of the foregoing discussion in Sections 2.4.2–2.4.6 was to introduce some salient anatomic features of the stems of grain crops, herbaceous biomass crops, plants producing bast fibers, woody monocotyledons, and the wood of trees, using examples from two species or species groups for each. This survey provides an introduction to a subset of the enormous variety of plant fiber that may be suitable for processing into composites. By way of summary of this discussion, a comparative table containing data on chemical composition of representative plant stems and dimensions of ulimate fibers (individual anatomic elements) has been prepared (Table 2.3). Additional comparative data on properties and composition of some natural fibers are given in Tables 8.3 and 8.4. Data such as these are useful in the evaluation of differing sources of lignocellulosic biomass, but some cautions must be observed. Many variables may enter into the collection of such data, including species variation, variation between and within plants of the same species, age of plant, conditions under which the plant was harvested, stored, and processed, and variation in methods of chemical analysis and measurement of fiber dimensions. Due to factors such as these, it is important to evaluate the properties of specific materials entering a manufacturing process on an ongoing basis in order to assure product quality.

2.6   Cellular Solids Revisited

Our brief survey of the anatomical structure of plants providing sources of lignocellulosic raw material clearly demonstrates their cellular nature. Since plant cell wall substance has a nearly constant density of about 1.5 g/cm^3, the variation in stem density between species is directly attributable to anatomical variation. This is significant in that density is positively correlated with the physical properties that govern material behavior during the manufacturing and subsequent use of lignocellulosic composites. Equally important is the observation that plant cell walls are organic polymer composites in-and-of themselves. From a materials science perspective, cellular lignocellulosic substances provide a thoroughly fascinating topic of investigation and a number of technological challenges toward their efficient utilization. These materials are also extremely versatile and useful from a practical standpoint, thanks in large measure to their fundamental cellular nature.

References

1. *Webster's New World Dictionary of the American Language*, 2nd College Edition. New York: The World Publishing Company; 1970.
2. Stenius P. editor. *Forest Products Chemistry*. Helsinki, Finland: Fapet Oy/Finnish Paper Engineers' Association/TAPPI; 2000.
3. Fengel D, Wegener G. *Wood: Chemistry, Ultrastructure, Reactions*. New York: Walter de Gruyter & Co; 1983.
4. Brown RM Jr, Saxena IM. Preface. In: Brown RM Jr, Saxena IM, editors. *Cellulose: Molecular and Structural Biology*. Dordrecht, The Netherlands: Springer; 2007. pp xiii–xv.

5. Sjöström E. *Wood Chemistry: Fundamentals and Applications*. New York: Academic Press; 1981.
6. Franz G, Blaschek W. Cellulose. In: Dey PM, editor. *Methods in Plant Biochemistry, Volume 2, Carbohydrates*. San Diego: Academic Press, Ltd; 1990. pp 291–322.
7. Albersheim P, Darvill A, Roberts K, Sederoff R, Staehelin A. *Plant Cell Walls: From Chemistry to Biology*. New York: Garland Science, Taylor & Francis Group, LLC; 2011.
8. Kondo T. Nematic ordered cellulose: its structure and properties. In: Brown RM Jr, Saxena IM, editors. *Cellulose: Molecular and Structural Biology*. Dordrecht, The Netherlands: Springer; 2007. pp 285–305.
9. Li Q, Renneckar S. Supramolecular structure characterization of molecularly thin cellulose I nanoparticles. *Biomacromolecules*, 2011;12:650–659.
10. Zugenmaier P. In: Timell TE, Wimmer R, editors. *Crystalline Cellulose and Cellulose Derivatives*. Berlin: Springer Verlag; 2008.
11. Gibson LJ, Ashby MF, Harley BA. *Cellular Materials in Nature and Medicine*. Cambridge, UK: Cambridge University Press; 2010.
12. Tsuchikawa S, Siesler HW. Near infrared spectroscopic monitoring of the diffusion process of deuterium-labeled molecules in wood. In: Stokke DD, Groom LH, editors. *Characterization of the Cellulosic Cell Wall*. Ames, IA: Blackwell Publishing; 2006. pp 123–137.
13. Groom LH, So C-L, Elder T, Pesacreta T, Rials TG. Effects of refiner pressure on the properties of individual wood fibers. In: Stokke DD, Groom LH, editors. *Characterization of the Cellulosic Cell Wall*. Ames, IA: Blackwell Publishing; 2006. pp 227–240.
14. Salmén L. The cell wall as a composite structure. In: Bristow JA, Kolseth P, editors. *Paper Structure and Properties*. New York: Marcel Dekker; Inc.; 1986. pp 51–71.
15. French AD, Johnson GP. Cellulose shapes. In: Brown RM Jr, Saxena IM, editors. *Cellulose: Molecular and Structural Biology*. Dordrecht, The Netherlands: Springer; 2007. pp 257–284.
16. Wakelyn PJ, Bertoniere NR, French AD, Thibodeaux DP, Triplett BA, Rousselle A. Cotton fibers. In: Lewin M, editor. *Handbook of Fiber Chemistry*, 3rd ed. Boca Raton, FL: CRC Press, Taylor & Francis Group; 2007. pp 521–666.
17. Batra SK. Other long vegetable fibers: abaca, banana, sisal, henequen, flax, ramie, hemp, sunn, and coir. In: Lewin M, editors. *Handbook of Fiber Chemistry*, 3rd ed. Boca Raton, FL: CRC Press, Taylor & Francis Group; 2007. pp 453–520.
18. Rowell RM, Stout HP. Jute and kenaf. In: Lewin M, editor. *Handbook of Fiber Chemistry*, 3rd ed. Boca Raton, FL: CRC Press, Taylor & Francis Group; 2007. pp 405–520.
19. Atalla RJ, Brady JW, Matthews JF, Ding S-Y, Himmel ME. Structures of plant cell wall celluloses. In: Himmel ME, editor. *Biomass Recalcitrance: Deconstructing the Plant Cell Wall for Bioenergy*. Chichester, UK: Blackwell Publishing Ltd; 2008. pp 188–212.
20. Liese W. *The Anatomy of Bamboo Culms*. Beijing, PR China: International Network for Bamboo and Rattan; Report No. 18. 1998.
21. Pettersen RC. The chemical composition of wood. In: Rowell RM, editor. *The Chemistry of Solid Wood*. Washington, DC: American Chemical Society; 1984. pp 57–126.
22. Laine C. *Structures of Hemicelluloses and Pectins in Wood and Pulp*. Espooi, Finland: Helsinki University of Technology; 2005.
23. Boerjan W, Ralph J, Baucher M. Lignin biosynthesis. *Annual Review of Plant Biology*, 2003;54: 519–546.
24. Ralph J, Lundquist K, Brunow G, Lu F, Kim H, Schatz PF. Lignins: natural polymers from oxidative coupling of 4-hydroxyphenyl-propanoids. *Phytochemistry Reviews*, 2004;3(1–2):29–60.
25. Brunow G. Methods to reveal the structure of lignin. In: Hofrichter M, Steinbüchel A, editors. *Lignin, Humic Substances and Coal*. Weinheim: Wiley-VCH; 2001. pp 89–116.
26. Ralph J. *Poplar Lignin Models*. USDA Agricultural Research Service, Madison, WI [cited 2011 November 22, 2011]; 2001. Available at: http://ars.usda.gov/Services/docs.htm?docid=10443 (last accessed 16 August 2013).
27. Jurasek L. Toward a three-dimensional model of lignin structure. *Journal of Pulp and Paper Science*, 1995;21(8):J274–J279.

28. Willför S, Sundberg A, Hemming J, Holmbom B. Polysaccharides in some industrially important softwood species. *Wood Science and Technology*, 2005;39:245–258.
29. Willför S, Sundberg A, Pranovich A, Holmbom B. Polysaccharides in some industrially important hardwood species. *Wood Science and Technology*, 2005;39:601–617.
30. Bismarck A, Mishra S, Lampke T. Plant fibers as reinforcement for green composites. In: Mohanty AK, Misra M, Drzal LT, editors. *Natural Fibers, Biopolymers, and Biocomposites*. Boca Raton, FL: CRC Press, Taylor & Francis Group; 2005. p 875.
31. Esau K. *Anatomy of Seed Plants*, 2nd ed. New York: John Wiley & Sons, 1977.
32. Niklas KJ. *Plant Biomechanics*. Chicago: The University of Chicago Press; 1992.
33. Mohanty AK, Misra M, Drzal LT. *Natural Fibers, Biopolymers, and Biocomposites*. Boca Raton, FL: Taylor & Francis; 2005.
34. Pickering KL. *Properties and Performance of Natural-Fibre Composites*. Cambridge, UK: Woodhead Publishing, Ltd; 2008.
35. Pande H. Non-wood fibre and global fibre supply. *Unasylva* [serial on the Internet], 1998;49(193): Available at: http://www.fao.org/docrep/w7990e/w7990e00.htm (last accessed 16 August 2013).
36. Energy USDo. Perlack RD, Stokes BJ, editors. *U.S. Billion-Ton Update: Biomass Supply for a Bioenergy and Bioproducts Industry*. Oak Ridge, TN: ORNL/TM-2011/224 Oak Ridge National Laboratory; 2011. p 227.
37. Sarath G, Dien B, Saathoff AJ, Vogel KP, Mitchell RB, Chen H. Ethanol yields and cell wall properties in divergently bred switchgrass genotypes. *Bioresource Technology*, 2011;102(20):9579–9585.
38. Heaton E, Voigt T, Long SP. A quantitative review comparing the yields of two candiate C4 perennial biomass crops in relation to nitrogen, temperature and water. *Biomass and Bioenergy*, 2004;27(1): 21–30.
39. Kaack K, Schwarz K-U, Brander PE. Variation in morphology, anatomy and chemistry of stems of *Miscanthus* genotypes differing in mechanical properties. *Industrial Crops and Products*, 2003;17: 131–142.
40. Munder F, Furll C, Hempel H. Processing of bast fiber plants for industrial application. In: Mohanty AK, Misra M, Drzal LT, editors. *Natural Fibers, Biopolymers, and Biocomposites*. Boca Raton, FL: Taylor & Francis; 2005. pp 109–157.
41. Killmann W, Fink D. 1996 *Coconut Palm Stem Processing Technical Handbook*. Available at: http://www.fao.org/docrep/009/ag335e/AG335E00.htm (last accessed 16 August 2013).
42. Han JS, Rowell JS. Chemical composition of fibers. In: Rowell RM, Young RA, Rowell JK, editors. *Paper and Composites from Agro-Based Resources*. Boca Raton, Florida: CRC Press, Inc/Lewis Publishers; 1997. pp 83–134.
43. Rials TG, Wolcott MP. Physical and mechanical properties of agro-based fibers. In: Rowell RM, Young RA, Rowell JK, editors. *Paper and Composites from Agro-Based Resources*. Boca Raton, FL: CRC Press, Inc./Lewis Publishers; 1997. pp 63–81.
44. Wiselogel AE, Agblevor FA, Johnson DK, Deutch S, Fennell JA, Sanderson MA. Compositional changes during storage of large round switchgrass bales. *Bioresource Technology*, 1996;56:103–109.
45. Ai J, Tschirner U. Fiber length and pulping characteristics of switchgrass, alfalfa stems, hybrid poplar and willow biomasses. *Bioresource Technology*, 2010;101(1):215–221.
46. Ververis C, Georghiou K, Christodoulakis N, Santas P, Santas R. Fiber dimensions, lignin and cellulose content of various plant materials and their suitability for paper production. *Industrial Crops and Products*, 2004;19:245–254.
47. Mohanty AK, Misra M, Hinrichsen G. Biofibres, biodegradable polymers and biocomposites: an overview. *Macromolecules*, 2000;276:1–24.
48. Shi J, Shi Q, Barnes HM, Pittman CU Jr. A chemical process for preparing cellulosic fibers hierachically from kenaf bast fibers. *BioResources*, 2011;6(1):879–890.
49. Sudo, S. Some anatomical properties and density of the stem of coconut palm (*Cocos nucifera*), with consideration for pulp quality. *IAWA Journal*, 1980;1(4):161–171.
50. Rowell RM. Natural fibres: types and properties. In: Pickering KL, editor. *Properties and Performance of Natural-Fibre Composites*. Cambridge, UK: Woodhead Publishing; 2008. pp 3–66.

3

Wood as a Lignocellulose Exemplar

3.1 Introduction

In Chapter 2, we introduced a number of plants as diverse, yet illustrative, sources of lignocellulosic material that may be useful for manufacturing selected natural fiber composites. In
this chapter, we will focus on wood as a teaching example or exemplar representative of many
of the characteristics of lignocellulosic materials in general. Certainly, wood has some unique
attributes that set it apart from other materials, and its uniqueness and availability make it by
far the most widely used lignocellulosic industrial feedstock. However, there are well-studied
attributes of wood that are also characteristic of other plant-derived materials. Thus, our study
of fundamental material properties of wood will enable the student to extend this knowledge
to other plant fibers.

3.2 Wood as a Representative Lignocellulosic Material: Important Physical Attributes

Of all the physical attributes of lignocellulosic materials that might be considered, the way in
which these materials interact with water, their density and specific gravity, and their fundamental mechanical properties are the most basic. An understanding of these characteristics of
wood will, by extension, provide insight into material behavior and utilization of a variety of
plant-based materials. As we have seen, lignocellulosic materials are fundamentally composed
of cellulose, hemicelluloses, and lignins. However, just as the specific chemical composition
and amount of these natural polymers may vary between plant species, so too will the particular
moisture interactions, density, specific gravity, and mechanical properties. With this in mind,
let us consider these attributes of lignocellulosic materials, using wood as our exemplar.

3.3 Moisture Interactions

Moisture interactions are significant considerations for the utilization of materials of all kinds.
While there are some materials that are seemingly impervious to moisture, most are nonetheless

Introduction to Wood and Natural Fiber Composites, First Edition. Douglas D. Stokke, Qinglin Wu and Guangping Han.
© 2014 John Wiley & Sons, Ltd. Published 2014 by John Wiley & Sons, Ltd.

affected, often adversely, by the presence of moisture. Moisture is contributory to various kinds of deterioration processes. It is also responsible for changes in material properties as the amount of moisture in the material varies. For example, some of the wood properties affected by moisture include mass, volume, and linear measures of dimensions, insulation value, heating value, electrical properties, acoustical properties, strength properties, and susceptibility to biological degradation. Interestingly, most of these changes occur within a specific range of moisture content, known as the hygroscopic range. It is within this band of moisture content, often assumed as zero to about 30% moisture on a dry basis, that changes in physical properties are observed. As moisture exceeds the approximate value of 30%, physical properties (other than mass) remain constant. Thus, a firm understanding of how wood interacts with moisture is essential to effective utilization.

3.3.1   Moisture Content

The moisture content of wood is almost without exception defined on a dry basis, as shown in Equation 3.1:

$$M = \left(\frac{W_M - W_o}{W_o} \right) 100 \qquad (3.1)$$

where

M = dry basis moisture content, %,
W_M = mass of wood plus moisture, that is, "wet weight" of wood, and
W_o = oven-dry mass of wood.

The determination of wood moisture content by weighing, oven drying, and then weighing again is known as the gravimetric method. Oven drying is carried out at 103°C $\pm$ 2°C, as specified in ASTM D-4442 [1]. Oven-dry mass is determined when the sample reaches approximate constant mass, that is, when there is no appreciable change of mass upon re-weighing at 2-hour drying intervals. Guidelines for determining approximate constant mass are given in the standard methodology [1].

Algebraic rearrangement of Equation 3.1 to solve for W_M yields Equation 3.2:

$$W_M = W_o \left(1 + \frac{M}{100} \right) \qquad (3.2)$$

Notice that $W_M = W_o + [W_o(M/100)]$, with emphasis on the fact that in the portion of this equation within brackets, the mass of water contained in wood is expressed relative to the dry mass of wood fiber.

The dry basis moisture content, M, is the usual and ordinary way of specifying the moisture content of both wood and other natural fiber. One notable exception is when moisture content is calculated for fuel value calculations. In this case, the wet-basis moisture content, M_{wet}, is used (Equation 3.3):

$$M_{wet} = \left(\frac{W_M - W_o}{W_M} \right) 100 \qquad (3.3)$$

3.3.2   Hygroscopicity

A hygroscopic material is one that is able to attract moisture from the surrounding atmosphere. This is a reversible interaction that is dependent upon the temperature, moisture content of air (commonly expressed as relative humidity), and the degree of hygroscopicity (water-attracting capacity) of the material. The hygroscopic nature of wood and other lignocellulosic materials is attributed primarily to the presence of hydroxyl (–OH) groups that are part of the natural organic polymers comprising the cell wall. Thus, it is mainly the carbohydrate-based fraction of the cell wall, hemicelluloses, and cellulose, which are responsible for wood's hygroscopicity. Moreover, it is primarily the hemicelluloses that dictate hygroscopicity, due to their short, branched, open structure, and their location on the surface of microfibrils, which results in availability of the OH groups for moisture sorption. Cellulose, though rich in OH content, is less hygroscopic than might be inferred from its chemical structure, due to the inter- and intra-molecular bonding within cellulose microfibrils, which precludes moisture uptake in the crystalline regions of cellulose. Lignin is regarded as either low in hygroscopicity, or as hydrophobic, that is, water-repelling.

The uptake and release of hygroscopically bound water occurs via the formation and breakage of hydrogen bonds between wood polymers and water molecules. Water, in vapor form, is attracted to and bound by wood's OH groups, known as sorption sites (Figure 3.1). This is a dynamic process that varies with temperature and relative humidity of the surroundings. This process is fundamentally driven by a free energy gradient, and as such, is a process that seeks equilibrium. Given sufficient time, this equilibrium of wood moisture content with the surrounding atmosphere will occur.

3.3.2.1   Equilibrium Moisture Content

The equilibrium moisture content (EMC) of wood or other hygroscopic materials may be defined as the moisture content that a material will assume if exposed to a given temperature and

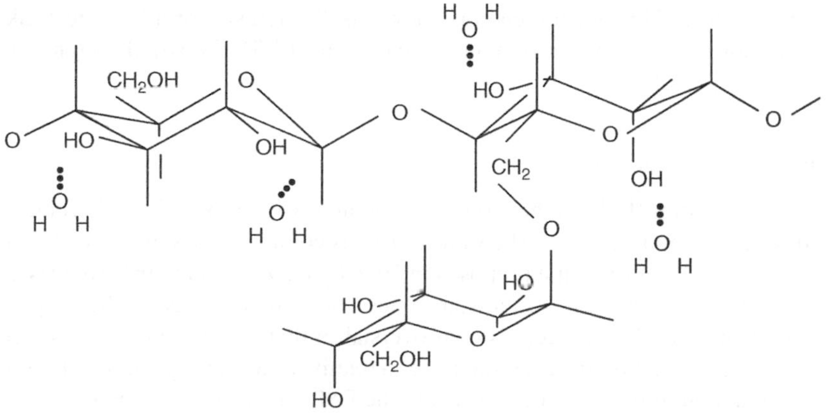

Figure 3.1   *Moisture sorption on a fragment of wood galactoglucomannan hemicellulose. Dots represent hydrogen bonds between water and –OH groups on the hemicellulose. Reproduced from [2] with permission of John Wiley & Sons, Inc © 2011.*

humidity condition for a sufficient period of time (Table 3.1). The EMC of wood is furthermore affected by a number of factors, such as the direction of moisture sorption (desorption or loss of moisture, versus adsorption or uptake of moisture), temperature, temperature history of the material (i.e., heat treatment of the material), mechanical stress, species factors such as extractives content and wood density, and chemical treatment. Needless to say, other lignocellulosic materials also experience moisture equilibrium and resultant impacts on the material properties.

EMC of wood may be estimated by reference to tabulated data such as that shown in Table 3.1 or by calculation using the Hailwood–Horrobin model (see Section 3.3.2.5). In either case, the EMC estimates are based on actual sorption data collected for Sitka spruce (*Picea sitchensis*).

The EMC conditions for wood vary with geographic region, due to obvious regional differences in annual climate variations. Since fluctuations in moisture content leads to dimensional change (shrinkage and swelling) of wood and wood-based products, problems due to dimensional change may be minimized by effective moisture management strategies. Ideally, wood fiber-based products should be conditioned to the average moisture content expected in use. Kiln drying of lumber is based on the establishment of target EMC conditions in the dry kiln by control of dry and wet bulb temperatures, that is, control of temperature and relative humidity. The EMC conditions in the kiln are varied as the wood moisture content is reduced throughout the drying process. Readers interested in further information on lumber drying are referred to the Dry Kiln Operator's Manual [4].

3.3.2.2   *Sorption Hysteresis*

A sorption isotherm is a plot of EMC (or MC) versus the relative humidity (RH or H) of the surrounding air. Often, a single, sigmoid curve is used to represent sorption data. In actual fact, however, there is some difference in the EMC of wood, depending on the "direction" of sorption, that is, whether moisture is being lost or gained by the wood (Figure 3.2). Here, it is seen that the EMC at a given H, upon initial desorption or loss of moisture from a green, water-saturated state, is greater than the EMC of wood at the same value of H upon subsequent adsorption of moisture. This phenomenon is known as "hysteresis" or a lag-effect. Skaar [5] shows ratios of adsorption to desorption moisture contents (A/D) for wood, ranging from 0.8 to 0.9, with an average of 0.85.

3.3.2.3   *Fiber Saturation Point*

The fiber saturation point (FSP) is the moisture content at which the cell walls contain maximum bound water, but no free water. The value of FSP is commonly assumed as 30% moisture content [3]. The concept of FSP, first proposed in 1906 by H.D. Tiemann [6] provides a useful, yet controversial concept relative to moisture interactions in wood and other hygroscopic, lignocellulosic materials. The concept is controversial in that experimental data suggest that bound water coexists with liquid (free) water over a relatively wide range of moisture contents [7]. In addition, although it is common to equate the EMC of wood at a vapor pressure ratio, h, equal to one (i.e., 100% relative humidity, H), this is not entirely true. Stamm [8] asserts that a sharp upward break in wood adsorption curves near $h = 1$ is due to condensation of capillary or liquid-free water. Thus, the true FSP is estimated to occur at h approximately

Table 3.1 Moisture content of wood in equilibrium with given temperature and relative humidity conditions.

Temperature		Moisture content (%) at various relative humidity values																	
°C	°F	5	10	15	20	25	30	35	40	45	50	55	60	65	70	75	80	85	90	95
−1.1	30	1.4	2.6	3.7	4.6	5.5	6.3	7.1	7.9	8.7	9.5	10.4	11.3	12.4	13.5	14.9	16.5	18.5	21	24.3
4.4	40	1.4	2.6	3.7	4.6	5.5	6.3	7.1	7.9	8.7	9.5	10.4	11.3	12.3	13.5	14.9	16.5	18.5	21	24.3
10	50	1.4	2.6	3.6	4.6	5.5	6.3	7.1	7.9	8.7	9.5	10.3	11.2	12.3	13.4	14.8	16.4	18.4	20.9	24.3
15.6	60	1.3	2.5	3.6	4.6	5.4	6.2	7	7.8	8.6	9.4	10.2	11.1	12.1	13.3	14.6	16.2	18.2	20.7	24.1
21.1	70	1.3	2.5	3.5	4.5	5.4	6.2	7	7.7	8.5	9.2	10.1	11	12	13.1	14.4	16	17.9	20.5	23.9
26.7	80	1.3	2.4	3.5	4.4	5.3	6.1	6.9	7.6	8.3	9.1	9.9	10.8	11.7	12.9	14.2	15.7	17.7	20.2	23.6
32.2	90	1.2	2.3	3.4	4.3	5.1	5.9	6.7	7.4	8.1	8.9	9.7	10.5	11.5	12.6	13.9	15.4	17.3	19.8	23.3
37.8	100	1.2	2.3	3.3	4.2	5	5.8	6.5	7.2	7.9	8.7	9.5	10.3	11.2	12.3	13.6	15.1	17	19.5	22.9
43.3	110	1.1	2.2	3.2	4	4.9	5.6	6.3	7	7.7	8.4	9.2	10	11	12	13.2	14.7	16.6	19.1	22.4
48.9	120	1.1	2.1	3	3.9	4.7	5.4	6.1	6.8	7.5	8.2	8.9	9.7	10.6	11.7	12.9	14.4	16.2	18.6	22
54.4	130	1	2	2.9	3.7	4.5	5.2	5.9	6.6	7.2	7.9	8.7	9.4	10.3	11.3	12.5	14	15.8	18.2	21.5
60	140	0.9	1.9	2.8	3.6	4.3	5	5.7	6.3	7	7.7	8.4	9.1	10	11	12.1	13.6	15.3	17.7	21
65.6	150	0.9	1.8	2.6	3.4	4.1	4.8	5.5	6.1	6.7	7.4	8.1	8.8	9.7	10.6	11.8	13.1	14.9	17.2	20.4
71.1	160	0.8	1.6	2.4	3.2	3.9	4.6	5.2	5.8	6.4	7.1	7.8	8.5	9.3	10.3	11.4	12.7	14.4	16.7	19.9
76.7	170	0.7	1.5	2.3	3	3.7	4.3	4.9	5.6	6.2	6.8	7.4	8.2	9	9.9	11	12.3	14	16.2	19.3
82.2	180	0.7	1.4	2.1	2.8	3.5	4.1	4.7	5.3	5.9	6.5	7.1	7.8	8.6	9.5	10.5	11.8	13.5	15.7	18.7
87.8	190	0.6	1.3	1.9	2.6	3.2	3.8	4.4	5	5.5	6.1	6.8	7.5	8.2	9.1	10.1	11.4	13	15.1	18.1
93.3	200	0.5	1.1	1.7	2.4	3	3.5	4.1	4.6	5.2	5.8	6.4	7.1	7.8	8.7	9.7	10.9	12.5	14.6	17.5
98.9	210	0.5	1	1.6	2.1	2.7	3.2	3.8	4.3	4.9	5.4	6	6.7	7.4	8.3	9.2	10.4	12	14	16.9
104.4	220	0.4	0.9	1.4	1.9	2.4	2.9	3.4	3.9	4.5	5	5.6	6.3	7	7.8	8.8	9.9			
110	230	0.3	0.8	1.2	1.6	2.1	2.6	3.1	3.6	4.2	4.7	5.3	6	6.7						
115.6	240	0.3	0.6	0.9	1.3	1.7	2.1	2.6	3.1	3.5	4.1	4.6								
121.1	250	0.2	0.4	0.7	1	1.3	1.7	2.1	2.5	2.9										
126.7	260	0.2	0.3	0.5	0.7	0.9	1.1	1.4												
132.2	270	0.1	0.1	0.2	0.3	0.4	0.4													

Source: Reproduced from [3].

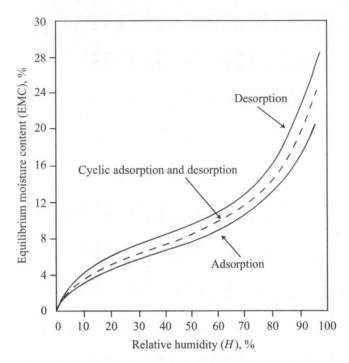

Figure 3.2   *Desorption and adsorption isotherms for wood, illustrating sorption hysteresis. Adpated from [3].*

equal to 0.995. In practical terms, however, FSP is ordinarily taken to occur at $h = 1$. Stamm presented adsorption moisture contents for 17 species of unextracted wood at ambient temperatures ranging from 10°C to 55°C. At $h = 1$, the moisture content, that is, FSP values ranged from 21% (in Western redcedar, *Thuja plicata*, a wood with a high-extractive content) to 33% (in white spruce, *Picea* sp.). Other techniques for determining FSP, such as the pressure plate technique, tend to produce estimates that are higher than those found with the common saturation over water vapor method [9].

Although FSP is commonly taken to be the EMC at $h = 1$ (or thereabouts), it may alternatively be defined in terms of water potential [10], or as the moisture content at which a change in an arbitrary physical property is detected. Above FSP, most wood mechanical properties are constant. As wood is dried below the FSP, wood properties change, with most mechanical properties increasing as moisture content decreases within the hygroscopic range (FSP to oven dry). However, since bound water likely coexists with free or liquid water in the vicinity of the FSP, thus, properties tend to show a change at a moisture content slightly below the sorption FSP. Accordingly, Siau [7] stated "… Tiemann [6] defined FSP as the moisture content at which the cell walls are saturated with no free or capillary water in the lumens. It appears to be an experimental impossibility to clearly separate these two kinds of water, in fact they seem to overlap with no clear line of distinction between them. It may be more appropriate to define FSP as the change in the slope of a $\delta P/\delta M$ relationship where P is any physical property

Table 3.2   *Intersection moisture content, M_p, for selected species.*

Species	M_p, %
Ash, white	24
Birch, yellow	27
Chestnut, American	24
Douglas-fir	24
Hemlock, Western	28
Larch, Western	28
Pine, loblolly	21
Pine, longleaf	21
Pine, red	24
Redwood	21
Spruce, red	27
Spruce, Sitka	27
Tamarack (larch, Eastern)	24

Source: Reproduced from [3].

(or its logarithm) which may be measured." When this approach is taken, the term used to describe the moisture content at which the measured property changes is not designated as FSP; rather, this is the Intersection Moisture Content, M_p. Representative values of M_p are shown in Table 3.2. Values for M_p are generally less than typical FSP values determined from sorption data.

With advanced techniques, contemporary researchers have kept the FSP concept as an issue for consideration within the discipline of wood science. For example, using nuclear magnetic resonance (NMR) relaxation times, Telkii et al. [11] found FSP values of 35 ± 4 and $45 \pm 7\%$ for pine (*Pinus sylvestris*) and spruce (*Picea abies*) sapwood, respectively. It is interesting to consider how a relatively simple concept such as FSP has piqued scientific curiosity for over a century.

Sidebar 3.1   Fiber Saturation Pernt?

The FSP of wood is said to average about 30% moisture content, with variations around that figure resulting from a variety of species and individual sample variations [3]. Even with such caveats, an interesting, if not tongue-in-cheek, analysis of FSP was offered by the eminent Professor Dr. John Siau:

" . . . In conclusion, it can be stated that fiber saturation point is, at best, an elusive concept although a very useful one for understanding the influence of bound water on physical properties. Bound and free water appear to coexist over a significant range of water potentials or partial vapor pressures. Furthermore, measured values depend on many factors such as the technique used, the size of the specimen, the structure of the wood, the extractives content, the temperature, and the static pressure. Nevertheless, FSP still enjoys the precise distinction of being called a *point*. Bent [12] discusses similar problems in thermodynamics with the Curie point, the lambda point, the ferroelectric point, and others. Bent's remedy for this anomalous situation is to coin a new scientific term, namely the *pernt*, defined in the following way. '*Pernt*, a transitive noun; a meandering point shaped like a small period; a point that is both here and there; a point that is intermediate due to its uncertainty.' There we have it. FSP could be more appropriately called the *fiber saturation pernt*." (Siau, 1995, p. 85) [7]

It seems to me that Dr. Siau, a precise and thoughtful man, was no doubt amused by the notion of FSP when he wrote this. A meandering *pernt* as opposed to a precise point? Having had the pleasure of knowing Dr. Siau, I don't find his humor too surprising.

—D.D.S.

3.3.2.4   *Bound Water: Sorption Isotherms*

The FSP is a concept based on bound water sorption phenomenon. Five types of sorption isotherms have been described, with three most applicable to the wood-water system. According to Skaar [13], "Type I characterizes sorption of a vapor on a substrate in which only a single layer of the sorbent (vapor) is found on the sorbate (substrate). It may also be considered that the vapor forms a hydrate with the substrate. In either case the attraction between the sorbent and the sorbate is greater than that between the sorbent molecules themselves in the liquid state. A type III sorption isotherm is obtained when several layers of sorbent are formed on the sorbate, that is, multilayered sorption occurs. It may also be viewed as a solution of the sorbent in the hydrated sorbate. The attraction of the sorbed molecules to the sorbate is considered to be minimal in this case, compared with the type I condition. The type II isotherm is typical of the moisture sorption isotherms of wood and many other natural hygroscopic polymers such as cotton, wool, cereal grains, etc. It is typically sigmoid in shape, and appears to be a composite of types I and III." Figure 3.3, "taken from Simpson [14] shows a typical type II moisture sorption isotherm for wood, together with its two components, the hydrated water (type I) and the dissolved water (type III) as analyzed by the Hailwood–Horrobin sorption theory. Similar curves of type I and type III are obtained assuming monolayer and multilayer sorption, respectively, based on a surface sorption model such as the Dent modification of the BET sorption theory."

Many biologically derived materials of both animal and plant origin have similar sorption properties, despite chemical differences. Examples include animal protein materials such as wool, silk, hair, and keratin. Plant materials include cotton, bamboo, rice straw, grains, and nuts.

3.3.2.5   *Sorption Models*

Many equations have been proposed to model sorption isotherms of biological materials; one review showed 77 isotherm equations applied to biologicals, including wood and other fibrous materials [15]. Though appearing in various forms, many of these equations are mathematically equivalent [16]. Some of the equations are empirical (based on experiment or observation rather than theory), some are theoretical, and some are semi-empirical.

William Simpson of the US Forest Products Laboratory applied a number of sorption equations to sorption isotherm data for wood [13, 14]. A number of models fit data for wood, but all have limitations. For example, calculation of thermodynamic parameters, for example, heat of sorption, and so on, are not always satisfactory. In all, about 18 sorption equations have been applied to cellulosic materials.

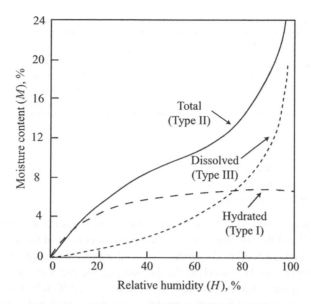

Figure 3.3 *Types I, II, and III sorption isotherms, representative of hygroscopic behavior of biological materials. Wood exhibits this sorption behavior, wherein Types I hydrated or (monolayer adsorption of water) and III dissolved or (multilayer adsorption) combine to yield the total or gross sorption characteristic represented by Type II. Reproduced from [13] with permission of Springer Science + Business Media © 1988.*

Most sorption theories include the following assumptions:

1. Bound water in wood consists of two components: one weakly bound, one strongly bound.
2. Water sorption is considered either a surface phenomenon or a solution phenomenon. In either case, it is assumed that sorption sites either on the surface or throughout the volume of the sorbate (here, wood) are equally accessible to water.

We will briefly consider here two surface models (BET) and (Dent), and a sorbate (solution) model (Hailwood–Horrobin). Details on the derivation of these models are given for BET and Dent by Skaar [13], and for Hailwood–Horrobin, by Siau [7].

The Brunauer-Emmett-Teller (BET) theory is the most widely used model for describing the sorption of gasses onto surfaces. It assumes that thermodynamic properties of secondary layers of water are essentially the same as liquid water. In the case of wood–water interactions, this assumption is invalid. This theory is also based on an assumption of one to many layers of water molecules per sorption site. The model is adequate for vapor pressure ratios <0.4. The BET model was modified by a researcher named Dent to provide a better description of moisture sorption isotherms for wood.

The Dent model more adequately represents sorption of water by wood, in that it is assumed that the thermodynamic properties of the first and second layers of water differ. The properties of all secondary layers are assumed identical, an assumption also made in the BET theory. Both the Dent and the Hailwood–Horrobin models are of the general form given in Equation 3.4:

$$\frac{h}{m} = A + Bh + Ch^2 \tag{3.4}$$

Table 3.3   *Parameters for the calculation of EMC with the Hailwood–Horrobin model (Equation 3.5).*

Parameter	For T in °C	For T in °F
W	$349 + 1.29T + 0.0135T^2$	$330 + 0.452T + 0.00415T^2$
k	$0.805 + 0.000736T - 0.00000273T^2$	$0.791 + 0.000463T - 0.000000844T^2$
k_1	$6.27 - 0.00938T - 0.000303T^2$	$6.34 + 0.000775T - 0.0000935T^2$
k_2	$1.91 + 0.0284T - 0.0000904T^2$	$1.09 + 0.0284T - 0.0000904T^2$

where

h = vapor pressure ratio,
m = fractional moisture content, and
A, B, and C = fundamental constants.

The Hailwood–Horrobin model [17] is a solution theory that has been applied to wood and textiles. The cell wall and its associated water is assumed to consist of three chemical species, the dry cell wall, the hydrated cell wall, and the dissolved water. The model appears as equation [6] in the Dry Kiln Operators Manual (DKOM) [4], despite the fact that the name, Hailwood–Horrobin, does not appear anywhere in the DKOM text. The EMC tables in the DKOM and the Wood Handbook are derived from this model, using sorption data for Sitka spruce. The Hailwood–Horrobin model is the most-widely used sorption model for wood today. EMC tables provided by the Forest Products Laboratory and Table 3.1 in this text are based on this model, a form of which appears in Equation 3.5:

$$\text{EMC} = \frac{1800}{W} \left[\frac{kh}{1 - kh} + \frac{k_1 kh + 2k_1 k_2 k^2 h^2}{1 + k_1 kh + k_1 k_2 k^2 h^2} \right] \tag{3.5}$$

where

EMC = equilibrium moisture content, %,
h = humidity ratio,
T = temperature, and
W, k, k_1, and k_2 = parameters as defined in Table 3.3.

3.3.3   States of Water in Wood

Moisture contained in wood may be in the states of matter described as frozen, liquid, or vapor. Liquid water, or free water, is held by capillary or tension forces in the cavities or spaces in wood, such as the cell lumens, pit chambers, and microvoids within the cell wall. Vapor may be found in any of these spaces as well. In addition to these "ordinary" states of water, moisture that is adsorbed, that is, bound water held by hydrogen bonds, represents a fourth "phase" of water. Bound water (sorbed, adsorbed, hydroscopic water) has thermodynamic properties that are intermediate between the frozen and liquid states.

According to Skaar [13], p. 53–54, "Sorbed water in the cell wall of wood is analogous to the frozen or solid state of ordinary water in that it has a lower enthalpy than liquid water. However, its enthalpy increases with increasing wood moisture content up to fiber saturation, above which it is essentially the same as that of liquid water. The enthalpy of capillary water in the cell cavities is slightly lower than that of ordinary liquid water, for one or both of two

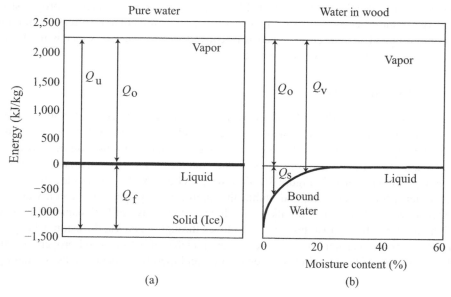

Figure 3.4   *Relative energy levels for various states of water: (a) pure water; (b) water contained in the woody cell wall. Q_o = heat of vaporization, Q_u = heat of sublimation, Q_f = heat of fusion, Q_v = heat of sorption, Q_s = differential heat of sorption. Reproduced from [13] with permission of Springer Science + Business Media © 1988.*

reasons. First, the vapor pressure (p_c) of capillary water is lower than (p_0) that of ordinary liquid water because of the curvature of the air–water meniscus in a capillary. Second, the presence of water-soluble extractives in the capillary water further reduces the vapor pressure."

Because enthalpy is a thermodynamic measure of the total energy or heat content, the foregoing analysis tells us that adsorbed water in the cell wall has total energy that is less than that of liquid (free) water. If we were to attempt to convert a given mass of frozen water, bound water, and liquid water to the vapor phase, we would find that the amount of energy input required would vary for each starting phase. This may be visualized by observing the energy states for frozen, liquid, and bound water, as shown in Figure 3.4, here presented on a mass, rather than molar basis, the latter a common basis for measures of enthalpy. The quantities depicted here for pure water may be understood as the energy input required to convert a given mass of ice to liquid water (heat of fusion, Q_f) and the energy to change the liquid phase to vapor (heat of vaporization, Q_o), with the sum of the two represented by the heat of sublimation (Q_u). Q_u is the energy needed to change phase directly from solid to vapor.

Bound water is analogous to frozen water in that the energy required to change "phase" to vapor is the sum of the heat of vaporization, Q_o, and the differential heat of sorption, Q_s. The major difference here is that Q_s varies with moisture content, that is, it is not a constant quantity, as is the comparable Q_f. At very low levels of moisture content, water in the cell wall is held tightly as a monomolecular layer of water molecules sorbed directly onto cell wall polymers. Removal of these water molecules requires a relatively high input of energy, as compared to the multilayers of water that accumulate as cell wall moisture content increases. Water adsorbed in a multilayer (think of water molecules stacked one on top of another), though held by hydrogen bonds, are not in as close proximity to the cell wall polymers as are monolayer water molecules, and are thus held by weaker forces. Consequently, progressively lower energy input is needed to remove bound water as moisture content increases within the hygroscopic range. The total energy needed to convert bound water to vapor is the heat of sorption, Q_v, analogous to Q_u for pure water.

3.3.4   Capillary or Free Water

Bound water is held within the cell wall by hydrogen bonding to the cell wall polymers. Free water is contained in the spaces created by the cell wall and is held by tension or capillary forces. It has higher enthalpy (total energy) than bound water, and thus requires less energy input to evaporate it from wood than does bound water, as seen by comparison of Q_o (free water) and Q_v (bound water) in Figure 3.4. Free water is liquid water, in this case, held within the cell lumens, tiny pit cavities, and within the microcapillary space in the cell wall. This free water moves within wood via bulk flow, which is descriptive of both capillary flow and hydrodynamic flow. In contrast, bound water moves through wood via diffusion. Changes in the amount of free water held in wood has no appreciable affect on wood properties, other than the total mass of the wood plus water system, and therefore also on density (mass per unit volume).

3.3.5   Shrinking and Swelling due to Moisture Flux

One of the important physical characteristics of wood affected by moisture flux is its dimensions. The shrinkage and swelling of wood induced by variation in moisture content is a commonly observed phenomenon. However, it is not always appreciated, except by those who regularly work with wood, that dimensional change occurs only with moisture flux within the hygroscopic range. That is, wood only shrinks and swells as moisture content varies in the moisture content range of fiber saturation point to oven dry (zero percent moisture content, in practical terms). This is because bound water sorbed onto cell wall polymers occupies space between those polymers, with addition or removal of bound water causing movement of polymers further or closer together, with concomitant breakage or formation of inter- and intra-molecular hydrogen bonds on the wood polymers. Free water, represented by moisture in addition to the maximum bound water content defined by FSP, does not affect the cell wall in this way. Free water is merely liquid water that adds mass to the wood–water system. It is the bound water interactions that are integral to dimensional change of wood. In addition, bound water content variation is also responsible for the changes observed in almost all other physical and mechanical properties within the hygroscopic range.

The structural anisotropy of wood results in anisotropic behavior with respect to dimensional change. Longitudinal shrinkage of normal, mature wood, is very small, typically on the order of 0.1% from FSP to oven dry. Transverse shrinkage, that is, in the radial or tangential directions, is considerably more, as shown in Table 3.4. As a "rule of thumb" the tangential shrinkage for any given species is roughly twice radial, that is, "$T = 2R$."

Shrinkage is assumed to be linear within the hygroscopic range, therefore, the shrinkage percent expected with any given change in moisture content within this range may be estimated with Equation 3.6:

$$S' = \frac{S \Delta M}{\text{FSP}}$$

(3.6)

Table 3.4 *Average shrinkage percentages, green (M ≥ FSP) to oven dry (M ≅ 0%) for selected species of North American woods.*

Species	Shrinkage,[a] %, from green[b] to oven-dry moisture content		
	Radial	Tangential	Volumetric
Hardwoods			
Ash, white	4.9	7.8	13.3
Aspen, quaking	3.5	6.7	11.5
Cottonwood, Eastern	3.9	9.2	13.9
Hickory, Shagbark	7	10.5	16.7
Maple, red	4	8.2	12.6
Oak, Northern red	4	8.6	13.7
Yellow-poplar	4.6	8.2	12.7
Softwoods			
Douglas fir, interior North	3.8	6.9	10.7
Fir, balsam	2.9	6.9	11.2
Hemlock, Eastern	3	6.8	9.7
Pine, Eastern white	2.1	6.1	8.2
Pine, Southern yellow (loblolly)	4.8	7.4	12.3
Spruce, red	3.8	7.8	11.8
Redwood, young growth	2.2	4.9	7

Source: Reproduced from [3].
[a]Expressed as percentage of original green dimension.
[b]"Green" moisture content means $M >$ fiber saturation point, FSP, commonly assumed as 30% moisture content. Oven dry is zero percent moisture content, in practical terms.

where

S' — shrinkage percentage associated with given moisture content change, ΔM,
ΔM = change in moisture content occurring within the hygroscopic range, that is, a moisture content difference that occurs between FSP and oven-dry condition, and
FSP = fiber saturation point, generally assumed to be 30% moisture content.

As an example, find the volumetric shrinkage percent associated with drying quaking aspen from 50% to 5% moisture content (Equation 3.7). From Table 3.4, we see that $S_v = 11.5\%$. Here, $\Delta M = 30 - 5 = 25\%$, noting that no dimensional change occurs with moisture loss above the assumed FSP of 30% moisture content. Thus:

$$S'_v = \frac{S_v \Delta M}{\text{FSP}} = \frac{11.5\% \times 25\%}{30\%} = 9.58\% \tag{3.7}$$

The change in linear dimension or volume associated with moisture flux in the hygroscopic range may be determined as shown in Equation 3.8:

$$\Delta D = D_i \left(\frac{S'}{100} \right) \tag{3.8}$$

where

$\Delta D =$ change in dimension (linear or volumetric, the latter designated ΔV) associated with a given change in moisture content within the hygroscopic range (ΔM),

$D_i =$ initial linear dimension (alternatively, initial volume V_i), and

$S' =$ shrinkage percentage as defined in Equation 3.6.

The final dimension of wood following shrinkage resulting from moisture content variation in the hygroscopic range may be found with Equation 3.9:

$$D_f = D_i - \Delta D \tag{3.9}$$

where

$D_f =$ final linear dimension (alternatively, volume, V_f) following shrinkage associated with given ΔM.

As an example, consider the volume of wood remaining if we start with 100 m^3 of quaking aspen at 50% moisture content, and dry the wood to a final moisture content of $M = 5\%$. In Equation 3.7, we estimated $S'_v = 9.58\%$. Therefore

$$V_f = V_i - V_i \left(\frac{S'_v}{100} \right) = 100\,\mathrm{m}^3 - 100\,\mathrm{m}^3 \left(\frac{9.58\%}{100} \right) = 90.42\,\mathrm{m}^3 \tag{3.10}$$

We see in Equation 3.10 that in this example, 9.58 m^3 of aspen (ΔD_v) vanishes into thin air simply by drying the wood!

3.4   Density and Specific Gravity of Wood

3.4.1   Density of Wood

3.4.1.1   *Density Defined*

Density of any material is mass per unit volume. Density is typically represented by the Greek letter, rho (ρ), and is expressed in units such as pounds/cubic foot (pcf), grams/cubic centimeter or kilograms/cubic meter. A generic expression for density is given in Equation 3.11:

$$\rho = \frac{m}{V} \tag{3.11}$$

where

$\rho =$ material density,

$m =$ mass of the object, and

$V =$ volume of the object.

Mass, m, is technically a measure of the amount of an object that does not vary with its location on earth or in space. In contrast, the weight of an object, Wt, is the mass multiplied by the acceleration due to gravity, g, that is, $Wt = mg$. Thus, an object with mass of 1 kg has a weight of about 9.8 Newtons on the surface of the earth, wherein 1 Newton = 100 grams force. However, despite this technical difference in the definition of the terms, it is common practice within industry and daily life to use the terms mass and weight interchangeably. We will endeavor in this text to use the term mass to refer to the amount of a material. However, we will nevertheless risk some confusion by using, as is commonly practiced in wood science, the capital letter W to refer to mass, since M and m are so commonly and ordinarily used within wood science and forest products to refer to wood moisture content percent or fractional moisture content, respectively. This being the case, wood density is expressed as shown in Equation 3.12:

$$\rho_M = \frac{W_M}{V_M} \qquad (3.12)$$

where

ρ_M = density of wood based on mass and volume at moisture content M,
W_M = mass of wood plus water at moisture content M, and
V_M = volume of wood at moisture content M.

3.4.1.2 *Effect of Moisture on Wood Density*

Defining density simply as mass per unit volume is insufficient with respect to a hygroscopic and absorptive material such as wood. Variation of moisture content within the hygroscopic range affects both the volume of the wood and the total mass of the wood plus water system. Wood volume is maximized at the FSP. Addition of free water above FSP does not affect wood volume, but the mass of the wood–water system is obviously changed. Thus, when defining density with respect to wood, it is important to know and specify the moisture content at which both mass and volume are determined. Often, these quantities are measured on the same material or sample at the same time, thus, the moisture content for both mass and volume determination is the same. However, this need not be the case. Accordingly, we may define wood density as shown in Equation 3.13:

$$\rho_{M_1,M_2} = \frac{W_{M_1}}{V_{M_2}} \qquad (3.13)$$

where

ρ_{M_1,M_2} = density of wood based on mass at moisture content, M_1, and volume at moisture content, M_2,
W_{M_1} = mass of wood plus water at moisture content, M_1, and
V_{M_2} = volume of wood at moisture content, M_2.

wherein it is common for $M_1 = M_2$, but $M_1 \neq M_2$ is also possible or permissible.

Some common bases for determination of wood density are mass and volume when "green" ($M \geq$ FSP), oven-dry mass and volume when green, or oven-dry mass and volume at 12% moisture content. The latter two determinations are commonly used for determination of wood specific gravity.

3.4.2   Specific Gravity of Wood

3.4.2.1   *Specific Gravity Is a Density Index*

Specific gravity (SG or *G*) is material density divided by the density of water, thus yielding a unitless density index. In the SI system, this is very convenient, given that the density of water is approximately 1 g/cm^3 (1,000 kg/m^3) at room temperature. In the Imperial System, water density is 62.4 lb/ft^3. As discussed in the previous section, for hygroscopic materials such as wood and other lignocellulosic substances, the moisture content at which both material mass and volume are determined must be known and specified for measures of density and specific gravity to be completely meaningful, given that wood volume and mass of the wood plus water system both vary with moisture content, with said variation dependent upon moisture flux within or in excess of the hygroscopic range.

With respect to the density of wood used in the determination of specific gravity, it is *always* based on *oven-dry wood mass* and wood volume at specified moisture content. In other words, the specific gravity of wood is, by definition, calculated using the oven-dry mass of wood. The volume base used for wood-specific gravity determination may vary and must be specified. We may thus define specific gravity mathematically as follows (Equation 3.14):

$$G_M = \frac{\rho_{o,M}}{\rho_w} = \frac{W_o}{V_M \times \rho_w} \tag{3.14}$$

where

G_M = specific gravity of wood based on volume at moisture content *M*,
$\rho_{o,M}$ = density of wood based on oven-dry mass and volume at moisture content *M*,
ρ_w = density of water,
W_o = oven-dry mass of wood,
V_M = volume of wood at moisture content *M*, and
M = moisture content of wood, percent, oven-dry mass basis.

3.4.2.2   *Volume Bases for Wood-Specific Gravity*

For practical reasons, reference data for wood-specific gravity are generally available for volume at either "green" condition (i.e., wood volume when cell walls are fully saturated with water, theoretically when $M \geq FSP$) or volume at 12% moisture content. Notice that we may readily convert from specific gravity, a unitless index, to density, mass per unit volume, by multiplying wood specific gravity by the density of water. When so doing, it is important to also report the moisture conditions measured or assumed for wood volume.

Tabulated data for wood specific gravity are given as G_{green} and G_{12}, which are read as follows:

G_{green} is the "green volume-base specific gravity of wood" or the "specific gravity of wood based on green wood volume." Oven-dry mass is always in the calculation of wood specific gravity, therefore, the moisture content at which the mass of wood is determined need not be stated; it is given by definition. Green volume is that volume associated with a moisture content that is greater than or equal to the fiber saturation point. Since wood volume is maximized at FSP, G_{green} will be the smallest G_M possible for any given wood sample, or

Table 3.5 Specific gravity of selected species of North American woods, based on green volume and volume at M = 12% (green volume is volume at M ≥ FSP).

Species	Specific gravity, G_M	
	G_{green}	G_{12}
Hardwoods		
Ash, white	0.55	0.60
Aspen, quaking	0.35	0.38
Cottonwood, Eastern	0.37	0.40
Hickory, Shagbark	0.64	0.72
Maple, red	0.49	0.54
Oak, Northern red	0.56	0.63
Yellow-poplar	0.40	0.42
Softwoods		
Douglas fir, interior North	0.45	0.48
Fir, balsam	0.33	0.35
Hemlock, Eastern	0.38	0.40
Pine, Eastern white	0.34	0.35
Pine, Southern yellow (loblolly)	0.47	0.51
Spruce, red	0.37	0.40
Redwood, young growth	0.34	0.35

Source: Reproduced from [3].

for the tabulated average for any given species. This value, G_{green}, is often termed the "basic specific gravity."

G_{12} is read as the "specific gravity of wood based on volume at 12% moisture content." This moisture-basis G is given in tabular data as it is a moisture content that approximates EMC of wood in a number of applications, particularly for use as framing lumber. Example data for wood specific gravity are shown in Table 3.5.

The specific gravity at any volume basis may be estimated if specific gravity at one moisture basis is known and the volumetric shrinkage is known at both moisture bases. The relationship is shown in Equation 3.15:

$$G_{M_2} = G_{M_1} \left(\frac{100 - S'_{v_{M_1}}}{100 - S'_{v_{M_2}}} \right)$$

(3.15)

where

G_{M_2} = specific gravity to be estimated at volume associated with moisture content M_2,
G_{M_1} = known specific gravity based on volume at moisture content M_1,
$S'_{v_{M_1}}$ = volumetric shrinkage, green to moisture content M_1, and
$S'_{v_{M_2}}$ = volumetric shrinkage, green to moisture content M_2.

3.4.2.3 *Specific Gravity of Solid Cell Wall Substance*

Interestingly, although the specific gravity (and hence, density) of wood varies significantly between tree species, it has been found that the cell wall specific gravity of woody plants is nearly constant. This is the specific gravity of the cell wall material without any voids, that is, the specific gravity of solid cell wall substance. The specific gravity of oven-dry wood solid cell wall substance, G_0^w, has been measured via fluid displacement by various researchers who report values ranging from 1.44 to 1.55 [7]. Polar and nonpolar fluids penetrate wall substance variously, so measurements are dependent upon the probe liquid used. In addition, there is some variation in density of bound versus free water, accounting for some of the variation seen in measures of wall specific gravity. Nevertheless, a value of G_0^w between 1.51 and 1.53 is typically assumed. Siau [7] used 1.53, and this is the value we will assume in this text. Since this value is approximately constant, the significant variation in wood specific gravity is due to variation in total porosity, or void volume of wood owing to the variability of wood anatomy between and within species.

3.4.2.4   Cell Wall Porosity

The specific gravity of wood is substantially less than that of solid cell wall substance due to the sum of all voids contained in wood. These voids include "macropores" (e.g., cell lumens and pit openings) and "micropores" or "microvoids" which are the tiny spaces within the cell wall itself (i.e., within and between cellulose, hemicelluloses, and lignin). The porosity or fractional void volume of wood (v_a) may be calculated as follows (Equation 3.16):

$$v_a = 1 - G_M \left(\frac{1}{G_0^w} + 0.01M \right) \tag{3.16}$$

where

G_M = specific gravity of wood based on volume at moisture content M,
G_0^w = oven-dry volume-base SG of solid cell wall substance, and
M = percent moisture content of wood, dry weight basis.

Alternatively,

$$v_a = 1 - G_M (0.653 + 0.01M) \tag{3.17}$$

if it is assumed $G_0^w = 1.53$.

The maximum moisture content of wood, M_{max}, may be estimated by Equation 3.18:

$$M_{max} = \left(\frac{100}{G_M} \right) - 65.3 \tag{3.18}$$

The size of voids in wood varies considerably. Macrovoids (cell lumens) typically range from 10 to >300 μm diameter with lengths of <0.5 to 3.5 mm. Pit openings are 0.02–4 μm in softwoods and 5–170 nm in hardwoods. Microvoids within the cell wall are approximately 0.3–60 nm, the smallest of which are the size of [some] nonpolar liquid or water molecules. Microvoid volume is 1.2–4.8% of the total void volume of wood [7]. It is interesting to note that water molecules are about 3 angstroms (0.3 nm) in diameter, with an angle, θ, between the two hydrogen atoms of about 105–109°. Lest we think, however, that water is always the same, a recent review shows 46 distinct models of water and its structural properties [18].

Microvoid volume has much to do with the enormous internal surface area within the cell wall. Internal surface area consists of "permanent" and "transient" surfaces, the latter created by the interaction with swelling solvents (polar liquids). Stamm and Millett [19] estimated the internal surface area of a 1 cm cube (6 cm^2 external surface area) of *Pinus lambertiana* (sugar pine) having a density of 0.36 g/cm^3. Permanent internal surface was calculated as 0.11 m^2 with transient $= 170$ m^2. Consider as a visualization exercise that 0.11 m^2 is 0.11 m $\times$ 1 m and 170 m^2 is 1 m $\times$ 170 m; one could think of the latter as a running lane for a competitive sprint on a running track. All of this surface area would be contained inside the 300,000 to 500,000 fibers making up that tiny 1 $\times$ 1 $\times$ 1 cm cube of wood. It makes one think of the micro- or nano-world in whole new light!

The point to be made in the foregoing discussion is that it is important to realize that moisture interaction in wood is a function of its combined chemical, supramolecular (ultrastructure) and microstructural (anatomical) architecture, each of which is truly remarkable. One of the joys of science is to contemplate the remarkable complexity of even the most seemingly simple things observed in everyday life.

3.5   Wood: A Cellular Solid

We have seen (Chapter 1) that cellular solids are substances "made up of an interconnected network of solid struts or plates which form the edges and faces of cells" [20]. Wood is inherently a natural cellular solid. One of the fundamental metrics useful in characterizing cellular solids is their relative density, the ratio of bulk material density to the density of the cell wall struts or plates defining the material. This ratio, ρ/ρ_s may be estimated for wood. In order to do so for a given wood species, we need two things: An estimate of wood density (ρ_{M_1,M_2}) and an estimate of the density of the cell wall material comprising the wood (ρ_s). Wood density values for many species of trees are readily available from published sources, thus, we may rely on tabulated data for this parameter, such as that summarized in Table 3.5. In Section 3.4.2.3, we noted that G_o^w is approximately 1.53. An estimate of density of wall material may be obtained by multiplying specific gravity by the density of water, with the caveats noted in the following section.

3.5.1   Relative Density of Wood

The literature values for the specific gravity of solid cell wall substance, G_o^w, are based on oven-dry mass and oven-dry volume. For the purpose of determining relative density, however, we need an estimate of wood cell wall density based on volume at either "green" condition or at 12% moisture content, given that these are the conditions for which reference data for whole wood is readily available. We shall select the 12% moisture content basis for the purpose of this discussion.

Since wood substance increases in volume with an increase in moisture content within the hygroscopic range, the specific gravity of solid cell wall substance, based on volume at 12% moisture content, G_{12}^w, will be a somewhat smaller number than that based on oven-dry volume (refer to Equation 3.14 to see why this is true). If we assume, for the purpose of the present discussion, $G_{12}^w = 1.50$ (as compared to $G_o^w = 1.53$), then we may further assume that the density of cell wall substance, ρ_s, will be 1.50 g/cm^3, based on oven-dry mass and volume at 12% moisture content, recalling that we may convert from the unitless index of specific gravity to density by simply multiplying by the density of water. Although the chemical composition of cell walls in woody- and nonwoody plants may differ substantially, the cell wall density of natural fibers throughout the plant kingdom is approximately constant at 1.5 g/cm^3 [21,22].

Table 3.6   *Selected physical properties of some commercially important woods of North America.*

Species	G_{12}	$\rho_{0,12}$ (g/cm^3)	Relative density[a] ($\rho_{0,12}/\rho_s$)	Compression parallel (kPa $\times$ 10^3)	Compression perpendicular (kPa $\times$ 10^3)
Ash, white	0.60	0.60	0.40	51.1	8
Aspen, quaking	0.38	0.38	0.25	29.3	2.6
Birch, yellow	0.62	0.62	0.41	56.3	6.7
Elm, American	0.50	0.50	0.33	38.1	4.8
Hickory, Shagbark	0.72	0.72	0.48	63.5	12.1
Maple, silver	0.47	0.47	0.31	36	5.1
Maple, sugar	0.63	0.63	0.42	54	10.1
Oak, Northern red	0.63	0.63	0.42	46.6	7
Oak, white	0.68	0.68	0.45	51.3	7.4
Douglas fir, coast	0.48	0.48	0.32	49.9	5.5
Fir, white	0.39	0.39	0.26	40	3.7
Larch, Western	0.52	0.52	0.35	52.5	6.4
Pine, Eastern white	0.35	0.35	0.23	33.1	3
Pine, loblolly	0.51	0.51	0.34	49.2	5.4
Pine, ponderosa	0.40	0.40	0.27	36.7	4
Pine, Western white	0.35	0.35	0.23	34.7	3.2
Redwood, old-growth	0.40	0.40	0.27	42.4	4.8
Redwood, young-growth	0.35	0.35	0.23	36	3.6
Spruce, white	0.36	0.36	0.24	35.7	3
Spruce, Sitka	0.40	0.40	0.27	38.7	4

Source: Values for G_{12} and compression strength are from the Reference 3.
[a] For the calculation of relative density, the density of solid cell wall substance, ρ_s, is assumed as 1.5 g/cm^3.

With ρ_s thus assumed as 1.5 g/cm^3, we may now get to the point of this discussion and easily calculate *relative density* using wood specific gravity from readily available reference tables, such as those found in the Wood Handbook [3]. For example, aspen, with a specific gravity $G_{12} = 0.38$ (see Table 3.5), hence, wood density ($\rho_{0,12}) = 0.38$ g/cm^3, has a relative density, ρ/ρ_s of $0.38/1.5 = 0.25$. Wood density and relative density are thus summarized in Table 3.6 for selected wood species.

Notice that because wall density is assumed constant regardless of species, the variations in relative density are due exclusively to variation in the porosity (i.e., cell lumen and micropore volume as a percent of bulk material volume) of the different species of wood. Figure 3.5 illustrates some of the structural variation of wood encountered at the anatomical level for woods with differing relative densities. This visual representation dramatically demonstrates the affect of wood anatomy on porosity and relative density.

3.6   Mechanical Properties

3.6.1   Compression Strength

There are many useful mechanical properties of wood that may be measured and applied to design and analysis of materials and structures made from wood and wood-based products. Among these are bending properties (chiefly moduli of elasticity and rupture), impact strength,

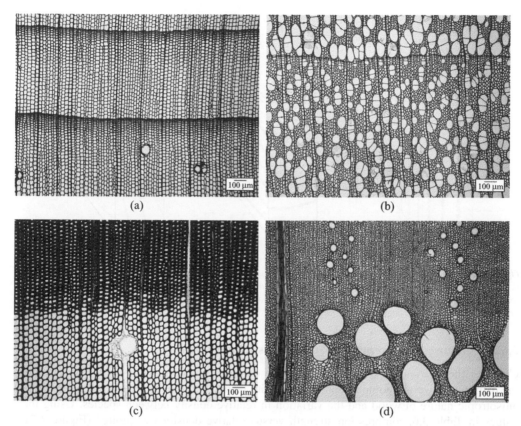

Figure 3.5 Cross-sections of (a) white spruce (Picea glauca), (b) aspen (Populus tremuloides), (c) loblolly pine (Pinus taeda), and (d) Northern red oak (Quercus rubra), with relative densities of 0.24, 0.25, 0.34, and 0.42, respectively. Since ρ_s is approximately constant between species, variation in relative density is ρ_s attributable to variation wood porosity, a function of wood anatomic structure. Light micrographs by D.D. Stokke.

shear strength, tension strength, side hardness, and compression strength. The structural composition of wood, from ultrastructural to anatomic levels, dictates that wood properties vary with direction of measurement, meaning that wood is an anisotropic material. This characteristic may illustrated by compression strength, as shown in Figure 3.6. Here, we see that compression is typically evaluated by testing of samples with three different orientations, longitudinal (Figure 3.6a, "along the grain") and in two transverse loading directions, radial and tangential (Figure 3.6b and c, "perpendicular to the grain"). Note that loading in the radial direction (Figure 3.6b) means that load is applied to the tangential face and that loading in the tangential direction results from load application to the radial face (Figure 3.6c). See Figure 2.31 for a refresher on the three principal planes of wood structure. Compression strength perpendicular to the grain is typically reported as a single number (e.g., Table 3.6), the average of radial and tangential loading results.

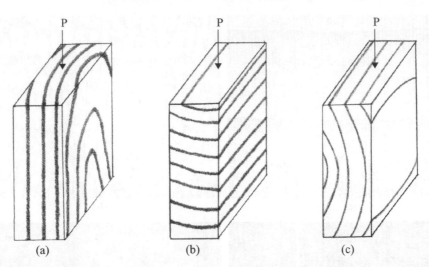

Figure 3.6 *Three principal modes of compression strength testing of wood, an anisotropic material. (a) Load, P, imposed parallel to the grain (longitudinal direction); and perpendicular to the grain, b and c. (b) Load in the radial direction, or 90° mode as assessed relative to annual ring orientation; (c) Load in the tangential direction, or 0° mode. Artist's renderings by Alexa Dostart.*

3.6.2 Compression Strength of Wood versus Relative Density

Consider the relationship of relative density to compression strength. In Table 3.6, observe the disparity in wood's compression strength with respect to grain direction, reflective of the anisotropic nature of wood and the variation in relative density between species. Using the values in Table 3.6, compression strength versus relative density was plotted (Figure 3.7). In this figure, a linear relationship of strength to relative density is observed, illustrating the importance of relative density as an indicator of properties in these natural cellular solids. Notice the similarity of this relationship to that shown in Figure 1.7, in which we observed the correlation of relative density to Young's modulus in some synthetic materials.

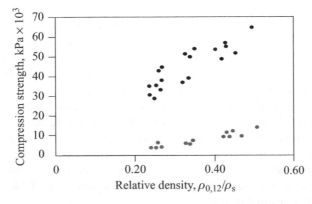

Figure 3.7 *Compression strength of wood parallel (•) and perpendicular (•) to the grain versus relative density for the wood species listed in Table 3.6.*

3.6.3   Mechanical Properties in Context

To place this discussion of compression strength versus relative density in a practical context relevant to the focus of this book, we will note that compression strength perpendicular to the grain is an important physical characteristic governing behavior of wood furnish during thermocompression processes used for manufacturing of the majority of adhesive-bonded wood composites. Relative density is thus an indicator of likely consolidation response of a given material.

Other mechanical properties, such as bending strength and stiffness, are also significant determinants of consolidation behavior of lignocellulosic materials. Whether the mode of material deformation is compression or bending, the fact that wood responds as a viscoelastic material is of primary importance, and will be discussed in Section 4.4, wherein viscoelasticity is treated in context of thermocompressive consolidation.

3.7   Wood Is the Exemplar: Extending Principles to Other Plant Materials

In this Chapter, we have focused on wood as a teaching example or exemplar representative of many of the characteristics of lignocellulosic materials in general. Specifically, our attention has been on moisture interactions, density and specific gravity, and the relationship of relative density to compression strength. The goal of seeking a basic comprehension of these concepts is to provide insight into material behavior and utilization of a variety of plant-based materials. Though the specific values for EMC, FSP, ρ, G, and strength will vary between different types of plant material or fiber, the essential concepts will prove useful in their understanding and application.

References

1. ASTM International. D-4442. Standard test methods for direct moisture content measurement of wood and wood-base materials. *Annual Book of ASTM Standards*. West Conshohocken, PA: ASTM; 2010.
2. Shmulsky R, Jones PD. *Forest Products and Wood Science, An Introduction*, 6th ed, Wiley-Blackwell. 2011.
3. USDA Forest Products Laboratory. *Wood Handbook: Wood as an Engineering Material*. FPL-GTR-190, Madison, WI: USDA Forest Products Laboratory; 2010. p 508.
4. Simpson WT, editor. *Dry Kiln Operator's Manual, Agriculture Handbook No. 188*. Madison, WI: US Department of Agriculture, Forest Service, Forest Products Laboratory; 1991.
5. Skaar C. *Water in Wood*. Syracuse, NY: Syracuse University Press; 1972.
6. Tiemann HD. Effect of moisture upon the strength and stiffness of wood, Bulletin 70. US Department of Agriculture, Forest Service; 1906 p 144.
7. Siau JF. Wood: Influence of moisture on physical properties. Blacksburg, VA: Department of Wood Science and Forest Products, Virginia Polytecnic Insitute and State University; 1995.
8. Stamm AJ. *Wood and Cellulose Science*. New York: The Ronald Press Company; 1964.
9. Hoffmeyer P, Engelund ET, Thygesen LG. Equilibrium moisture content (EMC) in Norway spruce during the first and second desorptions. *Holzforschung*, 2011;65: 875–882.
10. Stone JE, Scallan AM. The effect of component removal upon the porous structure of the cell wall of wood. II. Swelling in water and the fiber saturation point. *Tappi*, 1967;50: 496–501.
11. Telkki V-V, Yliniemi M, Jokisaari J. Moisture in softwoods: fiber saturation point, hydroxyl site content, and the amount of micropores as determined from NRM relaxation time distributions. *Holzforschung*, 2013;67: 291–300.

12. Bent HA. *The Second Law*. New York: Oxford; 1965.

13. Skaar C. In: Timell TE, editor. *Wood–Water Relations*. Berlin: Springer-Verlag; 1988.

14. Simpson WT. Sorption theories for wood. Symposium on Wood Moisture Content and Humidity Relationships, Virginia Tech University and US Forest Products Laboratory; 1979. pp 36–46.

15. VandenBerg C, Ruin S. Water activity and its estimation in food systems: theoretical aspects, In: Rockland LB, Stewart GF, editor. *Water Activity: Influences on Food Quality A Treatise on the Influence of Bound and Free Water on the Quality and Stability of Foods and Other Natural Products*. New York: Academic Press; 1981. pp 1–61.

16. Boquet R, Chirife J, Iglesias HA. Technical note: on the equivalence of isotherm equations. *International Journal of Food Science and Technology*, 1980;15(3):345–349.

17. Hailwood AJ, Horrobin S. Absorption of water by polymers in terms of a simple model. *Transactions of the Faraday Society*, 1946;42B:84–102.

18. Guillot B. A reappriasal of what we have learnt during three decades of computer simulations on water. *Journal of Molecular Liquids*, 2002;101:219–260.

19. Stamm AJ, Millett MA. Internal surface of cellulosic materials. *Journal of Physical Chemistry*, 1941;45:43–54.

20. Gibson LJ, Ashby MF. *Cellular Solids: Structure and Properties*, 2nd ed. Cambridge, UK: Cambridge University Press; 1997.

21. Gibson LJ, Ashby MF, Harley BA. *Cellular Materials in Nature and Medicine*. Cambridge, UK: Cambridge University Press; 2010.

22. Rowell RM. Natural fibres: types and properties. In: Pickering KL, editor. *Properties and Performance of Natural-Fibre Composites*. Cambridge, UK: Woodhead Publishing; 2008. pp 3–66.

4

Consolidation Behavior of Lignocellulosic Materials

4.1 Introduction

The majority of the reconstituted, adhesively bonded wood products produced today are formed under thermocompression processes in which a "mattress" of wood furnish (e.g., strands, flakes, particles, or fibers), blended with adhesive resins and other additives, is subjected to heat and pressure in the process of forming the final product. The application of heat and pressure to moisture-containing wood or other lignocellulosic material results in consolidation, or compaction and compression. Consolidation behavior is a term used to describe the combined effects of temperature, pressure, and moisture on the composition and structure of lignocellulosic material. Consolidation has a significant influence on the material properties of the resulting composites.

We have seen that the lignocellulosic cell wall is a multicomponent, organic polymer system. Since plant cell walls are composed of natural polymers, it is therefore logical to apply concepts of polymer theory to the understanding of how these materials behave during manufacturing processes and the resulting properties of wood- and fiber-based composite products. As an introduction to these concepts, we will consider some of the properties of synthetic polymers.

4.2 Synthetic Crystalline and Amorphous Polymers

Polymers are molecules that are composed of repeating chemical units. A synthetic polymer is one that is manufactured, or synthesized, by a chemical process devised by humans. Synthetic polymers play a significant role in our modern world. Often, we identify these polymers as "plastics." Plastics are used in a myriad of applications, including packaging and containers, toys, tools, automobile parts, computer housings, and sporting goods, to name just a few. The versatility of plastic polymers results from the ability to form and mold them, commonly under heat and pressure.

Introduction to Wood and Natural Fiber Composites, First Edition. Douglas D. Stokke, Qinglin Wu and Guangping Han.
© 2014 John Wiley & Sons, Ltd. Published 2014 by John Wiley & Sons, Ltd.

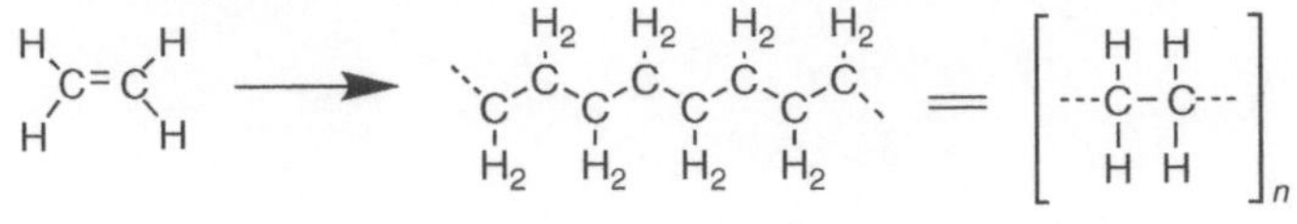

Figure 4.1   *Polymerization of ethylene monomers forms the thermoplastic polymer, polyethylene. The degree of polymerization is represented by* n.

In Chapter 2, we considered the polymer nature of the three major organic constituents of lignocellulosic materials, that is, cellulose, hemicelluloses, and lignins. These are natural polymers that are produced by the cells of organisms as part of their normal life processes. As polymers, these natural materials share similarities with synthetic plastics. We have seen that the structure of cellulose is both crystalline and amorphous, producing a significant influence on its physical characteristics. Hemicelluloses are likewise crystalline and amorphous, with lignin typically described as an amorphous polymer. Accordingly, we would like to expand on the concept of crystalline and amorphous polymers via a brief discussion of these attributes as expressed in some synthetic plastics, with a view toward the application of resultant polymer theory to natural polymers.

4.2.1   Polyethylene

Polyethylene is a ubiquitous synthetic plastic used in products of everyday life. Milk jugs, toys, and tool handles are, but a few, examples of products made from polyethylene. As a thermoplastic polymer, it melts at elevated temperature, allowing it to be easily formed and molded, remaining in its formed shape as a rigid material upon cooling.

Polyethylene is formed by the polymerization, or linkage, of many individual ethylene (or ethene) monomers (Figure 4.1).

Under the appropriate conditions of chemical synthesis, the ethylene molecules link end-to-end to form a long, straight chain polymer of polyethylene. In so doing, the ethylene loses its double bond, with the resulting polyethylene macromolecule having only single bonds along the carbon chain backbone. Given the linear structure, it is easy to imagine that the individual polymers can "stack" one on top of the other to form a closely packed, regular structure in the bulk material. This is, in fact, a reasonable representation of the structure of high-density polyethylene (HDPE), which is a crystalline polymer. Crystalline plastic polymers are optically opaque. At room temperature, HDPE is opaque, but as it is melted, it becomes clear, indicating that the individual macromolecules comprising the HDPE have separated from one another to some degree, thus disrupting the crystalline structure. As the crystalline structure is altered, the material becomes optically clear, but upon subsequent cooling, the material will once again become crystalline and opaque.

4.2.2   Polystyrene: Isotactic, Syndiotactic, and Atactic

Polystyrene is another synthetic polymer found in everyday, modern life. It is a hard, clear solid used in disposable drinking cups, CD cases, toys, computer housings, and other consumer products. It is also foamed to make insulation.

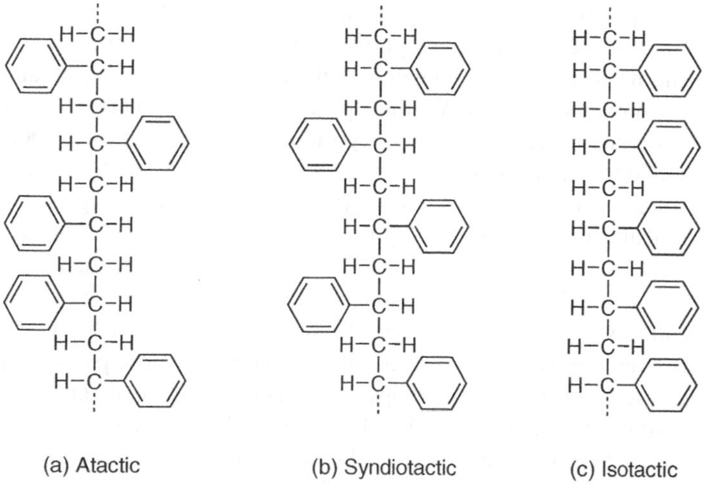

(a) Atactic   (b) Syndiotactic   (c) Isotactic

Figure 4.2   *(a) Atactic, (b) syndiotactic, and (c) isotactic polystyrene.*

The polystyrene found in commercial products is invariably of a type known as atactic polystyrene (Figure 4.2a).

Notice that polystyrene is composed of a linear carbon backbone, exactly like that found in polyethylene. The difference between polyethylene and polystyrene is obvious; at certain points along the carbon chain, hydrogen atoms are substituted with a phenyl group. In atactic polystyrene, the phenyl substituents are randomly distributed, such that the polystyrene macromolecules are unable to pack closely in the bulk material. As a result, the rigid polystyrene material is amorphous in structure, resulting in an optically transparent material.

Due to the relative ease of chemical synthesis, atactic polystyrene is, by far, the most common type of polystyrene in use. However, two other types may be formed, that is, syndiotactic and isotactic polystyrene (Figures 4.2b and 4.2c). In the syndiotactic variety, the phenyl substituents alternate along the carbon backbone, forming a semicrystalline, translucent material. The phenyl substituents are all arranged on one side of the backbone in the isotactic polystyrene, allowing the polymer chains to pack closely together in the bulk material. This results in a highly crystalline, opaque material. Of course, other material properties vary between the three types of polystyrene, but it is easiest to "visualize" the continuum of optical transparency to translucence to opacity as related to the molecular structure of the three varieties. Some physical properties of polymers tend to increase with degree of crystallinity. These include density, stiffness, strength, toughness, and heat resistance.

4.2.3   Degree of Crystallinity, Revisited

In a theoretical sense, the example of atactic and isotactic polystyrene present a nice, neat illustration of how crystalline or amorphous structure affects a tangible physical property of a polymer system. Our discussion thus far implies that a material may be completely crystalline or completely amorphous, or perhaps somewhere in between, as represented by the semicrystalline, syndiotactic polystyrene. In fact, it is well known that no polymer is completely crystalline, that is, no polymer has a degree of crystallinity equal to 100%. However, the relative amount of crystallinity (or lack thereof) is significant. We have already seen in Sections 2.2.3.2 and 2.2.3.4 that the natural organic polymer, cellulose, has both crystalline and amorphous regions as part of its ordinary structure. We thus speak of the degree of crystallinity of cellulose as a measure of its relative degree of order (crystalline nature) versus disorder (amorphous nature). Recall that cotton cellulose is in excess of 90% crystalline, with wood pulp cellulose in the range of 70–85%. The same is true of synthetic polymers. With regard to the optically opaque HDPE, typical degree of crystallinity is on the order of 92%, making it a (primarily) crystalline material. Though similar in basic chemical structure, low-density polyethylene (LDPE) is approximately 55% crystalline and optically translucent. Syndiotactic polystyrene may have a crystallinity on the order of 50% [1], whereas isotactic polystyrene approaches 100% crystallinity. Another way to express this idea is to say that those parts of a polymer that are not crystalline are amorphous, or perhaps somewhere in between (i.e., semicrystalline) in terms of structural order versus disorder. Therefore, when we consider the behavior of amorphous polymers when subjected to elevated temperature, we are, in essence, considering the effect of temperature on the "amorphous parts" of the polymer in question. We must also bear in mind that in some cases (HDPE, as an example), thermal softening and melting of the bulk polymer will change it from a largely crystalline solid to largely amorphous liquid or semiliquid, with a return to the largely crystalline solid form upon subsequent cooling.

4.2.4 Thermal Softening of Amorphous Polymers: Glass Transition Temperature, T_g

Empirical measurements on structure–property relationships and the response of synthetic amorphous polymers to heat, diluents (diluting agents), and other factors have led to development of polymer theory. One fundamental concept of polymer theory is that of the glass transition temperature associated with a specific type of polymer. The glass transition temperature, or T_g, of an amorphous polymer is the temperature at which a polymer undergoes a change in the slope of the specific volume versus temperature curve. More simply, this may be understood as the temperature at which a material changes from a glassy (brittle) state to a rubbery state. Accordingly, the T_g generally corresponds to an abrupt decrease in material stiffness, or modulus. If this is indeed the case, then T_g should be evident in a plot of modulus versus temperature, as shown in Figure 4.3 for an amorphous polymer.

Several features of this relationship are evident in the regions delineated by vertical lines in Figure 4.3. At relatively low temperature, a glassy region represents a state in which the material in question is brittle (glassy) and/or stiff (high modulus). The temperature at which this condition exists can vary widely between materials. Some materials must be cooled to very low temperatures, whereas others are glassy at room temperature. Still others remain glassy at extremely high temperatures. Regardless of absolute temperature involved, the idea here is that if temperature is increased, there will come a point in an amorphous polymer at which the material will begin to soften, corresponding to a decrease in modulus. The T_g is generally defined as slightly greater than the point of inflection between the glassy and transition regions on a plot of temperature versus modulus. T_g is a fundamental material property for any given polymer, and is a significant determinant of how that material responds to heating.

Further ramping of temperature above T_g results in a significant decrease in modulus, corresponding to a transition region, followed by a modulus-temperature plateau described

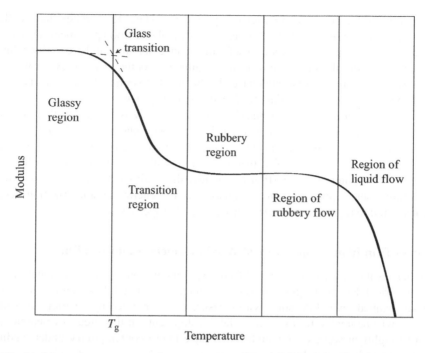

Figure 4.3 *Modulus versus temperature for an amorphous, uncrosslinked polymer. Note the location of the glass transition temperature, T_g. Adapted from [2] with permission of The Technical Association of the Pulp and Paper Industry © 1984.*

as a rubbery region. Rubbery and then liquid flow (melting) of the material results from greater temperature. Temperature increase beyond this point would cause thermal degradation (oxidation, pyrolysis/charring, burning). In the case of wood polymers, thermal degradation will likely occur before they exhibit liquid flow as depicted at the far right of Figure 4.3.

4.3 Glass Transition Temperature of Wood Polymers

The natural organic structural polymers comprising wood or other lignocellulosic materials may be understood through polymer theory. In particular, the concept of glass transition has direct application to the material properties of the polymers comprising plant cell walls. As a practical matter, lignocellulosic raw materials are routinely subjected to elevated heat and pressure during composite manufacturing processes. Thus, the T_g of wood or lignocellulosic polymers is of fundamental importance for the understanding of material behavior during composite consolidation. Throughout the following discussion, please continue to bear in mind that while wood will serve as our exemplar, the concepts presented may be also applied to understanding of other plant materials used as industrial feedstock.

Accordingly, we may ask if it is possible to evaluate T_g of wood polymers. The short answer is yes, with the caveat that those polymers must first be isolated from their native state. Though not trivial, it is possible to isolate wood polymers such that we may have reasonable confidence that the individual components (cellulose, hemicellulose, lignin) so acquired will be representative of the polymer in its relatively unaltered state. Researchers have collected empirical data or measurements on the T_g of chemically isolated wood polymers. This data has contributed to the development of models useful in the understanding and prediction of natural polymer behavior under the influence of heat and diluents. As the technology for measuring T_g has developed over time, researchers have also been able to conduct experiments on small solid wood specimens, resulting in data for *in situ* wood polymers.

The Latin term, *in situ*, means "in place" or not chemically isolated, in other words, the unmodified polymer(s) as found in wood that has not been chemically altered. It is often difficult to obtain data or information on the *in situ* polymers comprising complex natural materials, such as wood, due to the fact that many analytical techniques require at least some modification of the material in question to conduct the analysis. Nevertheless, in order to understand fundamental phenomenon, scientists seek to generate characterization data that may be considered to represent the *in situ* structure and properties.

4.3.1   Glass Transition Temperature of Wood Polymers: Empirical Data

Empirical data is the information derived from experimentation and measurement. The collection of empirical data on the glass transition temperatures of cellulose, hemicelluloses, and lignins have spanned several decades, with measurements obtained by a variety of methods. Although we will be unable to trace all of the developments in this area of research, it will be helpful to highlight some of the studies that have led to contemporary understanding of natural polymers in light of polymer theory.

4.3.1.1   Thermal Softening of Dry Wood Polymers

One of the pioneering studies concerning the thermal softening of cellulose, hemicellulose, and lignin was conducted by D.A.I. Goring in the early 1960s [3]. Materials studied included lignins from several wood species chemically isolated by various methods; three hemicellulose isolates; and several cellulose pulps. A device was built to contain a small sample of powdered material in a closed capillary tube, with a plunger to apply a compressive force to the sample in the tube. A method was devised to measure the force necessary for the plunger to compress a small amount of powdered material as it was heated in the closed tube. A key material parameter measured by this technique is the softening point. The softening point, T_s, was defined as the temperature at which the powder collapsed into a solid plug. Softening points for dry lignins and hemicelluloses, T_s^0 (where 0 represents dry samples), were in the range of 130–190°C, with higher softening temperatures observed for cellulose pulps. This "powder collapse" study and others like it generated empirical data on the thermal softening behavior of cellulose, hemicellulose, and lignin.

From the outset, it was recognized that T_s values obtained by the powder collapse method were an approximation of the glass transition temperature, T_g. Goring [3] stated, "It is proposed that T_s^0 as measured by the powder collapse method is within a few degrees of the glass transition temperature of the sample under study. It is further considered that the sorption of water or other solvents lowers the softening temperature in the same manner that a diluent will lower the glass transition temperature of an amorphous polymer." Accordingly, subsequent empirical studies quickly adopted the term, "glass transition temperature, T_g," in lieu of "softening point, T_s."

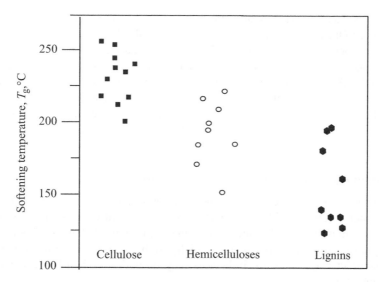

Figure 4.4   *Glass transition temperature of dry cellulose, hemicelluloses and lignins as determined by various researchers. Adapted from [4] with permission of The Technical Association of the Pulp and Paper Industry ©️ 1982.*

Back and Salmén [4] summarized a number of the early studies on thermal softening of wood polymers, referring to the empirical values as glass transition temperatures (Figure 4.4).

For each of the major organic constituents of wood, a range of T_g values was observed, owing to variation in the wood species or source, chemical isolation methods, and the techniques used to measure T_g. Nevertheless, it was evident from these studies that in general, the T_g of dry cellulose was greater than that of hemicellulose. This corresponds to our understanding of cellulose as a largely crystalline material (nevertheless having some amorphous regions), hemicellulose as a largely amorphous material (having some crystalline character), and lignin as an amorphous polymer.

Literature values for the T_g of cellulose ranged from 200°C to 250°C, with an assumed value of 220°C used for theoretical calculations [4]. The T_g of hemicellulose was summarized in the range of 150–220°C, with an assumed value of 180°C as the apparent mean T_g of hemicelluloses containing side groups (i.e., branched polymers) [4]. The T_g values of hemicelluloses were typically greater than those of lignins. However, the interpretation of lignin data was deemed somewhat more difficult in that it was expected that the cell wall fractionation process yielded a polymer with significantly lower molecular weight than the native, chemically unmodified *in situ* lignin. Thus, they assumed that the dry T_g of native lignin may be upward of 205°C. Despite the observed variation in the magnitude of temperature at which each of the major organic constituents of wood soften, it is evident that each of these natural polymers may be characterized by a glass transition.

The initial powder collapse methods for measuring softening temperature [3] soon gave way to more sophisticated instruments designed for measuring the properties of amorphous polymers. Thermal analysis instruments allow the study of material response under carefully controlled temperature conditions, that is, the temperature may be "ramped" (steadily increased) as material properties of interest are simultaneously measured and recorded.

Differential thermal analysis (DTA) and differential scanning calorimetry (DSC) are examples of this technology. Other methods allow the sample material to be subjected to various frequencies of vibrational or tortional energy. Given that temperature and frequency have similar effects on amorphous polymers, dynamic mechanical analysis (DMA) and dynamic mechanical thermal analysis (DMTA) represent additional advancement in the science and technology of polymer characterization. Although these more sophisticated instruments allowed for further study of the thermal properties of natural polymers and refinement of data, the fundamental findings of the early powder collapse methods were not refuted by them, but were, in essence, confirmed.

4.3.1.2   Diluent Effect of Moisture on T_g of Wood Polymers

The glass transition temperature indicates the polymer softening point with respect to temperature. Polymer softening within the plant cell wall may also be induced by moisture. Temperature and moisture actually interact to influence polymer behavior. This is important for hygroscopic materials such as wood, especially in the context of thermocompression in composite manufacturing. Moisture influences may be understood in the context of glass transition temperature.

A softening agent is a "diluent." Polar molecules such as water most effectively soften the largely hygroscopic structural components of wood or other lignocellulosic materials. In contrast, the diluents effective for most petrochemically derived synthetic polymers are nonpolar organic solvents. In either case, diluents effectively reduce the stiffness (modulus) of the materials on which they act. As a result, we should expect that addition of a diluent to a polymer system would lower the glass transition temperature. Another way to think of this is as a shift of the modulus versus temperature plot to the left.

Moisture sorption or uptake of water by the cell wall produces a diluent effect that is manifested as softening of the wall polymers. A schematic depicting the qualitative softening effects of moisture on the carbohydrate fraction of wood cell walls is shown in Figure 4.5.

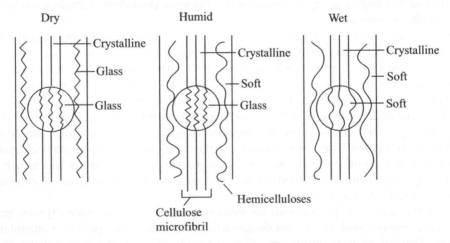

Figure 4.5   *Diluent effect of moisture on the carbohydrate fraction (cellulose and hemicellulose) of a lignocellulosic cell wall. Reproduced from [5] with permission of The Pulp and Paper Technical Association of Canada © 1985.*

The three moisture environments, represented as dry, humid, and wet, may be considered to produce equilibrium moisture content conditions as follows: Oven-dry or zero moisture content in the dry environment; moist in the humid environment, with the cell wall possibly having a moisture content somewhat below, but approaching the fiber saturation point (i.e., FSP, maximum bound or adsorbed water, but no free water); and moisture content above the fiber saturation point (containing free water) in the wet environment. In the dry condition, the cellulose and hemicellulose are glassy. In the intermediate, humid condition, the highly hygroscopic hemicelluloses become softened by the diluent effect of moisture, but it is assumed that neither the crystalline nor the amorphous regions of cellulose are so affected. Under wet conditions, the hemicelluloses remain softened, and the amorphous cellulose also becomes softened, that is, it assumes a rubbery state. In each of the conditions, crystalline cellulose is assumed to remain glassy. As the least hygroscopic component of the cell wall, lignin is omitted from this qualitative representation of moisture's diluent effect, even though it is known that moisture does have a limited softening effect on lignin.

4.3.1.3   *Diluent Effect of Moisture on T_g of the Carbohydrate Fraction*

Confining our discussion for the moment to the carbohydrate fraction of the cell wall, we find that empirical data (Figure 4.6) provide support for the qualitative model as just described (Figure 4.5).

This figure shows a steady decrease in T_g of hemicellulose from around 200°C at 0% moisture content to nearly 0°C at 30% moisture content. Amorphous or disordered cellulose is modeled with behavior similar to the hemicellulose component. This is in keeping with the prevailing theory of moisture sorption by cellulose occurring only in the amorphous

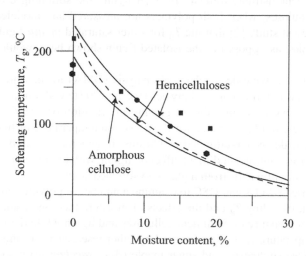

Figure 4.6   *Effect of moisture content on the softening temperature of xylan hemicellulose (solid lines) and the noncrystalline (amorphous) component of cellulose (dashed line). The two solid lines for hemicellulose were calculated by the method of Kaelble [6], assuming dry T_g of 220°C and 190°C (upper and lower lines, respectively). The circles, squares, and triangles represent empirical hemicellulose data from three different sources. Adapted from [4] with permission of The Technical Association of the Pulp and Paper Industry © 1982.*

regions, not in the crystallites. Using the same approach, Back and Salmén [4] modeled the relationship of cellulose softening versus temperature for degrees of crystallinity ranging from 0% (completely amorphous cellulose) to 80% (within the range for wood pulps cited in Section 2.2.3.4). However, they concluded that experimental data from various sources could not be reconciled in terms of demonstrating a glass transition for cellulose under moist conditions. Irvine [2] commented that although the T_g of cellulose in an aqueous system is "lowered substantially as the water content increases . . . [it is] difficult to determine the water content beyond which no further lowering of T_g occurs." Although amorphous regions exist within cellulose structure, it is presumed that the predominantly crystalline nature of cellulose precludes reliable modeling of its behavior under polymer theory applicable to amorphous polymers.

4.3.1.4   *Diluent Effect on T_g of Amorphous Hemicelluloses and Lignin*

Given the difficulties of modeling cellulose as an amorphous polymer, subsequent research focused on the effects of moisture content on glass transition of hemicelluloses and lignin, since these can be reliably modeled as amorphous polymers. Using a measurement technique known as differential thermal analysis (DTA), Irvine [2] studied the T_g of hemicelluloses from Eucalyptus and pine, as well as lignin from Eucalyptus (Figure 4.7). The T_g of dry hemicellulose was in the range of 160–180°C. Addition of moisture up to 20% of dry weight reduced the T_g of hemicelluloses to just above 40°C. Though not measured, it was specu-lated that water-saturated hemicellulose would have a T_g below room temperature. Dry milled wood lignin, with a T_g under 140°C, also exhibited a significant diluent effect of moisture, with the maximum softening (lowest T_g) plateauing at about 20% moisture content (Fig-ure 4.7). Contrary to these results, and as we shall observe in the following section, though water does have a demonstrable diluent effect on lignin, the softening effect is more sig-nificant for hemicelluloses, when both polymers are measured and modeled *in situ*. It was estimated in the present study [2] that the T_g for water saturated *in situ* lignin, that is, lignin in solid wood samples, as opposed to the isolated lignin shown here, would be in the range of 60–90°C.

Salmén [7] employed dynamic mechanical analysis (DMA) to measure the response of water-saturated solid wood to vibrations between 0.05 and 20 Hz with ramping of temperature from 20°C to 140°C, with the objective of gaining information on the T_g of *in situ* lignin. This study also suggested a broad range for the transition temperature of lignin, in the range of about 72–100°C under water-saturated conditions. In general, research shows that small amounts of water have a significant diluent effect on lignin, with broad temperature transitions reflective of the kinetics of a shift from a glassy to softened state.

Kelley et al. [8] used DMTA and DSC to examine a number of polymer properties of hemi-cellulose and lignin, including T_g and the effects of moisture thereon. Their work confirmed the diluent effects of moisture on both hemicelluloses and lignin while demonstrating transi-tions in similar temperature ranges as reported by other researchers. Using solid specimens of Sitka spruce (*Picea sitchensis*) and sugar maple (*Acer saccharum*) wood, they observed two transitions in both the DMTA and DSC scans, attributing them to the T_g of lignin and hemicellulose. One of the important findings reported in this paper was the ability to model the T_g of both lignin and hemicellulose by the Kwei equation.

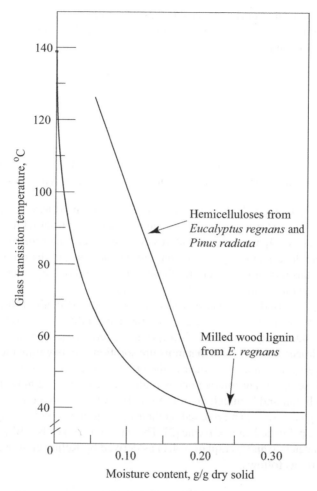

Figure 4.7 *Glass transition temperatures versus water content for hemicelluloses from a Eucalyptus and radiata pine and lignin from the Eucalypt. Extrapolation of the regression line representing hemicellulose gives a dry T_g of approximately 160°C. Note the significant diluent effect of moisture not only on the T_g of hemicellulose but also on that of lignin. Adapted from [2] with permission of The Technical Association of the Pulp and Paper Industry © 1984.*

4.3.2   Kwei Equation: Modeling T_g of Wood Polymers

Empirical data on physical phenomenon are valuable, in that their proper interpretation leads to insight into material properties and behavior. In addition, data may also be used to develop models to both explain observations and to predict material behavior under various conditions. For example, accurate modeling of changes in T_g of wood or wood components as moisture content is varied would be useful for the optimization of manufacturing processes. Within certain limits, the Kwei equation has proven to serve this purpose for the amorphous wood polymers, lignin, and hemicellulose.

The Kwei equation or model (Equation 4.1) was originally proposed as a means to model the T_g of a blend of two synthetic polymers capable of interacting with one another through the formation of hydrogen bonds [9]. The model is

$$T_g = \frac{(W_1 T_{g1} + W_2 T_{g2})}{(W_1 + kW_2)\,q\,W_1 W_2} \tag{4.1}$$

where

W_1 and W_2 represent the weight fractions of each of two polymer components;
T_{g1} and T_{g2} represent the glass transition temperature of each of the unblended polymers; and
k and q are experimentally derived parameters dependent on the polymer systems modeled.

This equation provided a good fit of experimental data for mixtures of certain synthetic polymers (specifically, mixtures of syndiotactic and isotactic poly(methyl methacrylate) with phenolic Novolac resins) [9]. Of particular interest, here is Kwei's interpretation of the qW_1W_2 term as representing the effect of hydrogen bonding interactions on glass transition temperature. The ability of this model to account for these important secondary interactions led to its application to the natural, organic, amorphous polymers of wood.

Kelley et al. [8] proposed the Kwei equation as an improved method to model moisture effects on hygroscopic wood polymers. They recognized that unlike other models based mainly on free volume considerations, this model incorporated a term to account for secondary interactions such as hydrogen bonds. Given that moisture sorption occurs within the lignocellulosic cell wall due to hydrogen bonding of water to the organic polymer constituents, a model that can account for these important interactions is highly logical. Accordingly, Kelley et al. [8] modeled the T_g of lignin and hemicellulose as a function of moisture content using the Kwei equation, with the parameters k and q based on their measurements on Sitka spruce and sugar maple wood and the published data of Irvine [2]. The interpretation of W_1, W_2, T_{g1}, T_{g2}, and the values for the parameters k and q (Equation 4.1) as derived by Kelley et al. were summarized by Wolcott et al. [10] as follows:

W = weight fraction for wood (W_1) and water (W_2);
T_{g1} and T_{g2} = dry glass transition temperature for wood polymers ($T_{g1} = 200°C$) and
       water ($T_{g2} = -137°C$);
k = adjustable parameter, 10 for lignin and 13 for hemicellulose; and
q = adjustable parameter, 585 for lignin and 355 for hemicellulose.

Figure 4.8 shows the effect of moisture content on the glass transition temperature of the amorphous structural polymers of the wood cell wall, lignin, and hemicellulose, as modeled by the Kwei equation (Equation 4.1) with the parameters listed above.

It is apparent from Figure 4.8 that softening of the amorphous polymers of wood occurs at moderate temperatures when the wood furnish moisture content is low. Notably, the majority of the reduction of T_g for both lignin and hemicellulose occurs below about 10% moisture content. This is significant in that the majority of the thermocompressive processes used to form adhesive-bonded wood-based composites are carried out with wood furnish moisture content below 10%, many of which are operated below 5%. Effective consolidation of wood materials to form composites is thus possible as a result of these phenomena.

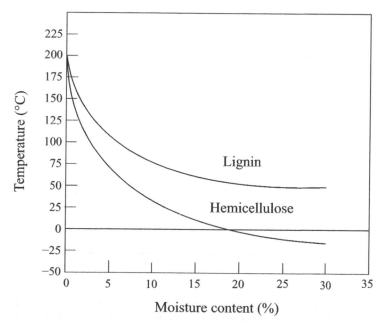

Figure 4.8 *Glass transition temperature versus moisture content for in situ wood lignin and hemicellulose as modeled by the Kwei equation. The diluent effect of moisture with respect to T_g is clearly demonstrable for both hemicellulose and lignin. Adapted from [10] with permission of the Society of Wood Science and Technology © 1990.*

4.4   Viscoelastic Behavior of Lignocellulosic Materials

4.4.1   Time–Temperature Superposition

The glass transition temperature is an important parameter that describes the behavior of thermoplastic, amorphous polymers when subjected to elevated temperature. This concept is relatively easy to understand in that we have everyday experience with softening or melting of materials as they are heated. Fundamentally, modulus/temperature relationship for amorphous polymers and the associated T_g are manifestations of the viscoelastic nature of such polymers. Interestingly, the same relationship exists between modulus and time. In Figure 4.9, we see that the "relaxation modulus" of an amorphous polymer is influenced by time in the same way as it is affected by temperature. Observe that at short time(s), the modulus is high. But, as time is extended, the modulus of the amorphous polymer decreases, showing the same net result as would be expected if the materials were heated. The fact that temperature and time have the similar effects on viscoelastic materials is called time–temperature superposition. As we have noted, diluents also affect modulus in a manner similar to temperature (and time).

The relationship depicted in Figure 4.9 is known as a "master curve" to describe an amorphous polymer as a viscoelastic material. A master curve is developed by collecting relaxation (or "creep") data over a series of temperatures, then using the time–temperature superposition principle to shift the information to span a wide range of log time. From an intuitive standpoint, the effects of temperature are typically easier to grasp than those of time. Nevertheless, it is important to recognize the similar effects of time and temperature. Viscoelastic materials

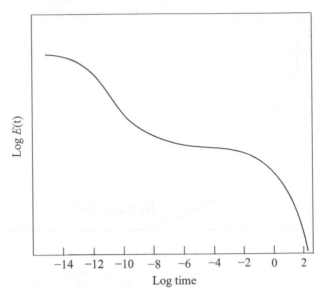

Figure 4.9   *Master curve for relaxation modulus, E(t), versus time for an amorphous polymer. Note the similarity of this curve to that of modulus versus temperature for an amorphous polymer, as shown in Figure 4.3. Adapted from [10] with permission of the Society of Wood Science and Technology © 1990.*

are elastic when the time span is short, but viscous when the time span is long. It is equally important to note that the definition of "short" and "long" with respect to time is dependent upon the T_g and diluent concentration.

4.4.2   Viscoelasticity in Mechanical Systems

An elastic material is one that recovers its size and shape after it has been deformed as a result of a mechanical load. A common example is the stretching of a rubber band. By pulling on the rubber band, a mechanical load is placed on it, causing the rubber band to stretch. When the rubber band is released, it snaps back, elastically, to its original size and shape. A viscous material is one that flows when placed under a load, resulting in a permanent deformation that is not recovered once the load is removed. A common example of this is taffy (a chewy candy) that is easily stretched, and remains so even after the force needed to stretch it is removed.

A viscoelastic material is one that is able to respond to a mechanical stress condition both viscously and elastically. Whether the response is viscous or elastic depends upon time, temperature, or diluent concentration, as well as the amount of imposed stress (force per unit area). Wood, as a composite of viscoelastic, amorphous polymers, exhibits such behavior. A plank of wood used as a bench may bend or deflect downward when someone sits on it, but will return to its original position once the person stands up, thus exhibiting elastic behavior. Viscous behavior is more difficult to observe under ordinary conditions, but its effects may be seen in old wood-framed structures. An example is the sagging roof of a building that has stood for decades or centuries. Over time, the natural wood polymers have undergone viscous flow under the imposed load of the roof, resulting in an observable permanent deformation.

4.4.3 Stress and Strain

Relevant to this discussion are the concepts of stress and strain. These are important parameters relative to the mechanical behavior of materials, defined as follows:

$$\text{stress} = \sigma = \frac{\text{force}}{\text{area}} = \frac{F}{A} \tag{4.2}$$

$$\text{strain} = \varepsilon = \text{unit deformation} \tag{4.3}$$

An example of the compression of a small wooden column will serve to illustrate these concepts.

Consider a solid column of sugar maple (*Acer saccharum*) wood, measuring 5.08 cm × 5.08 cm (2 in. × 2 in.) in cross-section by 20.32 cm (8 in.) long (Figure 4.10a). The cross-sectional area of such a sample, A, is 25.806 cm^2 (or 4 in.^{2}). These are the dimensions specified for a standard test of compression of wood parallel to the grain as given by ASTM D-143 [11].

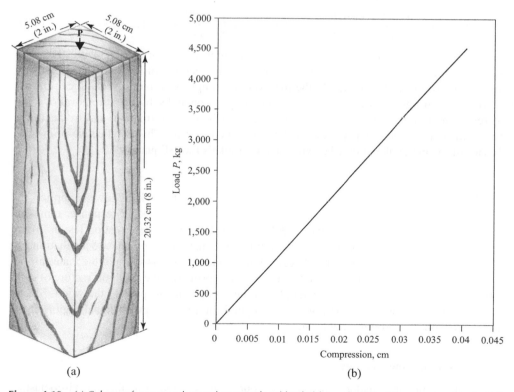

(a) (b)

Figure 4.10 *(a) Column of sugar maple wood prepared and loaded for evaluation of the compression strength parallel to the grain. Column drawn by Alexa Dostart. (b) Load versus compressive deformation for the column, using the example data described in the accompanying text.*

The column is placed on a firm surface (such as the table of a mechanical testing machine) and a uniformly distributed load or force, F, of 4,536 kg (10,000 pounds or lb) is applied to the top of the sample. The stress on the sample is given in Equation 4.4:

$$\sigma = \frac{F}{A} = \frac{4{,}536 \text{ kg}}{25.806 \text{ cm}^2} = 175.773 \text{ kg/cm}^2 (2500 \text{ lb/in.}^2 \text{ or psi}) \qquad (4.4)$$

Suppose that we measure the amount of compression (reduction of height or deformation) that occurs in the sample under this imposed stress, and find that the sample is compressed 0.04064 cm (0.016 in.) under the applied load of 4,536 kg. The strain is thus shown by Equation 4.5:

$$\varepsilon = \frac{\text{deformation}}{\text{original dimension}} = \frac{0.04064 \text{ cm}}{20.32 \text{ cm}} = 0.002 \text{ cm/cm} \qquad (4.5)$$

Note that strain is, by convention, reported as "centimeters per centimeter" (or inches per inch). The calculated strain in English units is 0.002 in./in.

The ratio of stress to strain provides the "modulus of elasticity in compression," which is also known as "Young's modulus" or Y (Equation 4.6). The word "modulus" means "measure of."

$$Y = \frac{\sigma}{\varepsilon} \qquad (4.6)$$

In our example, the Young's modulus is given in Equation 4.7:

$$Y = \frac{\sigma}{\varepsilon} = \frac{175.773 \text{ kg/cm}^2}{0.002 \text{ cm/cm}} = 87{,}886.5 \text{ kPa} \left(\frac{2{,}500 \text{ psi}}{0.002} = 1{,}250{,}000 \text{ psi} \right) \qquad (4.7)$$

This example of the compression of a wooden column serves primarily to illustrate the important concepts of stress and strain. Our plot of load versus compression is analogous to stress versus strain. In Figure 4.10b, the load versus compression plot reveals linear-elastic behavior. The straight regression line tells us that imposition and subsequent removal of load will result in complete recovery of compression, that is, the deformation, and by analogy, strain, will return to zero when load (stress) is returned to zero. This is reflective of elastic behavior of the material and may be expressed as Hooke's Law (Equation 4.8):

$$\sigma = \kappa \varepsilon \qquad (4.8)$$

Young's Modulus (Equation 4.6) is clearly a specific case of Hooke's Law as applied to compression of an elastic material. In the case of an elastic, Hookean response, removal of the load does not result in permanent deformation. Permanent deformation results from imposition of stress that exceeds the elastic limit of the material. Permanent deformation or damage occurs due to mechanical breakage, collapse, and/or viscous strain of the material. In the case of the maple column, permanent deformation is primarily the result of fracture of cell walls and compression of the bulk material into the porous lumen space of the wood. Since permanent strain occurs at stress/strain ratios beyond the linear or proportional region, it is said to be beyond the proportional limit.

4.4.4   A Trampoline Analogy

Perhaps the following analogy will further our intuitive understanding of viscoelastic responses. Let us think, for a moment, of a trampoline mat. When you step onto the mat, the trampoline deflects downward. The deflection represents an induced strain, ε, as a result of the force of your body exerted on the mat (P) over the sum of the area (A) of the bottom of your feet (imposed stress, $\sigma = P/A$). When you step off of the mat, thus removing the imposed stress, the trampoline returns to its original shape and position, meaning that the induced strain (ε) is completely, or elastically, recoverable.

A viscoelastic material also exhibits time-dependent deformation, or creep. Using our trampoline analogy again, consider what we might observe if we placed a long-term stress on it. To observe this effect, we place a large rock on the trampoline and just leave it there for a long time. We will assume that the mass of this rock is such that if we were to simply toss it on the trampoline, the rock would bounce up and the trampoline would return to its original shape and position. In fact, we will assume that this rock weighs exactly the same as you do, and furthermore, it has the same cross-sectional "footprint" that you do. In other words, if we placed the rock onto the trampoline and then quickly removed it, the response to the stress exerted by the rock would be completely elastic. But let us say we leave that rock stationary on the trampoline for a year. After sitting there one year, we remove the rock and find that the trampoline mat is now sagging down; it will not snap back up to its original, unloaded shape and position. What happened? The trampoline mat has undergone time-dependent deformation, also known as creep, resulting in a permanent deformation or set. This so-called creep behavior is characteristic of a viscoelastic material, attributable to molecular "slippage" of the polymers comprising the material as it is subjected to an ongoing stress. A time-dependent phenomenon such as this is called a rheological property.

Thinking back to the idea of time–temperature superposition for viscoelastic, amorphous polymers, we could extend the analogy further. Suppose that rather than leaving the rock on the trampoline for a year, we simply heat the trampoline somewhat above the glass transition temperature of the material from which it is made, causing it to soften. The net result we see of this heating—sagging or permanent deformation—is observed in less than a day, but is identical to that caused by a year of mechanical load. The net result is the same; time–temperature superposition yields identical observations.

4.4.5 Hysteresis

Viscoelastic materials are also characterized by hysteresis, that is, a "lag effect." This may be illustrated by considering the contrast between the stress–strain behavior of a completely elastic material and one that is viscoelastic, as illustrated in Figure 4.11. In the elastic material, stress and strain are proportional. The viscoelastic material, however, exhibits a hysteresis loop in its stress–strain behavior, meaning that the amount of strain induced by a given amount of stress is not constant, rather, it is dependent on whether the stress is being increased or decreased as strain is measured.

4.4.6 A Classic Model of Viscoelastic Stress Relaxation: Maxwell Body

Our discussion of glass transition, time–temperature superposition, and viscoelasticity has been intentionally presented in an inductive fashion. Using this approach, our aim is to lead the reader through a logical progression of concepts, all of which are intended to build an understanding of viscoelastic polymer systems. In so doing, we have injected at least one unconventional analogy (the trampoline) to support our inductive approach to the topic. At this point, we wish to introduce a more conventional conception of viscoelastic behavior as

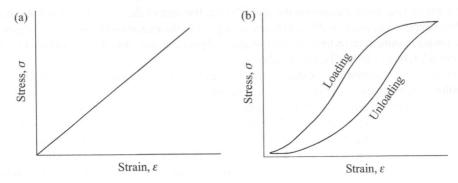

Figure 4.11 *Stress–strain diagrams for (a) purely elastic and (b) viscoelastic materials. Note the hysteresis loop in (b).*

represented by so-called spring and dashpot models. For those readers or instructors desiring additional background on the mathematical derivation or analysis of such models, any of a number of texts or references on the mechanics of materials or viscoelasticity of polymers may be consulted. Some specific examples include works by Brinson and Brinson [12], Lin [13], and Shaw and MacKnight [14].

Spring and dashpot models are mechanical analogues that are useful for understanding the behavior of materials. The two important elements in these models are "Hookean springs" and "Newtonian dashpots." A Hookean spring is an element (think of an ordinary spring) that responds to stress according to Hooke's Law, $\sigma = \kappa\varepsilon$ (Equation 4.8). In this case, σ and ε are analogous to the spring force and displacement, respectively, and the spring constant, κ, is analogous to Young's modulus (i.e., κ defines the stiffness of the spring). A Newtonian dashpot is an element that contains a viscous fluid, thus responding to a stress with a delayed or "damping" action. This may be visualized by thinking of a hydraulic shock absorber in an automobile, or a pneumatic door closer. The Newtonian dashpot is modeled as follows (Equation 4.9):

$$\sigma = \eta\dot\varepsilon \tag{4.9}$$

where σ is a strain rate (i.e., it is differentiated with respect to time), η is the viscosity of the fluid in the dashpot, and $\dot\varepsilon$ is a strain that is time dependent (as indicated by the dot above).

A Maxwell model or body consists of a Hookean spring and a Newtonian dashpot connected together in series (Figure 4.12). The Maxwell body mimics certain aspects of the response of a viscoelastic polymer to applied stress. The spring obviously responds elastically, as an analog to the behavior of many polymers under short-term instantaneous loading, including those in the lignocellulosic cell wall. The viscous response of the dashpot represents a time-dependent strain response that may be interpreted as molecular uncoiling or delayed movement within a complex polymer system. This, too, is a response observed in the lignocellulosic cell wall.

Within the Maxwell body, the ratio of viscosity to stiffness may be to represented as follows (Equation 4.10):

$$\tau = \frac{\eta}{\kappa}\sqrt{a^2 + b^2} \tag{4.10}$$

where τ has units of time.

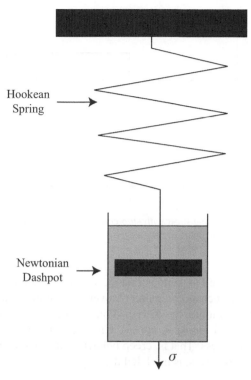

Figure 4.12　*Maxwell Model consisting of a Hookean Spring and a Newtonian dashpot.*

The stress on the body is equal to the stress in either the spring or the dashpot, that is, $\sigma = \sigma_s = \sigma_d$. The total strain is the sum of the strain in the spring and dashpot, that is, $\varepsilon = \varepsilon_s + \varepsilon_d$. It therefore follows from Equations 4.8 and 4.9:

$$\dot{\varepsilon} = \dot{\varepsilon}_s + \dot{\varepsilon}_d = \frac{\dot{\sigma}}{\kappa} + \frac{\sigma}{\eta} \tag{4.11}$$

Then, multiplying by κ and substituting with Equation 4.10:

$$\kappa \dot{\varepsilon} = \dot{\sigma} + \frac{1}{\tau}\sigma \tag{4.12}$$

If the strain is held constant, "stress relaxation" will be observed, that is, step-constant strain will result in stress that decreases with time, t. In this case, $\dot{\varepsilon} = 0$ and it can be shown that

$$\sigma(t) = \sigma_0 \exp\left(\frac{-t}{\tau}\right) \tag{4.13}$$

A significant observation resulting from Equation 4.13 is shown in Figure 4.13. Here, we have a graphic representation of stress relaxation over time (b), under step-constant strain (a). This tells us that the stress within a viscoelastic material will decrease with time when the strain is held at a constant level. This is, as we shall see, the situation when lignocellulosic material is compressed to a constant thickness during hot-pressing operations.

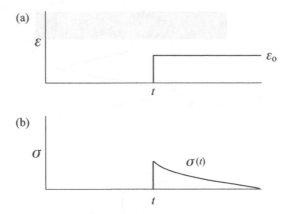

Figure 4.13   Stress relaxation: decreasing stress with step-constant strain in a Maxwell model of a viscoelastic material. (a) Strain, ε, versus time, t. (b) Stress, σ, versus t.

The Maxwell model serves well to demonstrate the stress relaxation behavior of viscoelastic materials. Other phenomena, such as creep and recovery, are more adequately represented by more complex models, for example, the Kelvin–Voigt model, which consists of a Maxwell body in parallel with a second Hookean spring. Characteristic of this model is that strain increases with step-constant stress. That is, creep behavior, or time-dependent deformation or strain under a constant load or stress, is modeled thereby.

4.4.7   Lignocellulosic Materials Are Viscoelastic

We may now summarize the attributes of viscoelastic materials: A viscoelastic material is one in which hysteresis is observed in a stress–strain diagram; stress relaxation occurs, characterized by decreasing stress resulting from step-constant strain; and in which creep occurs, that is, step-constant stress results in increasing strain. All of these observations are applicable to wood specifically and to lignocellulosic materials in general.

The viscoelasticity of lignocellulosic material is attributable to the multicomponent polymer system comprising the cellular solid. These attributes are especially manifested in thermocompression molding or manufacturing processes. In the following sections, we will seek to apply our understanding of viscoelastic behavior and thermal softening of cell wall polymers to manufacturing processes and resultant properties of the manufactured product. As a prelude to this discussion, we must first take a look at the basic mechanisms of heat and mass transfer.

4.5   Heat and Mass Transfer

4.5.1   Hot Pressing Parameters

Consider a hot pressing operation in which wood or lignocellulosic furnish (raw material) is blended with adhesive resin, formed into a "mattress" or "mat" of fibers, particles, flakes, strands, or veneer, and then subjected to heat and pressure to produce a densified panel product (Figure 4.14).

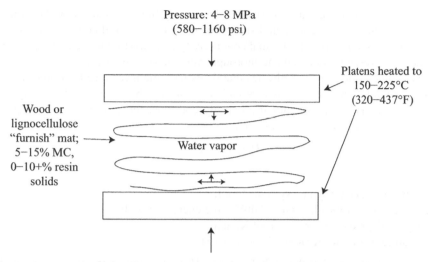

Figure 4.14 *Representative conditions for hot pressing of adhesive-bonded wood-based panel products.*

Typical industrial ranges for panel material and process conditions are as follows: Wood furnish at 5–15% mc with 0–10% adhesive resin solids, temperature of 150–225°C (302–437°F), pressure of 4–8 MPa (580–1160 psi), time 60–240 seconds. There is a caveat here, recognizing that most modern operations are not carried out in stationary platen batch presses as illustrated, but in continuous presses; nevertheless, this serves to illustrate the principles involved. In such a scenario, heat is first transferred from the press platens to the wood material (furnish) by conductive and convective mechanisms. Moisture at the surfaces of the furnish in contact with the platens is vaporized and driven toward the center of the mat, as well as toward the edges. Vaporization of moisture and simultaneous application of mechanical pressure due to closing of the press platens results in increased internal pressure on the mat. This manufacturing operation thus imposes transient conditions (i.e., changing with time) of temperature, moisture content, and pressure on the lignocellulosic furnish. The furnish, understood fundamentally as a natural, polymeric material, undergoes softening and compression or compaction, as cellulose, hemicellulose, and lignin respond to these conditions. Application of heat is thus important to allow formation of the final product. Heat is also necessary to induce polymerization or cure of the thermosetting adhesive resins typically used by industry. The entire process of thermomechanical softening and compaction of the wood furnish is described as consolidation. This process has also been termed viscoelastic thermal compression (VTC) [15–17].

4.5.2   Thermodynamics 101

In the preceding pages, we have discussed viscoelasticity, a subject that could readily fill a lifetime of work, in very rudimentary terms. Nevertheless, we hope this approach has been instructive and possibly a catalyst for further study. In similar fashion, following are some basic concepts relative to heat and mass transfer mechanisms as important fundamentals to the internal conditions within a mattress of wood furnish undergoing heating and compression. The concepts introduced here in somewhat cursory fashion are often the subject of entire courses or even careers of research. Perusal of University library holdings on the subject of thermodynamics and the closely related subject of mass transfer will confirm the enormous scope of the topic. With respect to thermal conductivity and moisture movement in wood, the text by Siau [18] presents a detailed and thorough, yet concise, treatment.

Thermodynamics is a branch of physics that deals with the mechanical action or relations of heat. Underlying many of the concepts embodied within the subject of thermodynamics are the fundamental principles of enthalpy, entropy, and free energy. The relationship of these parameters is expressed as follows (Equation 4.14):

$$H = G + TS \qquad (4.14)$$

where

H = enthalpy, or total energy in a system, J/mol;
G = free energy (sometimes known as Gibb's free energy), J/mol;
S = entropy, or degree of disorder, J/mol K; and
T = temperature in Kelvin, K (K = °C + 273.15).

Equation 4.14 informs us that the enthalpy term, H, is the sum of the free energy and the product of temperature and entropy. Free energy is understood as the difference between the total internal energy (enthalpy) and the unavailable energy, which is equal to the product of entropy and Kelvin temperature, that is, $G = H - TS$.

The entropy term, S, is a measure of "disorder" that always increases with temperature due to decrease in molecular order. One way to conceptualize entropy is to consider the difference between liquid water and water vapor. The water molecules in vapor are spaced further apart and undergo more random collisions with other molecules in the vapor state, when compared to the spacing and motion of water in liquid form. Similarly, ice has a higher degree of order than does liquid water, that is, ice has a lower entropy (less disorder) than liquid water.

Enthalpy may, in certain cases, consist of sensible heat, which is the product of specific heat and temperature difference. Specific heat, c, is a material property defined as the amount of energy required to raise the temperature of a given mass of material by one degree. If the material in question is raised multiple degrees of temperature, then the temperature difference becomes important in determining total energy, H. If, however, a change in phase of matter is involved (i.e., frozen to liquid or liquid to vapor), then enthalpy will consist of heats of fusion, sorption, and vaporization.

4.5.3   Thermodynamics of Water

4.5.3.1   Heats of Vaporization, Fusion, and Sublimation

The heat relations of water are necessarily governed by the relationship of H, G, T, and S. When water is heated or cooled, intuitively, we know that the "total energy in the system" will necessarily change. If a given mass of water is to be heated, we may consider that mass of water to be the system in question. If the water is simply warmed, the total energy will be expressed as the sensible heat, which is the product of the specific heat of water, $c = 4,185$ J/kg K at 15°C, and the temperature difference. That is, the sensible heat is the energy consumed or liberated to change the temperature of a given amount of a substance. But, if that water is to be evaporated, then there must be additional energy supplied to change the phase of the water from liquid to vapor.

In order to evaporate water, energy must be supplied by an increase in temperature or a decrease in pressure. The heat energy needed to evaporate water is called the latent heat of vaporization, Q_o. Latent heat is the energy consumed or liberated to change the phase of a substance. If the phase change is from liquid to vapor, then Q_o is sufficient to effect the phase change. If, however, the phase change is from solid (ice) to vapor, then the heat of fusion, Q_f, must be added to Q_o. The heat of fusion is the energy required to melt a given mass of ice. The sum of Q_f and Q_o is the heat of sublimation, Q_u.

Given that many hot-pressing manufacturing processes are operated at or above the boiling point of water, it is easy to understand why we should be cognizant of the heat quantity associated with evaporation, namely, Q_o. However, one may logically ask why the energy associated with a phase change from ice to liquid is part of this discussion. The reason for including Q_f has to do with the analogous behavior of ice and bound water in the lignocellulosic cell wall. Specifically, the heat of fusion for ice, Q_f, is an interesting analog to the quantity of energy needed to liberate bound water from the cell wall, the differential heat of sorption, Q_s.

4.5.3.2   *Bound Water: Differential Heat of Sorption*

Bound water, also known as adsorbed or sorbed water, is held by hydrogen bonds to the sorption sites (hydroxyl groups) that are intrinsic to the natural polymers of the lignocellulosic cell wall. Bound water exists in the hygroscopic range of "zero" percent moisture content (oven dry) to about 30%, with this upper limit representing the fiber saturation point. Although the energy states of water were discussed in Section 3.3.3, it will nonetheless be instructive to revisit these ideas here, and to build upon them with respect to rudimentary thermodynamics and the context of hot-pressing operations.

Bound water is held within the cell wall in a state analogous to frozen water in that the entropy (degree of disorder) of bound water is decreased relative to liquid water or water vapor, just as ice is less disordered (or, we might say, more highly ordered) than the other two states of matter. Liberation of bound water molecules from their hydrogen-bonded state requires an input of energy, just as the melting of ice requires energy. The amount of energy required to liberate bound water is a variable quantity, dependent upon the moisture content. The energy varies with wood moisture content, M, because at very low M, water is theoretically sorbed in a monomolecular layer. If this is the case, then the adsorbed water molecules are held, one-to-one, via hydrogen bonding directly to the cell wall substrate. Water molecules sorbed directly onto the wood polymers are tightly held. As moisture content increases, additional bound water is added by building up multilayer sorption as water molecules successively hydrogen bond to other water molecules. These additional layers of water are not directly sorbed onto the cell wall, and are therefore held by weaker bonds. Thus, as M increases, some of the water molecules are less tightly bound than others, with the result that less energy is required to liberate some of the water molecules from their hydrogen bound condition. This variation in energy required to remove bound water is known as the differential heat of sorption, Q_s.

Figure 4.15 graphically illustrates the analogy between water in the solid, liquid, and vapor phases, with water in wood, that is, bound, free (liquid), and vapor. Free water, which is liquid water held in the capillary space of cells (lumens, pit chambers, etc.) and water vapor in capillary space behave just like ordinary water. Bound water is analogous to ice in that it has lower entropy than liquid or vapor. However, unlike the constant value of Q_f, the heat of fusion of ice, the heat of sorption for bound water, Q_s, is variable. The analogy of Q_f and Q_s is

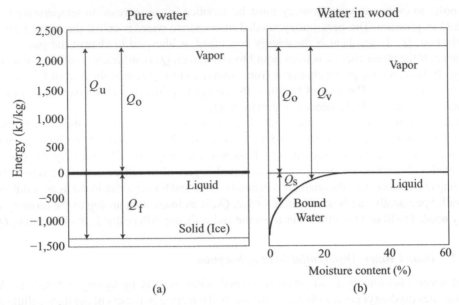

Figure 4.15   *Relative energy levels for various states of water: (a) pure water; (b) water contained in the woody cell wall. Q_o = heat of vaporization, Q_u = heat of sublimation, Q_f = heat of fusion, Q_v = heat of sorption, Q_s = differential heat of sorption. Reproduced from [19] with permission of Springer Science + Business Media © 1988.*

apparent in Figure 4.15, just as their difference is apparent. Note that the total energy to move bound water to vapor is the heat of sorption, Q_v, which is the sum of Q_o and Q_s. Q_v is clearly analogous to the heat of sublimation, Q_u, of "ordinary" water.

Because most industrial wood-based composites processes are operated with wood moisture contents in the lower half of the hygroscopic range, the quantities Q_s and Q_v are significant. Since heat transfer from the surface to the core of a furnish mat is largely dependent upon mass transfer and condensation of water vapor (steam), it is apparent that process energy required to effect this form of heat transfer will increase as mat moisture content decreases due, in part, to the differential heat of sorption, Q_s. That is, more energy is required to move water from the bound water state to the vapor state as the overall wood moisture content decreases. Water in the vapor state is the primary means by which mass transfer, and consequently convective heat transfer, occurs within the consolidating mat. As water vapor moves through the mat, it periodically condenses and in so doing, provides additional and significant amounts of heat transfer.

4.5.4   Mass Transfer: Moisture Vapor Movement

The fundamental driving force for moisture movement is a free energy gradient; moisture always moves to a region with lower free energy. This may be manifested as hydrodynamic flow, capillary flow, water vapor diffusion, or bound water diffusion. Mass flow of free water (via either hydrodynamic or capillary flow) is not generally considered relevant to wood composites manufacturing, since the furnish moisture content is typically well below the fiber saturation point. Although some liquid water may be present in a wood furnish mat (typically as the carrier droplets for adhesive resins), most water within the wood will be in the form of water vapor or water bound to the cell walls by hydrogen bonding. Thus, mass transfer within the wood substrate is limited by the rates of diffusion for these two "states" of water. During typical wood drying processes, bound water diffusion is the limiting factor with regard to moisture movement, as it is the slowest means by which water may move through wood. In hot pressing operations, bound water diffusion is not a significant mechanism of mass transfer [20]. Bound water diffusion is too slow a process to occur appreciably in the time intervals afforded industrial hot pressing operations.

Assuming a wood furnish mat consisting of discrete fibers, particles, strands, veneers, or the like, the movement of moisture between said particles is driven by a pressure gradient in that process temperatures are (typically) sufficient to produce evaporation of moisture from the wood, thus creating steam within the mat. Since the surface of the mat in contact with heated platens is the first to heat up, steam pressure drives moisture toward the center of the mat. Heat transfer accompanies the mass transfer of steam from surface to core.

4.5.5   Heat Transfer: Conduction

Heat transfer within or between materials may occur by one or more of three mechanisms, that is, conduction, convection, and radiation. In the context of heating and consolidation of wood materials in a hot press, radiation is insignificant relative to the first two means of heat transfer [20,21]. Conduction involves the transfer of heat energy from one body of material to another, as in the heat transferred directly from a heated metal platen in contact with a mattress of wood materials ready for consolidation into a composite. Heat transfer results from a difference in temperature, or temperature gradient, between the platen and the furnish. Heat is likewise transferred from discrete particles of furnish in contact with one another (Figure 4.16a).

Conduction is represented by Fourier's First Law (Equation 4.15) as:

$$q = -k\Delta T \tag{4.15}$$

where

q = heat flux or flow;
k = thermal conductivity, a material property; and
ΔT = temperature difference between two adjacent bodies.

In the case of heat transfer between a heated press platen and a mat of wood furnish, this may be represented (Equation 4.16) as shown by Kamke [22]:

$$\dot{q}_z = U \left(T_{\text{platen}} - T_{\text{material surface}} \right) \tag{4.16}$$

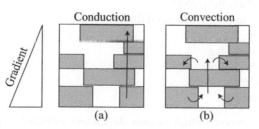

Figure 4.16   *Heat transfer in a hot press: (a) conduction; (b) convection. Radiation is omitted as it is considered insignificant in this instance.*

where

$\dot{q}_z$ = heat flux in the z (thickness) direction, J/s/m^2;
U = surface conduction heat transfer coefficient, J/s/°C/m^2; and
T = temperature, °C.

Heat conduction within the mat of wood furnish occurs from discrete wood element (particle, strand, fiber, etc.) to wood element. Heating through the thickness of the mat is crucial, and may be represented by the following form of Fourier's First Law (Equation 4.17):

$$\dot{q}_z = -k \left(\frac{dT}{dz} \right) \tag{4.17}$$

where

k_z = thermal conductivity of the mat in the z (thickness) direction, J/s/°C/m; and
z = distance in Cartesian coordinates, m.

The thermal conductivity of the mat, k_z, is dynamic during the pressing operation as a result of transient (time-dependent) conditions of moisture content and material density, the latter increasing as compaction pressure increases.

4.5.6   Heat Transfer: Convection

Within a consolidating mat, "gas convection is the main mass transfer mechanism... Gas convection is the bulk flow of the gas mixture along a total gas pressure gradient... Convective heat transfer is associated with convective gas flow through the void system of the mat, in combination with phase change of water" [23]. The main or primary mass transfer mechanism of gas convection is modeled according to Darcy's law, which includes a term to allow for heat transfer due to condensation of water vapor onto wood particle surfaces within the mat, that is, h_g, the enthalpy of the vapor. Condensation involves a substantial change in enthalpy. The latter component of heat transfer (condensation of vapor) may be more than ten times greater than that associated with convection alone under circumstances representative of internal mat conditions. Darcy's Law may be represented (Equation 4.18) as shown by Kamke [22]:

$$\dot{E}_{bulk} = \left(-\frac{K_z MP}{n_g RT} \right) \left(\frac{dP}{dz} \right) h_g \tag{4.18}$$

where

$\dot{E}_{bulk}$ = heat flux due to bulk flow, J/m^2/s;
K_z = gas permeability in the z (thickness) direction, m^2;
P = total gas pressure, Pa;
h_g = enthalpy of the gas, J/kg;
M = molecular weight, kg/mol;
n_g = viscosity of gas, Pa·s; and
R = gas constant, J/mol/K.

Heat transfer within the wood furnish is thus dependent primarily on the mechanisms of conduction and convection (refer again to Figure 4.16). Internal heating, combined with the diluent, or softening effect of moisture (vapor; steam) on the amorphous polymers comprising the wood, are important determinants of consolidation behavior during hot pressing. Heating throughout the mat is also necessary to allow for cure of thermosetting adhesive resins.

4.5.7   Internal Mat Conditions

4.5.7.1   *Temperature and Gas Pressure Within a Mat*

Internal conditions in a mat of wood or lignocellulose furnish undergoing hot-pressing may be monitored to provide data useful for understanding of material behavior, for modeling the consolidation process, or for optimizing the manufacturing parameters. Two of the key measurable variables are temperature and pressure. These may be monitored by insertion of thermocouples and pressure probes connected to a means of data collection and processing. A well-known system for making such measurements is the PressMAN™ Press Monitoring System [24]. This technology was developed by the former Alberta Research Council, now a part of Alberta Innovates Technology Futures, Alberta, Canada. The system is designed to provide control of laboratory pressing, based on the real-time temperature and pressure readings.

Using the PressMAN or similar technologies, researchers have gathered data on internal mat conditions [25–29]. Although of interest from a strictly observational perspective, the real aim of much of this data gathering is to develop and validate computer models of consolidation [21, 23, 30–33]. Computational modeling is a useful tool for understanding internal mat conditions, material response to the dynamic environment within a consolidating mat, and for evaluation and prediction of material properties. Models are useful for the improvement and optimization of manufacturing processes and the final products.

A representative example of measured mat temperature and pressure versus time is shown in Figure 4.17.

Although the absolute values for temperature will vary according to mat moisture content, platen temperature, furnish type, press closing time, platen pressure, and related variables, the trends shown in Figure 4.17 are typical in terms of general response. The surface, in

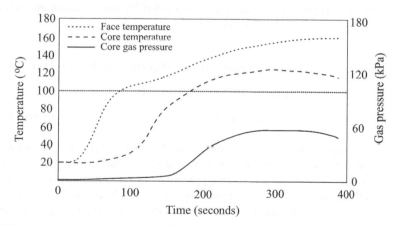

Figure 4.17   *Typical face (surface) and core (center) temperature and core gas pressure versus time for a laboratory medium density fiberboard panel during hot pressing. Press platen temperature was 210°C. Adapted from [28] with permission of the Society of Wood Science and Technology © 2005.*

contact with the heated platens, is the first to heat up as a result of conductive heat transfer. Conduction of heat from the surfaces to the core is accompanied by convective heat transfer as the surfaces, followed by the core, are heated above the boiling point of water. Although some gas pressure escapes from the unsealed edges of the mat, much of the heated gas and water vapor is driven toward the center of the mat, providing heat and mass transfer sufficient to heat the core in a relatively short period of time. Pressure builds rapidly in the core once the core reaches the boiling point of water. The internal mat pressure is generally sufficient to break the developing adhesive and/or autoadhesive bonds, that is, the board will "blow" (fracture internally) if the gas pressure is not allowed to dissipate slowly via a gradual decrease in applied platen pressure. Thus, press closing strategies (affecting initial heat transfer and mat density profile development) and press opening strategies (to avoid blows in the panel) are directly affected by the related phenomenon of mat temperature and pressure profiles.

4.5.7.2   *Mat Counterpressure*

Internal mat pressure results from the generation of heated air and steam. In addition, the wood material resists compressive deformation. The combined effects of internal gas pressure and the response of the furnish to compression results in mat counterpressure, or resistance to compression, with the latter contributor of primary importance. This counterpressure is generally greatest shortly after the press is closed to final position (i.e., final distance between platens), with the counterpressure decreasing as the closed press time increases (Figure 4.18).

In the case illustrated in Figure 4.18, the press was closed to stops, meaning that the platens were closed to a fixed final position or specified final thickness of the compressed mat. In the laboratory, this is typically accomplished by placement of solid metal stop bars of the appropriate thickness at the edges of the mat. When compressed in this manner to a constant thickness, the mat of wood furnish is subjected to step-constant strain, and the furnish undergoes stress relaxation in the same manner as that illustrated in Figure 4.13 for a viscoelastic material. Thus, the decrease in counterpressure as shown in Figure 4.18 is due to the viscoelastic stress relaxation in the wood furnish, as influenced by both mat moisture content (a) and press temperature (b), and aided by the concomitant thermal- and diluent-softening of the amorphous wood polymers.

4.6   Consolidation Behavior: Viscoelasticity Manifested during Hot Pressing

4.6.1   Response of a Viscoelastic Foam to Compression

Synthetic cellular elastic-plastic foam materials respond to compression in a manner illustrated by Figure 4.19.

Such a material undergoing compression exhibits three characteristic zones or regions as strain is increased. First, the material responds with a linear-elastic region in which the cell edges bend and the faces stretch. The horizontal plateau represents plastic yielding, in which the cells collapse. Cell collapse may be accompanied by fracture of the cell walls, or if the walls are sufficiently pliable, bending and buckling of the walls without fracture. Collapse may involve a combination of fracture and buckling. Once the cell cavities are collapsed, densification occurs as the cell wall material contacts that adjacent, collapsed cells. It has been demonstrated that the compression of wood may be understood to occur in this same manner [36].

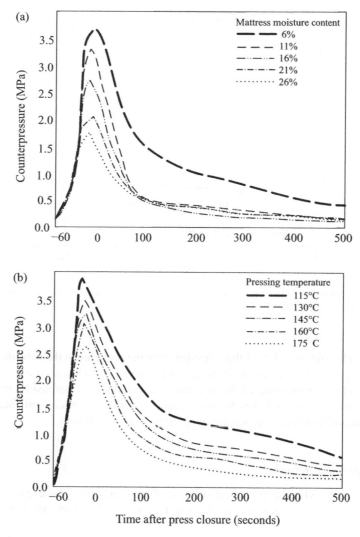

Figure 4.18 *Mat counterpressure (total resistance to compression) in laboratory hot-pressed particleboard. Time zero represents the point at which the platens reach final closure position (stops). (a) Effect of varying mat moisture content. (b) Effect of varying press platen temperature. Reproduced from [34] with permission of De Gruyter/Holzforschung © 1989.*

4.6.2   Viscoelastic Response of Lignocellulosic Material to Thermocompression

4.6.2.1   Cellularity: Intra- and Interparticle

Wood is conceptualized as a cellular, viscoelastic material. In the context of consolidation behavior, we may observe two levels of cellularity, namely, the interparticle voids consisting first, of the space between individual fibers, particles, strands, and so on, in a furnish mat, and second, the inherent or intrinsic cellularity of the material, which we may call the intraparticle voids (Figure 4.20).

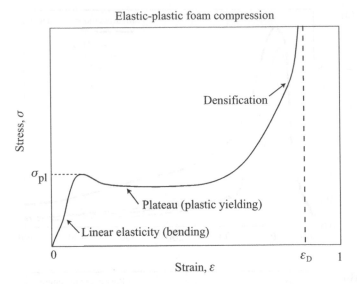

Figure 4.19   *Generalized stress–strain diagram for compression of an elastic-plastic foam. Reproduced from [35]* *with permission of Cambridge University Press © 2010.*

The interparticle voids consists of the space between the particles in the furnish mat. These voids are therefore largely dependent on the geometry of the particles, their placement in the mat, and the level of compression of the mat. Intrinsic cellularity is defined by the anatomic structure of the wood. This includes the void volume due to cell lumens, pit chambers, and other permanent and transient microcapillary space within the cell wall.

4.6.2.2   *Compressive Behavior*

As the furnish is softened and compressed, collapse of voids at both of these levels of structure results in compression behavior that is characteristic of a cellular elastic foam (Compare Figures 4.19 and 4.21).

In the specific case of compression of wood particles, the stress–strain diagram is interpreted as follows: The particle mat initially responds to compressive stress with linear-elastic strain as the particles shift relative to one another into available interparticle void space (Figure 4.21A). This is followed quickly by a plateau of stress over a relatively wide range of strain, during which period the mat has compressed sufficiently to produce continuous columns of particles from top-to-bottom within the mat. During this stage (Figure 4.21B), the wood deforms primarily by bending, thus further occupying the interparticle void space. When most of the interparticle void space has been eliminated from the mat, the stress level increases rapidly with diminishing amounts of strain as the particles are compressed, now reducing intraparticle void space, that is, deformation of the anatomic structure of the wood (Figures 4.20c and 4.21C). Stress increases steeply as the cell wall itself is compressed (Figure 4.21D). Comparison of Figures 4.21 and 4.19 show that wood mats do, indeed, behave like an elastic-plastic foam material when subjected to compression [37, 38].

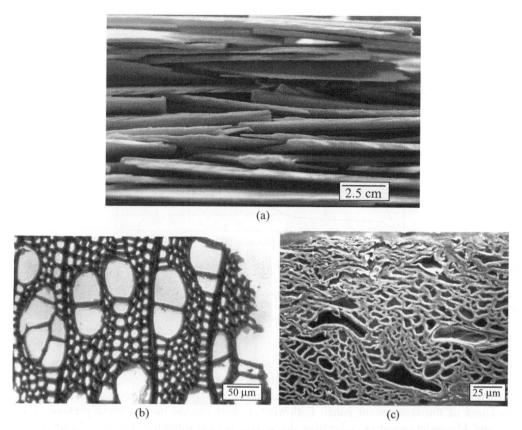

Figure 4.20 (a) Interparticle voids are apparent in this uncompressed mattress of aspen (Populus sp.) wood strands (b) Epi-fluorescence micrograph of the cross-section of an uncompressed aspen strand. Note the normal anatomic shape of the large pores and smaller fibers. (c) Scanning electron micrograph of the cross-section of an aspen strand extracted from the surface of a mattress of strands compressed under 3.45 Mpa (500 psi) and 204°C (400°F) at 5% moisture content. Note the consolidation-induced distortion of pores and fibers from their respective normal elliptical and round shapes, and the concomitant reduction of intraparticle void space. Photo and micrographs by D.D. Stokke.

Dai studied the stress–strain behavior of individual aspen (*Populus tremuloides*) flakes [39] and of carefully prepared six-flake columns of under compression [40], and found the response for the individual flakes and flake columns identical to the pattern shown in Figure 4.21. Simulation modeling of void volume in wood strand mats shows that voids in the compressed panel increase with strand thickness. Strand density is also an important determinant of void volume, whereas strand length and thickness had a less significant effect [41].

The consolidation of a wood furnish mat results from the viscoelastic response of wood to applied heat and pressure, in combination with the diluent effects of moisture on the wood poymers. During hot pressing, the pressure required to consolidate to a given thickness reduces as the mat undergoes thermoplastic softening and viscoelastic stress relaxation under step-constant strain. This is illustrated in Figure 4.22, in which we see a representation of the

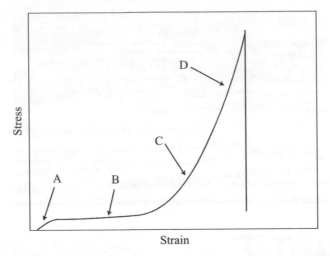

Figure 4.21   *Stress–strain diagram for a wood particle mat under compression. The zones labeled A–D are characterized as follows: (A) Linear-elastic compression; unrestrained particles slide past one another. (B) Particle contact from top to bottom of mat; particle bending begins. (C) Compression of particles; reduction of interparticle and intraparticle void space. (D) Densification; most cell lumens are collapsed, with compression of cell wall substance in progress. Reproduced from [22] with permission of Dr. Fred Kamke.*

thermal softening and viscoelastic relaxation of wood superimposed above a plot of pressure vs. press time for laboratory-pressing of a mattress of wood flakes.

Observe that mat (counter) pressure is maximized shortly after compression to the target thickness of the compressed panel. Following a short plateau of maximum pressure, the mat quickly undergoes stress relaxation due to the thermal softening and viscoelastic stress relaxation under step-constant strain. This is followed by release of platen pressure as the press is opened near the end of the pressing cycle. In this example, the press starts to open at around 300 seconds press time, causing a gradual reduction in mat pressure to zero. Notice that there

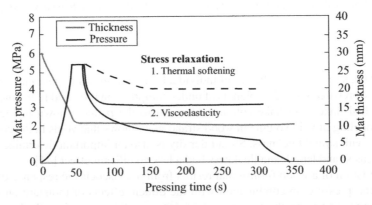

Figure 4.22   *Mat pressure and thickness versus press time representative of consolidation of wood flake mats. Stress relaxation following initial consolidation of the viscoelastic mat is evident. Reproduced from [27] with permission of Dr. Chunping Dai.*

is a slight increase in mat thickness as the press opens. This is attributable to the viscoelastic response known as springback.

4.6.2.3 Springback

Springback is the term used to describe the usually small increase in thickness of a compressed mat of wood furnish following release of platen pressure. Springback is a recovery of elastic strain within the viscoelastic furnish. Transient changes in temperature and moisture content throughout the mat influence the T_g of wood polymers. As the temperature within the furnish is raised beyond T_g, the polymers soften, resulting in a decrease in modulus within the mat. This produces densification of the wood furnish. By densification, we mean that the process causes the density (mass per unit volume) of the compressed material to exceed that of the input raw material. Adequate softening of the cell wall permits this densification to occur without undesirable fracture of the cell walls. Under this scenario, the cell wall polymers are repositioned or reoriented at a molecular level, consistent with their response to the significantly lower glass transition resulting from temperature and diluent effects.

Once the hot-pressing operation is completed, the material remains densified, resulting from unrecoverable viscous strain or permanent deformation of the viscoelastic furnish. Even so, some of the densification is reversed when the press is opened as a result of recoverable elastic strain in the material. In adhesive-bonded composites, springback is reduced by the development of adequate adhesive bond strength between furnish particles. The adhesive bonds counteract strain that would be recoverable, were it not for the adhesive strength.

4.6.3 Vertical Density Profile

4.6.3.1 Development of Vertical Density Profile

Hot pressed wood mats invariably develop a gradation in final panel density throughout the thickness of the panel. This variation is called "vertical density profile" (VDP), also known as vertical density gradient, density profile, or vertical density distribution. VDP is a function of raw material (furnish) and process variables.

Consider a lignocellulosic furnish consolidated in a cold-press, that is, a pressing operation conducted at room temperature using a cold-setting adhesive. If we assume a uniform furnish density, then theoretically, conditions could be such that the compressive stress will be constant throughout the mat. If this were indeed the case, then the resulting cold-pressed material would be characterized by a density through-the-thickness, or VDP, in which $\rho_f = \rho_i = \rho_c$, where ρ_f is the density of the face layer(s), ρ_i is the density of an "intermediate" layer, and ρ_c is density of the core or center of the mat (Figure 4.23).

In hot pressing operations, thermal and moisture-induced softening reduces wood particle compression strength and modulus perpendicular to the grain, as well as bending modulus. Thus, as the press closes, the layers of the mat in contact with the platens are weakened and therefore readily densified. Steam generated by heating of the surface layer is driven toward the mat core, increasing the core temperature and moisture content. The sublayers, that is, the intermediate and core regions of the mat, begin to weaken and densify, but at a different rate and degree than the surface layers. The general result of this dynamic process will be the development of a vertical density profile throughout the panel thickness in which

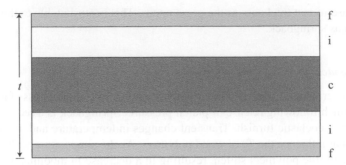

Figure 4.23 *Density (ρ) of a compressed wood composite may vary through the thickness (t) in zones identified as face, intermediate or core layers, here designated as f, i, and c. The corresponding densities may be represented as ρ_f, ρ_i, and ρ_c, respectively.*

$\rho_f > \rho_i > \rho_c$. In actuality, the variation between "layers" is graduated, as opposed to being identifiable in obviously discrete layers as suggested here.

The key determinants of vertical density profile were enumerated by Wang and Winistorfer [32] as follows:

1. Moisture condition. This refers to the initial moisture content of the mat, including variations in face versus core moisture content. Transient moisture content of the mat during the press cycle is also a significant factor.
2. Mat structure. The mat structure is determined by furnish particle geometry, particle distribution (as influenced by mat formation technique/technology) and inherent material density of the particles.
3. Pressing environment. Temperature, pressure and pressure cycle, and time (total and press closing time) are the key parameters that define pressing environment.

The interaction of moisture, mat structure and pressing environment produce dynamic variations in local conditions within the mat which impinge upon the viscoelastic response of the consolidating material. These interactions have the net effect of rendering a final product that has vertical density gradient that is predictable in form, and often, in degree.

4.6.3.2 *Measurement of Vertical Density Profile*

At one time, VDP was measured by finding the density of a sample of panel, followed by determination of layer density by tedious sanding off of thin layers of material, accompanied by frequent measurement of the thickness and mass of the remaining sample and calculation of layer density by difference. This gravimetric method eventually gave way to x-ray densitometry, which allowed the measurement of VDP in a much more efficient and accurate manner [32]. Commercial systems are now routinely used for industrial quality control, product development, and academic research [42].

A representative VDP is shown in Figure 4.24. The shape of the profile is typical of that seen for strand- or flake-, particle- or fiber-based composites.

Density is at a maximum just below the top and bottom surface planes of the composite. These zones, exceeding the average density of the composite as a whole, are deemed the faces. The extreme outer surface of the face is generally lower than the maximum, mainly due to overcure of adhesive resin in immediate contact with the heated press platens. The center or

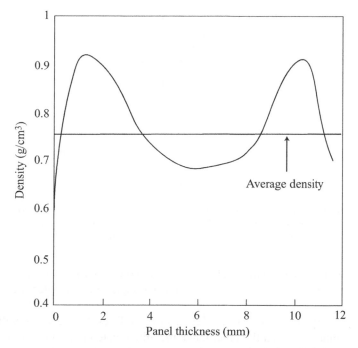

Figure 4.24 *Representative vertical density profile (VDP) for a hot-pressed, wood-based composite. Thickness of zero is interpreted as the top of the sample; the bottom plane of the sample is at about 12 mm thickness. Reproduced from [43] with permission of the Society of Wood Science and Technology © 2000.*

core of the composite has a density lower than the overall average. Analysis of the density profile could involve division of the thickness into arbitrary zones defined by absolute density values, for example, face, intermediate, and core. However, it is common practice to simply recognize the face zone, with $\rho_f > \rho_{\text{panel average}}$ and the core zone, where $\rho_c < \rho_{\text{panel average}}$.

4.6.3.3   *Implications of Vertical Density Profile*

There are two major implications of vertical density profile with respect to physical properties of hot-pressed composites. First, higher density of the faces improves bending strength and stiffness due to the positive relationship between density and most mechanical properties of wood and wood-based materials. Second, because the density of the core is lowest, this means that the core will generally exhibit the lowest mechanical properties as measured through the thickness. This includes tension strength perpendicular to the panel surface plane(s). As a result, tension strength perpendicular to the panel is often measured as an indicator of adhesive bond integrity. Accordingly, this metric is customarily referred to as internal bond (IB) strength.

4.7   Press Cycles

4.7.1   Effect of Press Closing Time on Development of Vertical Density Profile

With regard to pressing parameters, let's now consider the effect of press closing time (PCT), which is defined as the time required for the press platens to move from their initial, open

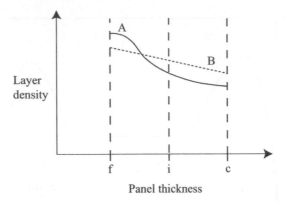

Figure 4.25   *Density versus panel thickness, expressed as face (f), intermediate (i), and core (c) layers for a wood-flake composite consolidated with "short" (A) or "long" (B) press closing times. The average panel density of types A and B is the same. Figure developed from concepts presented by Smith [44].*

position (loaded with furnish) to the final platen position (i.e., approximate product thickness) during one press cycle. Here, we will think in terms of a single-opening batch press to illustrate the concept.

If the initial press (platen) pressure is high, a short PCT results (here, we will interpret a short PCT as <1 minute). If, on the other hand, the initial platen pressure is low, a long PCT is seen (1–2 minutes). When PCT is short, the upper and lower layers of the mat are densified quickly and to a greater degree than if the PCT is long. This occurs because the press will close before there is sufficient time for appreciable transfer of heat and moisture from the face to the core. The face layers initially retain somewhat more moisture and are heated more rapidly by conduction from the platens, whereas core plasticization and densification is delayed. In contrast, a longer PCT results in a more uniform densification of faces and core (Figure 4.25). Given these observations, we may conclude that press cycle, particularly press closing time, may be used to control or modify VDP.

The net effect of PCT on properties is therefore seen in its influence on vertical density profile (VDP). Those materials with a relatively higher face layer density, that is, those manufactured with a short PCT, would be thus understood to have greater bending properties due to the positive correlation of density to strength and stiffness. Materials with a higher core density, that is, those manufactured with a long PCT, will conversely have greater internal bond strength (tensile strength perpendicular to the plane of the panel). Thus, for the examples shown in Figure 4.25, panel A will maximize bending strength, whereas B will maximize internal bond strength, assuming the same average panel density for each.

4.7.2   Effect of Mat Moisture Content on Face Density

Mat moisture content, particularly that of the face, combines with PCT to influence VDP. High initial face moisture content results in greater face layer density. Conversely, lower face moisture content results in lower face density. This is due to the diluent effect of moisture on the wood/lignocellulose polymers. Adhesive resin content has little direct effect on VDP, but final material properties are influenced by resin since greater resin content should result in greater interparticle bonding.

4.7.3   Effect of Furnish Density and Compaction Ratio on Composite Properties

Finally, we need to recognize the influence of furnish density. This is tied to the concept of compaction ratio, defined as the ratio of density of the final manufactured product to the density of the wood or lignocellulosic furnish (Equation 4.19).

$$CR = \frac{\rho_{final\ product}}{\rho_{raw\ material}} \qquad (4.19)$$

As an example, consider a product made from aspen wood flakes or strands. Let us assume a wood raw material density of 0.385 g/cm³ (24 pounds per cubic foot, pcf) and a typical final oriented strand board product density of 0.675 g/cm³ (42 pcf). Thus, the compaction ratio for this example is (Equation 4.20):

$$CR = \frac{0.675\ g/cm^3}{0.385\ g/cm^3} = 1.75 \qquad (4.20)$$

Assuming that the composite is adequately bonded, we should expect a substantial increase in most mechanical properties over those of the unmodified wood, given the direct correlation of mechanical properties with density. Some classic results on studies of laboratory panels made from veneer flakes (laboratory flakes made to exacting dimensional specifications) formed with random flake orientation, as reported by Hse [45] and Hse et al. [46], will serve to illustrate the point (Figures 4.26–4.28).

In Figure 4.26, the relationship between modulus of elasticity (MOE), a measure related to panel bending stiffness, and compaction ratio is illustrated.

Inspection of Figure 4.26 reveals that MOE increases by approximately the same factor as compaction ratio, that is, a compaction ratio of 1.5 results in MOE that is nearly 1.5 times greater than the MOE of boards with a compaction ratio (CR) of 1.0. This is borne out by the regression line (based on data in psi) calculated as MOE, psi = 123,574 + (545,429 CR), with $r = 0.83$. These results imply that low-density furnish may be used to advantage, given

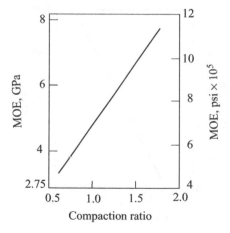

Figure 4.26   *Modulus of elasticity (MOE) in bending versus compaction ratio for laboratory panels made from Southern hardwood veneer flakes, bonded with phenolic resin. Original data were reported in pounds per square inch, psi. Note: 1 GPA = 1,000 MPa. Adapted from [46] with permission of the Forest Products Society © 1975.*

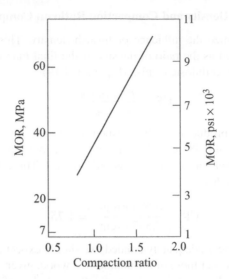

Figure 4.27 *Modulus of rupture (MOR) in bending versus compaction ratio for laboratory panels made from Southern hardwood veneer flakes, bonded with phenolic resin. Original data were reported in pounds per square inch, psi. Adapted from [46] with permission of the Forest Products Society © 1975.*

the positive influence of compaction ratio on MOE. Use of low density furnish has the benefit of keeping final product density and weight to manageable levels, while still assuring adequate stiffness and strength. The bending strength, or Modulus of Rupture (MOR), is highly correlated with both MOE and CR. Hse's results [45] show that MOR may increase by a factor of three as compaction ratio is increased by a factor of nearly 1.8, with a regression of MOR, psi $= -2,760 + (7,782 \text{ CR})$; $r = 0.93$ (Figure 4.27).

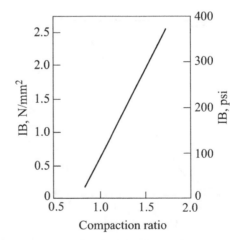

Figure 4.28 *Internal bond strength (IB) versus compaction ratio for laboratory panels made from Southern hardwood veneer flakes, bonded with phenolic resin. Original data were reported in pounds per square inch, psi. Note: 1 N/mm² = 1 MPa. Adapted from [46] with permission of the Forest Products Society © 1975.*

The effect of compaction on MOE and MOR also reinforce the observations and rationale regarding the influence of press closing time and vertical density profile on composite properties.

Figure 4.28 shows the relationship of internal bond strength (tension strength perpendicular to the panel surface) to compaction ratio.

Compaction ratio has a significant, positive effect on internal bond. In this example, as compaction ratio is increased by a factor of 1.5, IB increases by approximately four times, with the regression line in Figure 4.28 given by IB, psi $= -358 + (423\ \mathrm{CR})$, $r = 0.85$. Since the IB is a measure of core strength, the importance of compaction ratio to development of adequate IB should be carefully considered. High density furnish may preclude the opportunity to produce a panel with sufficient IB. Again, the advantage of using lower density furnish is seen. Indeed, species such as aspen (*Populus tremuloides*) and yellow poplar/tuliptree (*Liriodendron tulipifera*) are favored in North America as desirable for manufacture of oriented strandboard intended for structural applications. These species have wood densities, $\rho_{0,12\%}$, of 0.38 and 0.42 g/cm^3, respectively [47]. Wood raw material density in this range allows for adequate compaction ratio in the final product, without inordinately high final product density and panel weight.

4.8 Horizontal Density Distribution

4.8.1 In-Plane Density Variation

Horizontal density distribution (HDD) is the variation of composite density in the plane of the board or panel surface. This is readily observed by dividing a composite panel into successively smaller sampling units and measuring the density of each unit. For example, an interesting laboratory exercise with commercial wood-based panel composites, e.g., oriented strandboard (OSB), particleboard (PB), and medium density fiberboard (MDF) is as follows. First, determine the density of the entire panel for each product type. This value will represent the average product density. Then, divide each panel in half and determine density. Divide each half again, and determine the density of each quarter of the panel. This may be repeated, or the exercise may proceed to sampling small squares (5.1 cm × 5.1 cm or 2 in. × 2 in.) along original midlines of the panel, both lengthwise and crosswise. The results of such an exercise are shown in Table 4.1. This exercise provides an indication of variation in material density within the plane of a composite panel. In addition, the influence of furnish particle geometry (size) is readily apparent by comparing the coefficient of variation (COV) for each type of panel. OSB, consisting of the largest wood constituents (strands), has the greatest variation in density, with the fiber-based MDF having the least variability in HDD. Whereas vertical density profile results from a complex interaction of most raw material and process variables, HDD is primarily a function of particle or flake geometry [48]. Flake or wood element geometry, along with variables attributable to the mat formation process, produce within the mat a number of air spaces or voids, which we have previously called interparticle void space. This space, in totality within the mat, may be defined as the void volume fraction (VVF). In the case of wood strand materials, the VVF remaining in a consolidated panel is a significant determinant of the engineering constants of the product [49].

Table 4.1 *Density variation of three commercial wood composite panel products.*

Product	Sample	ρ_{test}, g/cm³	COV%	n
OSB	Half-sheet	0.620	–	1
	5 × 5 cm, along sheet length	0.616	5.1	20
	5 × 5 cm, across sheet width	0.614	7.4	10
PB	Half-sheet	0.759	–	1
	5 × 5 cm, along sheet length	0.735	1.5	20
	5 × 5 cm, across sheet width	0.737	2.2	10
MDF	Half-sheet	0.730	–	1
	5 × 5 cm, along sheet length	0.713	1.03	20
	5 × 5 cm, across sheet width	0.737	1.10	10

Original 122 × 244 cm (4 × 8 foot) panels were cut in half to 122 × 122 cm (4 × 4 foot) half sheet samples for density determination. Small 5 × 5 cm (2 × 2 in.) samples were then cut from the centerline of one of the half-sheets, along the length of the original sheet. Similar samples were also cut from the center to one edge perpendicular to the length of the original sheet. Data generated by students in the Forestry 485 class, Iowa State University, 2008. The coefficient of variation (COV%) is the standard deviation expressed as a percentage of the mean.

4.8.2 Thickness Swelling

Performance of wood and natural fiber composites in service is often measured in terms of mechanical strength and stiffness properties. Equally important is the dimensional stability of the product. Given that lignocellulosic substances are hygroscopic materials, it is expected that they will interact with moisture in the environment. One outcome of this interaction is dimensional change, both in the panel plane and perpendicular to the plane, the former measured as linear expansion (LE), with the latter quantified by thickness swelling (TS). In the case of flake- or strand-based wood composites, linear expansion both along and across the machine direction of the panel is primarily a function of flake alignment and density [50]. Horizontal density distribution (HDD) has a direct relationship to thickness swelling (TS) resulting from interaction of the wood composite material with moisture, and to internal bond strength (IB), particularly of panels subjected to cyclic water soak tests [51].

Thickness swelling variation in a panel can be appreciable (Figure 4.29). As illustrated in the figure, swelling at the edges of a panel may be considerably greater than that at the center. This is attributable both to HDD and to variation in residual stresses resulting from the

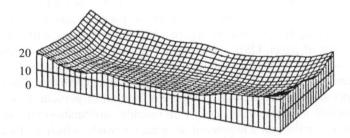

Figure 4.29 *Variation in thickness swelling of an industrial waferboard. The panel was cut into 1,012, 100 × 100 mm specimens. Each specimen was conditioned to equilibrium at 20°C and 65% relative humidity, followed by soaking in water for 24 hours, after which the percent thickness swelling was determined. Reproduced from [34] with permission of De Gruyter/Holzforschung © 1989.*

hot-pressing operation, the latter producing a greater springback effect (Section 4.6.2.3) near the panel edges [34].

References

1. Pellegrino M, Tavone S, Guerra G, DeRosa C. Evaluation by Fourier transform infrared spectroscopy of the different crystalline forms in syndiotactic polystyrene samples. *Journal of Polymer Science B: Polymer Physics*, 1997;35:1055–1066.
2. Irvine GM. The glass transitions of lignin and hemicellulose and their measurement by differential thermal analysis. *Tappi*, 1984;65(2):118–21.
3. Goring DAI. Thermal softening of lignin, hemicellulose, and cellulose. *Pulp and Paper Magazine, Canada*, 1963;64(2):T517–T527.
4. Back EL, Salmén L. Glass transitions of wood components hold implications for molding and pulping processes. *Tappi*, 1982;65(2):107–110.
5. Salmén L, Kolseth P, de Ruvo A. Modeling the softening behaviour of wood fibres. *Journal of Pulp and Paper Science*, 1985;11(2):J102–J107.
6. Kaelble DH. *Physical Chemistry of Adhesion Science*. New York: Wiley-Interscience; 1971.
7. Salmén L Viscoelastic properties of *in situ* lignin under water-saturated conditions. *Journal of Materials Science*, 1984;19:3090–3096.
8. Kelley SS, Rials TG, Glasser WG. Relaxation behavior of the amorphous components of wood. *Journal of Materials Science*, 1987;22:617–624.
9. Kwei, TK. The effect of hydrogen bonding on the glass transition temperature of polymer mixtures. *Journal of Polymer Science: Polymer Science Letters Edition*, 1984;22(2):307–313.
10. Wolcott MP, Kamke FA, Dillard DA. Fundamentals of flakeboard manufacture: viscoelastic behavior of the wood component. *Wood and Fiber Science*, 1990;22(2):345–361.
11. ASTM. D-143-09. Standard test methods for small clear specimens of timber. In: *Annual Book of ASTM Standards*. West Conshohocken, PA: ASTM International; 2010.
12. Brinson HF, Brinson LC. *Polymer Engineering Science and Viscoelasticity: An Introduciton*. New York: Springer; 2008.
13. Lin Y-H. *Polymer Viscoelasticity: Basics, Molecular Theories, Experiments and Simulations*, 2nd ed. Singapore: World Scientific Publishing Co Pte. Ltd; 2011.
14. Shaw MT, MacKnight WJ. *Introduction to Polymer Viscoelasticity*, 3rd ed. Hoboken, NJ: Wiley-Interscience; 2005.
15. Kamke FA. Sizemore H. III (inventors). Viscoelastic thermal compression of wood. Eagle Analytical Company, Inc., assignee. US Patent No. 7,404,422; 2008
16. Kutnar A, Kamke FA, Nairn JA, Sernek M. Mode II fracture behavior of bonded viscoelastic thermal compressed wood. *Wood and Fiber Science*, 2008;40(2):362–373.
17. Kutnar A, Kamke FA, Sernek M. Density profile and morphology of viscoelastic thermal compressed wood. *Wood Science and Technology*, 2009;43:57–68.
18. Siau JF. *Wood: Influence of Moisture on Physical Properties*. Blacksburg, VA: Department of Wood Science and Forest Products, Virginia Polytechnic Institute and State University; 1995.
19. Skaar C. In: Timmell TE, editor. *Wood-Water Relations*. Berlin: Springer-Verlag; 1988.
20. Thoemen H, Humphrey PE. Modeling the physical processes relevant during hot pressing of wood-based composites. Part I. Heat and mass transfer. *Hoz als Roh- und Werkstoff*, 2006;64(2):1–10.
21. Thoemen H, Humphrey PE. Modeling the continuous pressing process for wood-based composites. *Wood and Fiber Science*, 2003;35(2):456–468.
22. Kamke FA. Physics of hot pressing. In: Winandy JE, Kamke FA, editors. *Fundamentals of Composite Processing, FPL-GTR-149*. Madison, WI: U.S. Department of Agriculture, Forest Service, Forest Products Laboratory; 2004. pp 3–18.

23. Thoemen H. Simulation of the pressing process. In: Winandy JE, Kamke FA, editors. *Fundamentals of Composite Processing, FPL-GTR-149*. Madison, WI: U.S. Department of Agriculture, Forest Service, Forest Products Laboratory; 2004.

24. Anonymous. PressMAN Press Monitoring System. Alberta Innovates Technology Futures; [February 20, 2012]; Available at: http://www.albertatechfutures.ca/RDSupport/IndustrialBioproducts/EngineeredCompositePanels/PressMAN.aspx.

25. Kamke FA, Casey LJ. Gas pressure and temperature in the mat during flakeboard manufacture. *Forest Products Journal*, 1988;38(2):41–43.

26. Kamke FA, Casey LJ. Fundamentals of flakeboard manufacture: Internal-mat conditions. *Forest Products Journal*, 1988;38(2):38–44.

27. Dai C, Wang S. Press control for optimized wood composite processing and properties. In: Winandy JE, Kamke FA, editors. *Fundamentals of Composite Processing, FPL-GTR-149*. Madison, WI: U.S. Department of Agriculture, Forest Service, Forest Products Laboratory; 2004.

28. Garcia RA, Cloutier A. Characterization of heat and mass transfer in the mat during the hot pressing of MDF panels. *Wood and Fiber Science*, 2005;37(2):23–41.

29. Garcia PJ, Avramidis S, Lam F. Horizontal gas pressure and temperature distribution responses to OSB flake alignment during hot-pressing. *European Journal of Wood and Wood Products*, 2003;61(2):425–431.

30. Hata T, Kawai S, Sasaki H. Production of particleboard with steam-injection. *Wood Science and Technology*, 1990;24(2):65–78.

31. Humphrey PE, Bolton AJ. The hot pressing of dry-formed wood-based composites. Part II. A simulation model for heat and moisture transfer, and typical results. *Holzforschung*, 1989;43(2):199–206.

32. Wang S, Winistorfer PM. Consolidation of flakeboard mats under theoretical laboratory pressing and simulated industrial pressing. *Wood and Fiber Science*, 2000;32(2):527–538.

33. Zombori BG, Kamke FA, Watson LT. Simulation of the internal conditions during the hot-pressing process. *Wood and Fiber Science*, 2003;35(2):2–23.

34. Bolton AJ, Humphrey PE, Kavvouras PK. The hot pressing of dry-formed wood-based composites: Part VI. The importance of stresses in pressed mattress and their relevance to the minimisation of pressing time, and the variability of board properties. *Holzforschung*, 1989;43:406–10.

35. Gibson LJ, Ashby MF, Harley BA. *Cellular Materials in Nature and Medicine*. Cambridge, UK: Cambridge University Press; 2010.

36. Wolcott MP, Kamke FA, Dillard DA. Fundamental aspects of wood deformation pertaining to manufacture of wood-based composites. *Wood and Fiber Science*, 1994;26(2):496–511.

37. Lenth CA, Kamke FA. Investigations of flakeboard mat consolidation. Part I. Characterizing the cellular structure. *Wood and Fiber Science*, 1996;28(2):153–167.

38. Lenth CA, Kamke FA. Investigations of flakeboard mat consolidation Part II. Modeling mat consolidation using theories of cellular materials. *Wood and Fiber Science*, 1996;28(2):309–319.

39. Dai C, Steiner PR. Compression behavior of randomly formed wood flake mats. *Wood and Fiber Science*, 1993;25(2):349–358.

40. Dai C. Viscoelasticity of wood composite mats during consolidation. *Wood and Fiber Science*, 2001;33(2):353–363.

41. Li P, Dia C, Wang S. A simulation of void variation in wood-strand composites. *Holzforschung*, 2009;63(2):357–361.

42. Anonymous. QMS x-ray based measurement equipment. Knoxville, TN: Quintek Measurement Systems; [March 5, 2012]; Available at: http://www.qms-density.com/.

43. Winistorfer PM, Moschler WW, Wang S, DePaula E, Bledsoe BL. Fundamentals of vertical density profile formation in wood composites. Part I. *In-situ* density measurement of the consolidation process. *Wood and Fiber Science*, 2000;32(2):209–219.

44. Smith DC. Waferboard press closing strategies. *Forest Products Journal*, 1982;32(2):40–45.

45. Hse CY. Properties of flakeboards from hardwoods growing on Southern pine sites. *Forest Products Journal*, 1975;25(2):48–53.

46. Hse CY, Koch P, McMillin CW, Price EW. Laboratory-scale development of a structural exterior flake-board from hardwoods growing on Southern pine sites. *Forest Products Journal*, 1975;25(2):42–50.

47. Laboratory FP. *Wood Handbook: Wood as an Engineering Material*. Madison, WI: U.S. Department of Agriculture, Forest Service, Forest Products Laboratory; 2010.

48. Suchsland O. The density distribution of flakeboards. Michigan Agriculture Experiment Station, East Lansing, *Quarterly Bulletin*, 1962;45(1):104–121.

49. Wu Q, Lee JN, Han G. The influence of voids on the engineering constants of oriented strandboard: A finite element model. *Wood and Fiber Science*, 2004;36(2):71–83.

50. Wu Q. In-plane dimensional stability of oriented strand panel: effect of processing variables. *Wood and Fiber Science*, 1999;31(2):28–40.

51. Suchsland O, Xu H. A simulation of the horizontal density distribution in a flakeboard. *Forest Products Journal*, 1989;39(2):29–33.

5

Fundamentals of Adhesion

5.1 Introduction

Adhesives have been used to bond wood and other natural materials since antiquity, as evidenced by the fine veneered furniture of ancient Egypt and Rome. In today's world, the importance of adhesives can hardly be overestimated. Adhesives are used for almost all types of product found in everyday life, from packaging and postage stamps to computers, automobiles, and aircraft. Not surprisingly, adhesives play a key role in the efficient utilization of renewable natural materials, providing a means to effectively bond wood and other plant fiber to form products required by human society.

In this chapter, we will introduce fundamental concepts of adhesion science, with a view to understand the interplay of the complex mechanisms and interactions involved in the adhesion phenomenon. This knowledge forms a sound basis for the proper use of adhesives in manufacturing and the development of new and improved processes for bonding natural fiber.

5.2 Overview of Adhesion as a Science

5.2.1 A Brief History of Adhesion Science

Although adhesives in various forms have been used for thousands of years, "adhesion science" is a relatively new field of study, having developed from the 1920s. The paper by J.W. McBain and D.G. Hopkins, published in 1925, is recognized by many as representing the beginning of adhesion science [1]. In this paper, "mechanical interlocking" is proposed as an adhesion mechanism, based on studies of wood bonding. Interestingly, only four years after the publication of McBain and Hopkins' paper, Browne and Brouse [2] discounted the mechanical interlocking mechanism in favor of underlying mechanisms of "specific adhesion." In some sense, this debate has continued to the present. Also in the decade of the 1920s, Truax et al. [3] published work conducted at the US Forest Products Laboratory on the mechanical testing of wood joints as a means to select wood glues. These works, and others, signified the beginning of the earnest study of adhesion as a science.

Introduction to Wood and Natural Fiber Composites, First Edition. Douglas D. Stokke, Qinglin Wu and Guangping Han.
© 2014 John Wiley & Sons, Ltd. Published 2014 by John Wiley & Sons, Ltd.

Although there has been a significant accumulation of practical knowledge concerning adhesion, not only in the decades since the advent of adhesion science but also over the centuries since antiquity, knowledge gaps still remain. Fourche [4] has observed that "... fundamental knowledge about adhesion phenomenon is still very modest and fragmented... one has to admit that the mechanisms governing adhesion are still fairly imprecisely understood. Moreover, no single global theory or model can explain all the phenomena." However, although it is generally conceded that there is no *single* global theory of adhesion, there are a number of adhesion theories that may be used to develop an understanding of the forces and mechanisms involved in this important phenomenon. Together, these theories contribute toward the goal of adhesion science, which is "to predict adhesive bond strength from first principles" [5].

5.2.2   Adhesive Bonding

In order to approach adhesion as a science, we must have a working definition of adhesive bonding. Pocius [5] offers this: "Adhesive bonding is a method by which materials can be joined to generate assemblies. Adhesive bonding is an alternative to more traditional mechanical methods of joining materials, such as nails, rivets, screws, etc." Succinctly, adhesive bonding is the joining of materials to form assemblies.

Adhesive bonding enjoys certain advantages over mechanical means of joining materials, including minimization of stress concentrations and the absorption of applied mechanical load via the viscoelastic behavior of the adhesive. In addition, dissimilar materials that are otherwise difficult to fasten may be joined by the use of adhesives. For example, metal mechanical fasteners may be corroded by contact with lumber containing certain preservative salts. Thus, appropriate adhesives provide a superior alternative.

Disadvantages of adhesive bonding arise from the complex behavior of adhesives as they interact with the surface of materials to be joined, possibly leading to unpredictable performance attributes of the resulting bond. Furthermore, it is often difficult to establish whether proper or adequate bonding occurs within composite material systems. Nevertheless, adhesive bonding is used in ever-expanding applications as a means of choice to join materials together.

At the most basic level, adhesion science is directed toward the understanding of the materials to be bonded, the substance providing the adhesive action, and the structure resulting from the interaction of the two. These are known as the adherend(s), the adhesive, and the adhesive joint, respectively.

Sidebar 5.1   Etymology of Adhesion

'Adhesion' seems to have come into the English language in the seventeenth century. It is interesting that one of the earliest examples (1661) quoted by the Oxford Dictionary is its use in a scientific context by Robert Boyle. The word 'adhesion' comes from the Latin *adhaerere*, which means to stick to. Its use by Lucretius of iron sticking to a magnet anticipates the present technological application of the word. *Adhaerere* itself is a compound of *ad* (to) + *haerere*, where *haerere* also means to stick. Cicero uses the expression *haerere in equo* literally 'to stick to a horse' to refer to keeping a firm seat. (It follows from this that the word 'abhesion' sometimes encountered, supposedly meaning 'no adhesion' or 'release' is etymological nonsense and should be abandoned.)

—Packham [6]

5.2.3 Adherend

An adherend is any substance that is to be bonded together with an adhesive. A synonymous term for the adherend is substrate; accordingly, these two terms will be used interchangeably throughout this text. Substrates are generally solid materials whose physical structure and surface chemical characteristics dictate their performance in the development of an adhesive bond. In the case of wood substrates, the wood anatomical features (e.g., cell lumen size and distribution of cell types) and the surface chemistry, dictated predominantly by the amount and nature of chemical extractives (nonstructural molecules and polymers), are the major players determining the performance of the substrate with respect to adhesion.

5.2.4 Adhesive

Adhesives are generally polymeric materials that exhibit viscoelastic properties. Adhesives are capable of bonding to adherends and transferring mechanical load from the adherend to the adhesive. Though often liquids, adhesives may have other forms. Think, for example, of pressure-sensitive adhesive materials such as cellophane tape or certain postage stamps. Adhesives must effectively interact with the surfaces of substrates and then produce a bond that will resist deformation (breakage or de-bonding). As we will see, wetting of the adherend by the adhesive is the first prerequisite of adhesive bonding. Once an adhesive wets the surface of the substrate, the load-transferring capacity of the bond is developed by cooling, solvent loss, chemical reaction, or some combination of these curing mechanisms. Thus, the two primary functions of an adhesive are achieved, that is, to adequately bond to the surface of adherends and to contribute appropriate mechanical properties subsequent to bond cure.

5.2.5 Adhesive Joint

In its most basic conceptual form, an adhesive joint, also known as an adhesive bond, is an assembly consisting of two substrates and an adhesive. Conceived as such, it is a three-part system: substrate–adhesive–substrate. The substrates to be bonded may be similar (e.g., two pieces of solid wood) or dissimilar materials (e.g., a piece of wood and a fiberglass reinforcement).

Sidebar 5.2   The Adhint: Some Terms Just Do Not Stick!

An assembly consisting of two adherends and an adhesive is an adhesive joint or an adhesive bond. During the development of adhesion science, other terms have been proposed for this and other concepts. Bikerman, credited with originating the theory of weak boundary layers, referred to an adhesive joint as an "*adhint*"— "this is an abbreviation for 'adhesive joint'" [7]. It seems this terminology never really caught on. Perhaps, we should say that it just did not stick!

Although the basic concept of an adhesive joint is simple—two substrates, which may be practically identical to one another, plus an adhesive—the actual reality is more complex. In recent years, scientists have come to recognize the importance of the interface between substrate and adhesive. When we speak of an interface, we are indicating the point(s) or plane(s) of contact between the adhesive and adherend, with the distance scale in the range of atomic to molecular to minute anatomic. At this interfacial level, an interphase is also observed.

The interphase is a minute region in which the substrate and the adhesive interact such that they may modify one another. This modification is typically an alteration of chemical properties of each component, effected by mutual solvation or dilution. As an example, natural extractive chemicals at the surface of a lignocellulosic substrate may migrate into the adhesive, just as adhesive and its solvent carrier migrate into the substrate surface. The intermixing of these compounds modifies the properties of each, such that the newly developed "interphase" has chemical and physical properties somewhat different from the neat (unmodified) adhesive and the *in situ* substrate. The boundary layer thus created may, in fact, dictate overall adhesive bond performance. This concept will be explored further as we discuss adhesion theories.

Our model of an adhesive bond, adherend–adhesive–adherend, has now expanded to a model consisting of five parts, that is, adherend–interphase–adhesive–interphase–adherend. Each of the two interphases result from the chemical interaction of the adhesive and adherend. Each interphase may be further subdivided into three smaller parts or zones, namely, the adhesive interphase, the adhesive–adherend interface (sic) and the adherend interphase. Given that an adhesive bond is conceptually symmetrical around the bulk adhesive (adhesive film), the model thus consists of nine parts: bulk adherend, adherend interphase, adhesive–adherend interface, adhesive interphase, bulk adhesive, adhesive interphase, adhesive–adherend interface, adherend interphase, and bulk adherend.

5.2.6   Marra's Concept of Bond Anatomy

The idea of a nine-component conceptual model of an adhesive bond is not new. Marra's "nine-link bond concept" of the anatomy of a wood–adhesive bond, dating back at least to 1981, is the basis of our foregoing discussion [8, 9]. He described an adhesive bond in terms of the adhesive film, an intraadhesive boundary layer, the adhesive–adherend interface, the adherend subsurface, and the adherend proper. While the model has remained useful as a descriptor of an adhesive bond, there have been changes in the nomenclature appearing in the scientific literature subsequent to Marra's original concept. For example, we have used the contemporary term "inter*phase*" to describe what Marra calls the "adhesive/adherend inter*face*." Another important term to note is that of "boundary layer" which has a different connotation today as compared to Marra's usage. This provides an excellent example of the dynamic nature of vocabulary as applied to advancing science and technology. We have summarized in Table 5.1 the terminology that you may encounter in the adhesion literature of recent decades, particularly with respect to wood adherends. Shifts in terminology as exemplified here, therefore, behooves the student or practitioner in any scientific or technical field of endeavor to stay abreast of the current literature and technical advances in the field.

5.2.7   Contemporary Concept of Bond Anatomy

A contemporary representation of an adhesive bond, slightly modified from Marra's original concept [8, 9], is shown in Figure 5.1, and is patterned after the adaptation and terminology advanced by Frihart [10]. The caption to this figure carries the contemporary, generic nomenclature we have used in developing a verbal description of the zones or links comprising an

Table 5.1 *Terminology used to describe zones within an adhesive bond, attributed to Marra [8, 9]; contemporary, generic terminology as encountered in today's adhesion literature; and that used by Frihart [10], specifically pertaining to wood as an adherend (The column headed "Link(s)" refers to the zones or links depicted in Figure 5.1).*

Link(s)	Marra's terminology [8, 9]	Contemporary, generic terminology	Frihart's terminology, specific to wood adherends [10]
1	Adhesive film	Bulk adhesive	Bulk adhesive
2 and 3	Intraadhesive boundary layer	Adhesive interphase	Adhesive interphase
4 and 5	Adhesive–adherend interface	Adhesive–adherend interface	Adhesive–wood interface
6 and 7	Adherend subsurface	Adherend interphase	Wood interphase
8 and 9	Adherend proper	Bulk adherend	Bulk wood

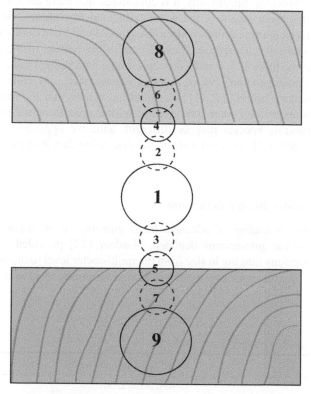

Figure 5.1 *A representation of adhesive bond "anatomy" as patterned after the original nine-link concept of Marra [8, 9]. Using contemporary, generic nomenclature, the links are identified, top to bottom, as follows: Bulk adherend [8], adherend interphase[6], adhesive–adherend interface[4], adhesive interphase[2], bulk adhesive[1], adhesive interphase[3], adhesive–adherend interface[5], adherend interphase[7], and bulk adherend[9]. Adapted from [10] with permission of Taylor © Francis Ltd © 2009.*

adhesive bond. By generic nomenclature, we mean reference to components of an adhesive bond without specifying the nature of the substrate or adherend.

A few observations regarding this concept of bond anatomy may be made here. Links 1, 8, and 9 represent the adhesive and substrates, assumed to be unmodified by one another, thus possessing the inherent properties of each. Links 6 and 7, representing subsurface layers of the substrate, are generally regarded as unmodified by the adhesive, but are in close enough proximity to the substrate surface to potentially have some structural and/or chemical modification resulting from surface preparation of the substrate. An example is seen in wood substrates, in which machining operations in preparation for adhesive bonding produce heat effects and/or mechanical damage to subsurface layers of the wood. All other links are assumed to exist because of some type of interaction and/or mutual modification of substrate or adhesive proper. Links 4 and 5 are shown as straddling the surface of the substrate, that is, at the "adhesive–adherend" or "adhesive–wood" interface. This is a reflection of current thinking with respect to wood bonding. Specifically, it is considered that there are different properties in the immediate subsurface layers of the wood (adherend interphases, links 6 and 7) as opposed to the interfacial planes (at the wood surface (links 4 and 5), both of which differ from the adhesive layers (adhesive interphases, links 2 and 3) just outside of the wood's surface. In Figure 5.1, notice that links 2, 3, 6 and 7 are represented with dashed lines. These links are "most vulnerable to malformation" [9], or in today's vernacular, may collectively represent potential "weak boundary layers" [7, 11]. We should also note that the nine links "appear to be mirror images about the center link. This is seldom true in actual practice because, for example, the substrate species may be different, adhesive application may vary, and surface qualities may differ. However in this general discussion they may be viewed as mirror images" [9].

5.2.8   Scale of Adhesive Bond Interactions

As we build our understanding of adhesive bond anatomy, it is important to consider the physical scale of the interactions depicted. Gardner [12] provided a comparison of wood–adhesive interactions ranging in size from the multi-meter level to the cell wall polymer scale of 10^{-8} to 10^{-9} m, thus representing physical size variations of 11 orders of magnitude (Table 5.2).

Table 5.2   *Orders of scale for wood–adhesive interactions.*

Scale	Wood–adhesive interaction
100–1 m	Glulam beam laminates
10^{-1} m or 10 cm	ASTM D2559 cycle delamination specimen [13]
10^{-2} m or 1 cm	ASTM D906 plywood shear specimen [14]
10^{-3} m or 1 mm	Polymer microdroplet on wood or cellulose fiber
10^{-4} m or 100 μm	Microscopic evaluation of wood–adhesive bondline
10^{-5} m or 10 μm	Diameter of bordered pit
10^{-6} m or 1 μm	Smallest resin droplets on medium density fiberboard furnish
10^{-7} m or 100 nm	Scale of cellulose nanocrystals
10^{-8}–10^{-9} m or 10–1 nm	Scale of wood cell wall polymers

Source: Adapted from [12] with permission of John Wiley & Sons, Inc © 2006.

Table 5.3   *Comparison of wood–adhesive interactions relative to length scale.*

Component	Length scale	
	μm	nm
Adhesive force	0.0002–0.0003	0.2–0.3
Cell wall pore diameter	0.0017–0.002	1.7–2.0
Phenol formaldehyde adhesive resin molecular length	0.0015–0.005	1.5–5.0
Diameter of particles that can pass through a pit	0.2	200
Tracheid lumen diameter	4–25	4,000–25,000
Glue line thickness	50–250	50,000–25,0000

Source: Adapted from [12] with permission of John Wiley & Sons, Inc © 2006.

It is conceivable that one might undertake an evaluation of glue bond performance at any of the spatial levels suggested by Table 5.2. The scale of analysis will depend upon the application of the knowledge sought. The industrial specialist will likely confine analyses to the meter (or greater) level down to about 10^{-4} m. Fundamental researchers may have greater interest in the submicrometer scales, but even so, the relationship to bond performance at larger scale is often of more practical application. Within the context of the link model of glue bond anatomy, our focus is on the millimeter to nanometer range.

Glue lines (adhesive film or bulk adhesive), at 50–250 μm in thickness (0.05–0.25 mm), are enormous compared to the scale of fundamental adhesive forces on the order of 0.2–0.3 nm (Table 5.3).

Uncured phenol formaldehyde (PF) adhesive molecules are in a size range that permits their interaction with wood substrates on a cell wall level. Solvent carriers for adhesive molecules (often water) are small enough to penetrate the most minute of cell wall pores, altering physical and/or chemical properties of the substrate in the process. These alterations or modifications likely vary at a molecular level, as a result of inconsistencies in surface characteristics of natural fiber substrates. In some sense, the nine-link glue bond model could, therefore, be expanded to a chain containing an expansive number of links of infinitesimally small size, each within a continuum of progressively smaller interphases.

Further appreciation for the scale of adhesive interactions at the submicrometer scale may be gained by studying Figure 5.2. Here, Frazier [15] diagrams the interaction of a cellulosic substrate (top of figure) with a hydroxymethyl resorcinol (HMR) coupling agent. HMR has been shown useful for improving the durability of wood–epoxy bonds [16]. Notice that the concept of the *interphase* proposed here ranges from 10^{-10} m or 0.1 nm (1 Angstrom, Å in older non-SI terminology) to 10^{-3} m or 1 mm. In this case, the interphase is conceptualized over a scale ranging seven orders of magnitude. One might consider each level of scale a tiny link in the adhesive bond chain, thereby expanding Marra's bond anatomy model exponentially.

Marra's chain link model and the contemporary adaptation of it, along with knowledge of the interphase, provide us with a foundation for adhesion theory. As we have noted, there is no single, global theory of adhesion thus far recognized as encompassing the adhesion phenomenon. In the following sections, we will describe the most commonly accepted theories of adhesion, with a view to applying the appropriate theories to bonding of wood and other porous, lignocellulosic substrates.

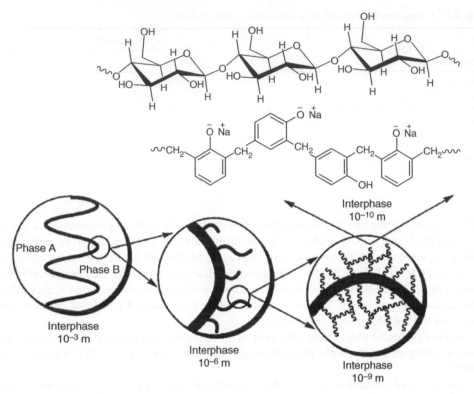

Figure 5.2   *Submicrometer adhesive interactions between cellulose (top) and a hydroxymethylresorcinol (HMR) coupling agent (middle). The interphases may be interpreted over a wide spatial scale (bottom). Reproduced from [15] with permission of Dr. Chip Frazier.*

5.3   Adhesion Theories

5.3.1   Overview of Adhesion Theories

Adhesion is a complex phenomenon involving physical and chemical interactions from molecular to macroscopic scales. As adhesion science has developed, a number of theories have been advanced to describe or model adhesive–adherend assemblies. Here, we will provide just a few examples of how these theories have been grouped in adhesion science literature. For example, Johns [17] listed adhesion theories as follows: Mechanical entanglement, adsorption/specific adhesion, diffusion/molecular entanglement, electrostatic/donor–acceptor interactions, and direct chemical bonding. Pizzi [18] categorized adhesion theories under the names of mechanical entanglement/interlocking theory, diffusion theory, electronic theory, adsorption/specific adhesion theory and covalent chemical bonding theory. Fourche [4] placed mechanical adhesion in a separate class from specific adhesion, the latter including the subcategories of the electronic model, diffusion model, thermodynamic adsorption or wetting model, rheological model, chemical adhesion model, and the model of weak boundary layers. Gardner [12] listed adhesion theories thus: Mechanical interlocking theory, electronic or electrostatic theory, adsorption (thermodynamic) or wetting theory, diffusion theory, chemical (covalent) bonding theory, and theory of weak boundary layers and interphases.

From this collection of literature, we see that although there is some variation in naming of the predominant theories, there is nevertheless convergence to a short list of theories advanced to describe adhesion as observed in various systems. We will use the following taxonomy in our discussion of adhesion theories:

- Mechanical adhesion
- Specific adhesion
  - Electronic theory
  - Diffusion theory
  - Adsorption theory
  - Covalent bonding theory
  - Weak boundary layers

It will be noted that we have elected to place "mechanical adhesion" and "specific adhesion" on equal classification footing, following the outline of Fourche [4] and other authors. As such, the other theories listed here are viewed as particular types of specific adhesion. Admittedly, the decision to separate mechanical adhesion from other adhesion mechanisms is somewhat arbitrary, but we believe it has merit, particularly with respect to porous substrates. In particular, we find this a useful taxonomy for pedagogy of adhesion theory and, therefore, a logical reason for outlining the material in this manner.

Our approach in the remainder of this chapter will be as follows: First, we will describe each one of the adhesion theories as categorized above. Second, we will discuss surface interactions, as these are significant in determining the degree of wetting of substrate by the adhesive system, with a view to providing an indication of the relative importance of the mechanisms as described in wood adhesion literature. Finally, we will look specifically at the factors that are at work in the bonding of wood and natural fiber substrates. If the reader is interested in further background on these theories, including mathematical models, reviews such as those given by Fourche [4] or Schultz and Nardin [19] are recommended as excellent starting points. The adhesion text by Pocius [5] also contains a more rigorous treatment of many of these topics.

5.3.2 Mechanical Adhesion

Mechanical adhesion or mechanical interlocking has long been thought to be a significant bonding mechanism on all porous substrates, for example, wood, paper, fabric, leather, and so on. Adhesive flow and penetration into surface cavities or pores, followed by cure (hardening) of the adhesive, results in an interlocked, solid system of substrate and adhesive. The scale of mechanical interactions ranges from nanoscale to microscale in systems as divergent as polymer-metal, polymer-polymer, or polymer-wood. The significance of surface roughness of adherends as a contributor to adhesion is complicated by the recognition of fractal surfaces in metals and polymers. Fractal surfaces are defined in terms of area according to the means by which the area is quantified [20]. Theoretically, surface area of adherends goes to infinity as the size of the probe used to measure the surface goes to zero. If this is the case, practically all adhesion phenomena could be viewed as a mechanical interaction ranging, in essence, from atomic to macroscopic scale.

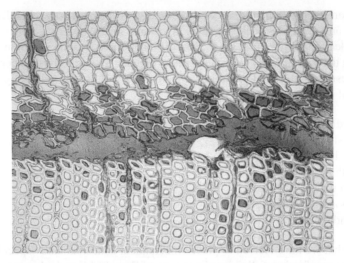

Figure 5.3   *Adhesive penetration into the surface layers of wood substrates. Notice the difference in penetration in earlywood (upper adherend) versus latewood (lower adherend). Reproduced from [21] with permisson of the Forest Products Society © 2007.*

Microscopic observation of adhesive joints containing substrates with porosity on the micrometer level provides a dramatic visualization and perhaps a more intuitive view of this mechanism. As shown in Figure 5.3, wood surfaces may be penetrated to a depth of few to several cell diameters. Cure of the adhesive within the wood cell cavities provides a mechanical linkage to the adhesive layer between the two substrates.

5.3.3   Specific Adhesion

We have already observed that the separation of mechanical and specific adhesion into distinct categories is to some degree, arbitrary. Schultz and Nardin [19] comment thus: "Among [the adhesion theory] models, one usually distinguishes rather arbitrarily between mechanical and specific adhesion, the latter being based on the various types of bonds (electrostatic, secondary, chemical) that can develop between two solids. Actually, each of these theories is valid to some extent, depending on the nature of the solids in contact and the conditions of formation of the bonded system. Therefore, they do not negate each other and their respective importance depends largely on the system chosen."

In reality, the bonding of porous substrates is likely dominated by mechanical interlocking, with other mechanisms of adhesion, that is, specific adhesion, also coming into play to varying degrees dictated by the substrate, adhesive, and conditions of bonding or manufacturing (clamping pressures, hot press environment, etc.). We could also take the view that mechanical interlocking operates primarily on a spatial scale that is significantly larger than the scale associated with specific adhesion (Table 5.4). Thus, specific adhesion devices at minute spatial scales are seen as contributory to the development of the mechanical interlocking phenomenon observed at a higher level of spatial order. Specific adhesion devices are described by electronic, diffusion, adsorption, and covalent bonding theories. Weak boundary layers are also included as a means of understanding adhesive bond performance in terms of failure mechanism.

Table 5.4   *Comparison of adhesion interactions relative to length scale.*

Category of adhesion mechanism	Type of interaction	Length scale
Mechanical	Interlocking or entanglement	0.01–1000 μm
Diffusion	Interlocking or entanglement	10 nm–2 mm
Electrostatic	Charge	0.1–1 μm
Covalent bonding	Charge	0.1–0.2 nm
Acid–base interaction	Charge	0.1–0.4 nm
Lifshitz-van der Waals	Charge	0.5–1.0 nm

Source: Adapted from [12] with permission of John Wiley & Sons, Inc © 2006.

5.3.3.1   *Electronic Theory*

The electronic theory suggests the formation of a double layer of electrostatic charge across the substrate–adhesive interface. Formation of this charge depends on material properties that allow electron transfer across the interface in conjunction with the intimate contact of smooth surfaces. The interactions thus developed are very weak and rather insignificant, so much so that this theory is generally the most controversial of all adhesion theories in that the evidence supporting it is most difficult to derive. Regardless, this mechanism or theory is not important for the bonding of wood or other natural, porous substrates. Pizzi [18] concluded, "Application of this theory to wood adhesion does not appear possible, and no experimental evidence of the existence of this contributory factor has ever been recorded for wood adhesion."

5.3.3.2   *Diffusion Theory*

The diffusion theory applies only when both adhesive and adherend are polymers that are, furthermore, compatible. This is based on the concept that the mutual attraction or autoadhesion within a homogenous polymer system is due to mobility of macromolecular chain ends and segments and the mutual solubility of said molecules. The consequence of such interactions is the disappearance of the interface, equated with the formation of an interphase. This may occur, for example, in synthetic thermoplastic polymers.

Since this adhesion mechanism occurs as interdiffusion of polymeric adhesives and adherends at the interface, it would appear that this mechanism is theoretically possible for wood adherends. In spite of this, the diffusion mechanism has been generally discounted in wood adhesion literature, with the exception of bonding of fiberboard without added adhesive via lignin autoadhesion [18]. However, Hansen and Björkman's study [22] of wood cell wall ultrastructure in terms of solubility parameters may provide a basis for further examination of this theory with respect to wood adherends. Gardner [12] has noted studies of wood thermoplastic matrix and molecular interpenetration of thermosetting polymers into woody cell walls as representative of the diffusion theory applied to wood substrates.

5.3.3.3   *Adsorption Theory*

Although there is no single, global theory of adhesion, the adsorption theory is, at present, the most prevalent, accepted model of adhesion [19]. This is particularly true for polymer-polymer adhesion, and is, therefore, directly applicable to the majority of adhesive–lignocellulosic substrate interactions of practical interest. The adsorption model rests on the premise that intimate contact of adhesive and adherend results in molecular interactions at the interface that are necessary, but not sufficient conditions for effective adhesion. van der Waals and Lewis acid–base interactions are the interfacial forces generally deemed of primary importance. These, then, represent the fundamental molecular interactions contributory to adhesion.

The adsorption theory is also known as the thermodynamic adsorption theory or simply as the wetting model. Wetting or spreading of a liquid adhesive onto a solid substrate is controlled by the surface energy characteristics of both liquid and solid. It is reasoned that once wetting of the substrate occurs, adhesive penetration, followed by cure of the adhesive, will produce an adhesive bond. Without thermodynamic adsorption, the subsequent steps of adhesive bond development cannot occur. Because of the importance of the wetting phenomenon to bond development in lignocellulosic adhesives, we will explore certain surface energy concepts further in Section 5.4.

5.3.3.4   Covalent Bonding Theory

The covalent bonding theory is based on the formation of primary chemical bonds between adhesive and substrate. Since covalent bonds are much stronger than secondary forces such as van der Waals interactions and hydrogen bonds (100–1,000 kJ/mol versus ≤50 kJ/mol), it might be reasoned that covalent bonds are required for good adhesion. However, the sheer number of secondary interactions between substrate and adhesive often produces more than adequate bond strength without the need for covalent bonds.

Do covalent bonds occur in lignocellulose-synthetic adhesive systems? Here is yet another topic in adhesion science on which the jury may still be out. One may find in the literature arguments both for [23] and against [18] the demonstration of covalent bonding in wood–adhesive systems. The structural polymers of wood, cellulose, hemicelluloses, and lignins certainly contain numerous potentially reactive hydroxylated substituents, particularly toward polymeric methylene diphenyl diisocyanate (pMDI), a wood adhesive resin. Although pMDI should be chemically reactive with wood, it readily undergoes a number of competing reactions, especially with water. Using solid state nuclear magnetic resonance (NRM), Zhou and Frazier [24], were unable to distinguish between polyurea and urethane bonds, attributable to reaction with water or cell wall polymers, respectively. Subsequently, Yelle et al. [25], using solution-state NMR, found covalent bonds between model pMDI and wood cell polymer model compounds. However, the same researchers were unable to demonstrate covalent bonds between model pMDI compounds and loblolly pine matchstick samples, leading to the conclusion that "covalent (i.e., wood carbamate) bond formation is not a probable mechanism for pMDI-wood bonding" [26]. Contemporary studies such as these continue to cast serious doubt on the occurrence of and need for covalent bonds in wood adhesion.

5.3.3.5   Weak Boundary Layers

The theory of weak boundary layers is also known as the Bikerman Model, so-named for the theory's originator [27]. While often categorized and discussed alongside adhesion theories, weak boundary layers is not an adhesion theory per se. Rather, it provides a framework for understanding adhesion in terms of failure. Central to this framework is the notion of a boundary layer or interphase.

The boundary layer is an interphase with properties that match neither the adhesive nor the adherend. In this interphase or weak boundary layer, there are agents other than the adhesive or adherends per se which may increase the probability of glue bond failure. "This concept of weak boundary layers is important because it introduces the notion of 'interphase' whose properties determine the joint strength" [4]. Thinking back to the contemporary chain-link analogy of glue bond anatomy (Figure 5.1), these interphases correspond to the links represented by dashed lines, that is, those links "most subject to malformation" [9].

Bikerman [7] originally proposed that when an adhesive joint is formed, air is usually present, thus air, adhesive, and adherend may combine to potentially form seven classes of weak boundary layers: Air, adhesive, adherend, air plus adhesive, air plus adherend, adhesive plus adherend, or all three. It is now recognized that other agents (e.g., contaminants, extractives) may also be present and have the potential to create a weak boundary layer.

The theory of weak boundary layers is based on observation of adhesion failure or failure propagation. When mechanical failure of an adhesive bond occurs, it often appears to be located either in the interphase or in the wood adherends. When adhesive joints fail in bonded wood materials, it is manifested as a cohesive failure of adherend(s). However, it is argued that when gluebond failure occurs, it is initiated within the weak boundary layer, but quickly proceeds to failure within the adherend(s). This is interpreted as follows: "When failure seems to be in adhesion—that is, the separation appears exactly along the adherend-adhesive interface—usually a cohesive break of a weak boundary layer is the real event" [7].

With respect to wood adherends, the classes of weak boundary layers that have been generally recognized include air (attributable to poor wetting of substrate by the adhesive), contaminants (in adhesive or adherend), or reaction products (resulting, e.g., from chemical interactions of air and adherend at the interface). An understanding of these classes of weak boundary layers is useful when troubleshooting adhesive joint failures.

Stehr and Johansson [11] proposed a distinction between "chemical weak boundary layers" (CWBL) and "mechanical weak boundary layers" (MWBL) for wood adherends (Figure 5.4). The segregation of WBL into CWBL and MWBL provides a conceptual model for understanding potential sources of adhesive joint failure when bonding wood and other lignocellulosic substrates.

With respect to CWBL, a distinction is made between deactivation of the adherend surface and actual chemical contamination. Deactivation refers to the degradation of glue bond performance when bonding aged wood surfaces, which have likely undergone oxidation, versus the adhesive joint performance when bonding freshly cut or machined wood surfaces. The deactivation phenomenon is not viewed as a CWBL, as this distinction is applied to chemical contamination, including migration or diffusion of wood extractives into the adhesive at the interface [11]. The authors consider an MWBL to consist "mainly of a layer of fibers of lower mechanical strength than the bulk wood." As an example, consider the wood surface damaged by machining processes as shown in Figure 5.5. Damage to the surface fibers in a case such as this could result in an MWBL.

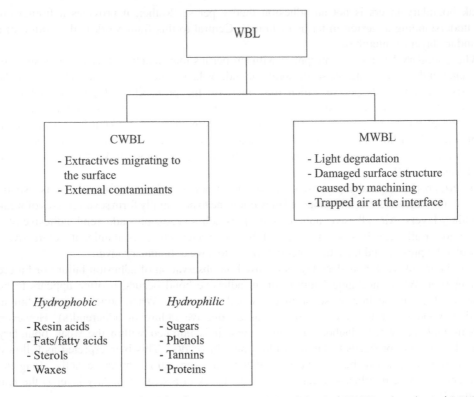

Figure 5.4   *A taxonomy of weak boundary layers (WBL), consisting of chemical (CWBL) and mechanical (MWBL) weak boundary layers. Adapted from [11] with permission of Taylor & Francis Ltd © 2000.*

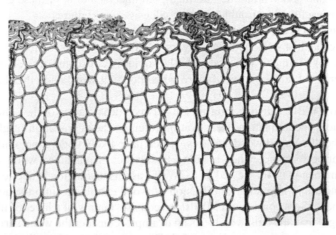

Figure 5.5   *Wood surface damaged by machining, followed by sanding with 80-grit sandpaper. Note crushing of earlywood tracheids at the surface. Light micrograph, cross-section of Douglas fir lumber. Reproduced from [21] with permisson of the Forest Products Society © 2007.*

5.4 Surface Interactions

5.4.1 Surface Interactions Are Critical Determinants of Adhesion

The theory of weak boundary layers does not represent an adhesion theory per se, but it is a useful means to describe adhesion performance in terms of glue bond failure mechanisms. To develop a more thorough understanding of adhesive bonding, we need to consider the physicochemical surface interactions of critical importance with respect to wetting of the adherend by the adhesive. It is often considered that wetting is the first prerequisite to adhesive bonding. Furthermore, wetting is an implicit assumption in the adsorption theory of adhesion. Poor wetting of adherends by an adhesive may be thought of as the development of a weak boundary layer at a critical adhesion interface. Surface interactions determine the degree of wetting expected when a liquid contacts a solid, and are thus primary determinants of adhesion performance.

Surfaces occur as a result of the interface between differing materials or states of matter. A solid wood substrate interfaces with the surrounding atmosphere at its surface. A liquid adhesive in an open container has surfaces at the interface with the wall of the container and at the top of the liquid where it contacts the atmosphere. When a liquid adhesive is spread onto a wood substrate, solid–liquid surfaces are created. Interactions at surfaces such as these are important determinants of adhesion. In the subsections that follow, we will discuss concepts of surface interactions that are important for a fundamental understanding of adhesion phenomenon. Instructors or students desiring a more thorough treatment of surface interactions and the work of adhesion are referred to Chapter 4 of Pocius [5].

5.4.2 Surface Energy of Liquids and Solids

Matter in various states—gas, liquid, or solid—has energy properties associated with it. Think, for example, of water vapor, liquid water, and ice. Water vapor, that is, water in gas form, has higher energy than liquid water, which in turn has higher energy than solid ice. The energy of the molecules within the bulk substance results in variations of intermolecular spacing, dictating the physical state of matter. The laws of thermodynamics inform us that physical systems tend to exist in states that minimize potential energy and maximize entropy or disorder.

Regardless of the state of matter, the molecules within the material in question have interactions with one another that either attract or repel. The molecules are attracted and repelled in such a way that potential energy is minimized. Bulk molecules (such as those on the inside of an adhesive droplet) experience uniform interactions that minimize the potential energy within the material (Figure 5.6).

Surface molecules (such as molecules in an adhesive droplet exposed to the surrounding atmosphere) encounter discontinuous interactions. In liquids, this results in higher energy at the surface. This surface energy, or surface tension, results from the cohesion of the molecules within the liquid. Insects, such as water striders, take advantage of this surface tension property of water as they rest or move along the surface of the water. A thin thread or string may also be suspended on the surface as a result of water's surface tension.

Surfaces result from the interaction of differing chemical substances or differing states of matter. As is often the case in adhesive systems, interactions occur between a solid substrate and a liquid adhesive. Surface energy is a significant determinant of how these solids and liquids interact. In the following subsections, we will discuss this important phenomenon.

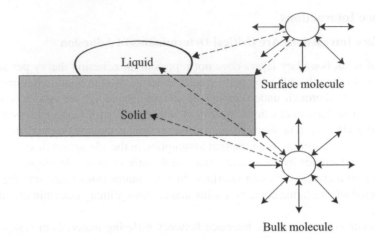

Figure 5.6   *Surface energy: Molecules within a bulk material and at surfaces (interface between two materials) interact with each other such that energy is minimized. Bulk molecules experience uniform interactions, as depicted by arrows, whereas surface molecules encounter discontinuous interactions resulting in higher surface energy, known as surface tension.*

5.4.2.1   *Surface Energy of Solids*

Bonds must be broken to create surfaces. The resulting "excess energy in the surface" from the breaking of bonds between two substances or phases is the surface energy, γ. The most fundamental concept of surface energy is conceptualized not for two substances, but for a single, uniform material. If an ideal, homogeneous material is separated such that two identical, uniform surfaces are produced, the work expended in so doing is called the work of cohesion, W_{coh} (Figure 5.7).

The work of cohesion is given by Equation 5.1:

$$W_{coh} = 2\gamma \tag{5.1}$$

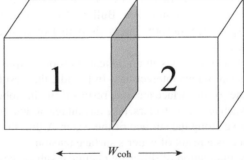

Figure 5.7   *Work of cohesion. If a homogeneous material is cleaved into two parts (1 and 2), with each new surface having the same cross-sectional area (shaded), the work expended will be twice the surface energy of the material, that is, $W_{coh} = 2\gamma$.*

where

W_{coh} = work of cohesion of a normalized area of two identical surfaces/materials in a
vacuum, and

γ = surface energy of the material.

Since the work needed to break a material into two surfaces is exerted over an area, the units associated with the surface energy defining the work may be expressed as free energy per unit area. Surface energy, γ, may therefore be expressed as Joules per square meter (J/m^2 or $J\ m^{-2}$).

With respect to adhesion, surface energy provides an indication of the degree of interaction expected when liquids or solids are brought into contact with one another. If we know the surface energy of adhesive and adherend, we will have an indication of their "compatibility" in at least one sense of practical importance.

The magnitude of surface energy depends not only on the specific substance(s) in question but also on the states of matter at the surface interface, for example, solid–vapor, solid–liquid, or liquid–vapor (S–V, S–L, or L–V). The surface energy associated with each is represented by γ_{SV}, γ_{SL}, and γ_{LV}, respectively.

Surface energy may be expressed either as energy per unit area, or as force per unit length (in the latter case, think of a thread suspended on water due to surface tension). If expressed in force per unit length, conventional SI units are Newtons per meter (N/m or $N\ m^{-1}$). From a thermodynamic standpoint, expression of surface energy as energy per unit area is preferred, in which case the SI units are Joules per square meter (J/m^2 or $J\ m^{-2}$). Since the values are very small numbers if reported in J/m^2, the units typically reported for surface energy of solids and liquids are millijoules per square meter (mJ/m^2), where a millijoule is 0.001 Joule.

Organic polymer surfaces are considered low energy surfaces, with typical surface energy values of $\gamma \leq 100$ mJ/m^2. Hard inorganic surfaces, such as metals or ceramics are high-energy surfaces with $\gamma \geq 100$s to 1000s mJ/m^2. As natural, organic polymers, wood and other lignocellulosic materials tend to have surface energy values of less than 100 mJ/m^2.

5.4.2.2 Surface Tension of Liquids

Surface tension is the surface energy of a liquid at a liquid–vapor interface, γ_{LV}. In this case, the vapor in question is the vapor of the associated liquid. So we could say that γ_{LV} is the "surface energy of a liquid against its vapor." Polar molecules have higher surface energy than nonpolar molecules. Typical surface energies of several liquids are shown in Table 5.5. It is apparent that water has a significantly greater surface energy than the other liquids shown in

Table 5.5 Surface tension of several liquids.

Liquid	γ_{LV} at 20°C, mJ/m^2
Water	73.05
Glycerol	63.40
Xylenes	28.37–30.10
Benzene	28.85
Acetone	23.70
Ethanol	22.75

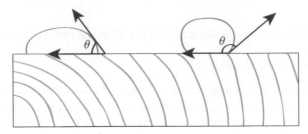

Figure 5.8   *Contact angle (θ) of sessile drops of liquid on a planar solid surface. The angle may be either obtuse (left) or acute (right), depending on the relative surface energies of liquid and solid.*

this table. It should be noted that water is the solvent-carrier of many of the adhesive systems employed in the wood-using industries.

5.4.2.3   *Contact Angle*

When a drop of liquid is placed on a solid surface, it is often observed that the liquid assumes a semi-spherical shape. Raindrops on a freshly waxed automobile form rounded "beads" of water that run over the surface, whereas raindrops falling onto a weathered wooden board often disappear quickly into the wood without first forming a noticeable bead. The difference in behavior of water on varying solid surfaces like these is attributable in large measure to the variation of surface energies of the solids in question.

The interaction of water or other liquids on a solid surface may be described in terms of a contact angle (Figure 5.8). The contact angle, θ, is the angle between the plane of the solid surface and the tangent of the drop's surface.

Contact angle is determined by the properties of the liquid and solid. Chief among these properties is the surface energy. Although there are more sophisticated means of predicting the behavior of a liquid drop on a solid surface, one simple and practical approach is to inspect and compare surface energy and surface tension values. For example, Table 5.6 shows surface energies of maple and red oak veneer and particles, as contrasted with the surface tension (surface energy) of water, glycerol, acetone, and ethanol.

In Table 5.6, we see that the surface energies of the wood materials, ranging from 44.33 to 53.73 mJ/m^2, are intermediate between the higher surface tension liquids, water and glycerol (73.05 and 63.40 mJ/m^2, respectively) and the lower surface tension liquids, acetone, and ethanol (23.70 and 22.75 mJ/m^2). The variation in surface energy between veneer and particles

Table 5.6   *Surface energy, γ_S, of selected wood materials and surface tension, γ_{LV}, of four liquids.*

Wood material	γ_S, mJ/m^2	Liquid	γ_{LV} at 20°C, mJ/m^2
Maple veneer	53.73	Water	73.05
Maple particles	45.54	Glycerol	63.40
Red oak veneer	50.12	Acetone	23.70
Red oak particles	44.33	Ethanol	22.75

Source: Wood surface energies from Table 5 of Gardner per the GVOC model [28].

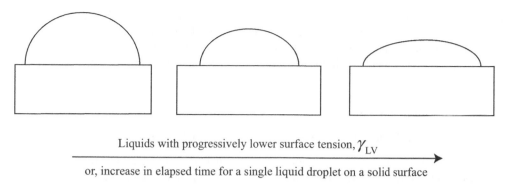

Liquids with progressively lower surface tension, γ_{LV}

or, increase in elapsed time for a single liquid droplet on a solid surface

Figure 5.9 *Repose of liquid drop(s) on a solid, planar surface, representative of liquids having decreasing surface tension, from left to right, on a surface with constant surface energy. The sequence could also represent a single drop of liquid with a constant surface tension as it spreads onto (wets) a solid surface over time.*

of the same species is attributed primarily to differences in the temperature at which the materials are processed during manufacturing. Although other factors are in play to determine the contact angle of the liquids on the solid wood materials, we may predict that those liquids with a larger surface tension than the wood surface energy will tend to form a drop with a large contact angle. In other words, this comparison suggests that water or glycerol will "bead up" on the wood materials. Conversely, those liquids with a surface tension smaller than the wood surface energy will form a small contact angle, that is, the liquids will rapidly spread over the wood surface without forming a noticeable bead. Thus, we would predict that the acetone and ethanol will be quickly absorbed into the wood, that is, we will observe "complete wetting" of these liquids on the wood surface. This is something you could easily test for yourself in the laboratory. Try it!

Contact angle may also be affected by time. In Figure 5.9, the three different drops depicted have significantly different contact angles, which may be interpreted in two ways, that is, as a time sequence, or as a surface tension sequence.

First, the variations in contact angle may be attributed to liquids in a decreasing progression of surface tension if the substrate is assumed identical for each case. That is, the sessile drop on the left is formed by a liquid with the largest surface tension, the sessile drop in the center by a liquid with an intermediate surface tension value, and the sessile drop at the right by a liquid of lowest surface tension. On the other hand, these three scenarios may represent a time sequence of a single drop, representative of dynamic contact angle. This would be the case if the liquid (type of liquid and total drop volume) and the substrate are identical in each of the three cases, with the drop on the left at time zero (immediately after placement onto the substrate), and the two other drops representing a time lapse over which the drop progressively spreads over the surface of the substrate. This is dynamic wetting behavior of liquid on solid.

Both sessile drop contact angle and dynamic contact angle are related to surface energy. In fact, the contact angle may be used to determine surface energy. The more common methods of measuring the surface energy of a liquid are the sessile drop, capillary rise, Wilhelmy plate, or duNouy Ring methods [29, 30].

Variations in contact angle, or contact angle hysteresis, may arise from advancing and receding of the liquid–solid interface or from microscopic "surface inhomogeneities" (roughness) of the substrate. Another cause of hysteresis is "heterogeneous contamination" of surfaces with low-energy impurities. In the case of wood adherends, such "impurities" or "contaminants" are typically hydrophobic extractives. Dynamic contact angle measurement (e.g., Wilhelmy plate or duNouy Ring methods) helps to account for these variations during data collection.

5.4.3   Wetting Phenomenon

5.4.3.1   *Adequate Wetting Is a Prerequisite to Effective Adhesion*

Regardless of adhesion mechanism, optimal adhesion is dependent upon effective contact of adhesive and adherend. Effective contact is, in turn, dependent upon the surface-wetting phenomenon as visualized by contact angles of liquid adhesives onto solid adherends. We might go so far as to say that adequate wetting of a substrate by the adhesive is the first prerequisite of adhesive bonding.

Beyond simply observing this phenomenon, wetting behavior may be modeled by Young's Equation, and the critical surface energy at which complete wetting occurs may be determined by a Zisman Plot. These mathematical and graphical tools are introduced as a means to further our understanding of the first prerequisite of adhesion.

5.4.3.2   *Young's Equation*

In the year 1805, Thomas Young proposed a mathematical relationship to explain the equilibrium of a drop of liquid on a solid surface [31]. Fourche [4] described this equilibrium as follows: "When a droplet of liquid, L, with its vapor, V, is at rest on a solid surface, S, it takes a configuration which minimizes the energy of the system and highlights the liquid–solid interactions." These interactions are represented visually in Figure 5.10, and the mathematical expression for the equilibrium is given by Young's Equation (Equation 5.2):

$$\gamma_{SV} = \gamma_{SL} + \gamma_{LV} \cos\theta \qquad (5.2)$$

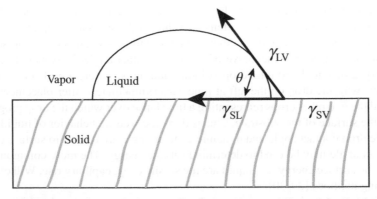

Figure 5.10   *Surface interactions of solid, liquid, and vapor interfaces. Symbols are defined with Equation 5.2 in the text.*

where

γ_{SV} = surface energy of the solid–vapor interface,
γ_{SL} = surface energy of the solid–liquid interface,
γ_{LV} = surface energy of the liquid and its vapor, and
   θ = contact angle formed by the liquid drop and solid.

This expression provides us with a theoretical picture of the equilibrium of a liquid drop on a surface. Although direct calculation of the contact angle is impeded by our inability to measure all of the parameters required, for example, γ_{SL} is a good example of a quantity that cannot be determined experimentally [32], we should nevertheless note here that a solution to Young's Equation would provide a criteria for wetting:

If $\theta = 0°$, spreading or complete wetting occurs.
If $\theta < 90°$, wetting is favorable.
If $\theta > 90°$, wetting is not favorable.

Another expression of Young's Equation is as follows (Equation 5.3):

$$\gamma_{LV} \cos \theta = \gamma_{SV} - \gamma_{SL} = \gamma_S - \gamma_{SL} - \pi_e \tag{5.3}$$

where

γ_S = surface energy of a solid measured in a vacuum, and
π_e = equilibrium spreading pressure of the adsorbed vapor of the liquid on a solid.

The equilibrium spreading pressure, π_e, is introduced because the surface energy of the solid-saturated vapor interface, γ_{SV}, is sometimes less than the surface energy of the solid as measured in a vacuum, γ_S. Thus, we could express π_e as follows (Equation 5.4):

$$\pi_e = \gamma_S - \gamma_{SV} \tag{5.4}$$

Now, π_e is very small and may be neglected for most polymer materials [19]. Recalling that complete spreading or wetting occurs if the contact angle, θ, is equal to $0°$, and that the cosine of $0° = 1$, we may define a condition for spontaneous wetting as given in Equation 5.5:

$$\gamma_S \geq \gamma_{SL} + \gamma_{LV} \tag{5.5}$$

This condition is sometimes presented as a spreading energy or spreading coefficient, S (Equation 5.6):

$$S = \gamma_S - \gamma_{SL} - \gamma_{LV} \tag{5.6}$$

If the spreading coefficient, S, is ≥ 0, then spontaneous wetting is indicated.

   Equations 5.5 and 5.6 inform us that complete wetting will occur if the surface energy of the solid exceeds that of the surface energies of the solid–liquid and liquid–vapor interfaces combined. In Section 5.4.2.3, we suggested that wetting could be predicted simply by comparing the surface energy of the solid, γ_S, with the surface tension of a liquid in equilibrium with its vapor, γ_{LV}. In essence, we were using the spreading criterion of Equation 5.5, assuming that we could ignore the experimentally unmeasured surface energy of the solid–liquid interface, γ_{SL}. In other words, we assumed a spontaneous wetting condition exists if $\gamma_S \geq \gamma_{LV}$. Although this approach will work in some cases, the fact of the matter remains that this comparison is based on an assumption that there are many factors determining wetting behavior that may be ignored. In many instances, such assumptions are neither justified nor accurate. As a result, the concept of critical surface energy is helpful here.

5.4.3.3   *Critical Surface Energy: Zisman Plot*

Recognizing a practical need to predict wetting behavior while simultaneously accounting for a number of difficult-to-determine parameters, Zisman and co-workers [33–35] proposed the concept of critical surface tension. Critical surface tension, γ_C, is an empirically derived surface tension at which complete wetting occurs. When we say that this factor is determined empirically, it means that the value is found by conducting experiments. In this case, the experiments involve the measurement of contact angle of several liquids on the same solid surface. The series of liquids used in such experiments are called the probe liquids. The cosines of the contact angles of the probe liquids are then plotted against the surface tension of the liquids. Extrapolation of the plot to the point at which the cosine of the contact angle is equal to 1 (i.e., $\cos \theta = 1$ and $\cos^{-1} \theta = 0°$) provides the critical surface tension, γ_C. Furthermore, as shown in Equation 5.7:

$$\cos \theta = 1 + b\,(\gamma_C - \gamma_{LV}) \tag{5.7}$$

where

$\theta =$ measured contact angle,
$b =$ the slope of the line,
$\gamma_C =$ critical surface energy, and
$\gamma_{LV} =$ liquid surface energy (surface tension).

Ideally, experiments for the determination of critical surface tension are conducted on a planar, nonporous solid surface. This is obviously not the case with wood surfaces. Nevertheless, such experiments can provide insight into surface characteristics and wetting behavior of wood surfaces. The data from one such investigation are presented here in order to illustrate how critical surface tension is determined (Table 5.7). In this experiment, never-dried Southern yellow pine heartwood samples were surface-inactivated by drying at 200°C for several hours. Following this treatment, the wood was probed with liquids having different, known, surface tension values. The contact angle of each probe liquid on the inactivated wood surface was measured and is shown in Table 5.7, along with the calculated cosine of contact angle.

Using data from Table 5.7, a Zisman Plot may be constructed (Figure 5.11). A Zisman Plot is simply a plot of the cosine of contact angle versus the surface tension of the probe liquid.

Table 5.7 *Surface tension of probe liquids and contact angle, θ, on the surface of heat-inactivated Southern yellow pine heartwood (the cosine of contact angle is also given).*

Probe liquid	Surface tension, mN/m	Contact angle, θ, degrees	Cosine θ
Water, deionized	72.8	100.8	−0.19
Glycerol	64.0	83.0	0.12
Formamide	58.3	55.1	0.57
Ethylene glycol	48.3	46.4	0.69
Bromonaphthalene	44.6	7.7	0.99

Source: Reproduced from [36] with permission of Dr. Milan Šernek.

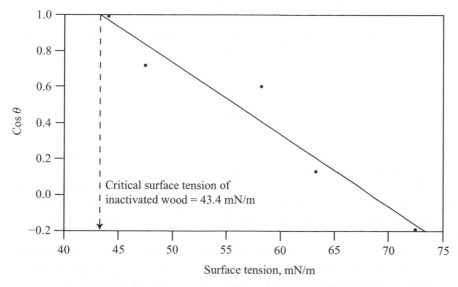

Figure 5.11 *Zisman Plot of the data given in Table 5.7. The critical surface tension, γ_C, is the surface tension value when cos θ equals 1. Reproduced from [37] with permission of Dr. Milan Šernek.*

The Zisman Plot allows for the estimation, by extrapolation, of the regression line to the point at which cos θ is equal to one, or by calculation if the regression equation is known. Here, the linear regression for this set of empirical data is given by cos $\theta = -0.0398\gamma + 2.7279$. Thus, for cosine $\theta = 1$, $\gamma_C = (1 - 2.7279)/ - 0.03982 = 43.4$ mN/m, as shown on the horizontal axis. This value represents the surface energy at which complete wetting of the solid substrate will occur.

As illustrated in the foregoing example (Table 5.7 and Figure 5.11), the Zisman Plot is useful for estimating the surface energy of a substrate at which complete wetting will occur. This estimate may be used to evaluate the relative effects of various wood surface treatments, for example, or to compare alternative adhesive formulations. As we shall see, there are some techniques that may be used to adjust both substrate surface energy and the surface tension of liquids. This is important to promote wetting, the first prerequisite of adhesion.

Sidebar 5.3 Surface Energy: There Is More to the Story

The surface energy of a solid material is a significant determinant of its interaction with a liquid. As we have considered some basic concepts of surface energy, we have done so with a view that the total surface energy is the key factor that determines wetting behavior. While this is true, it is also true that, as with most topics, there is more to the story if one cares to look. Surface energy is generally considered to be the sum of two main components, namely nonpolar and polar components. These are attributed to London dispersive forces and acid–base interactions, respectively [28, 32]. The magnitude of surface energy of natural materials, γ_S, found in much of the literature is often in agreement with the dispersive component of material surface energy, γ_S^D. The dispersive component, γ_S^D, is generally greater than 40 but less than 50 mJ/m^2 for lignocellulosic materials. Some references report total surface energy, γ_{tot}, with values generally in excess of 50 mJ/m^2 [38]. Regardless of whether we use γ_S or γ_S^D, these natural organic polymer surfaces are considered low-energy surfaces.

5.5   Work of Adhesion: Dupré Equation

In Equation 5.1, we observed that the work of cohesion is twice the surface energy of the homogeneous material in question, that is, $W_{coh} = 2\gamma$. This has been described thus: "An elastic material of unit cross-section is subjected to a tensile force. The material breaks, creating two new surfaces. [Since] the new surfaces are each made of the same material, then the total energy expended must be twice the surface energy of the material" [5]. This energy expended, or work of cohesion, is descriptive of the internal forces holding a single type of material together.

Within an adhesive joint, we minimally have two types of materials, the adhesive and adherend. Within the joint, these materials are in intimate contact. If they are subjected to a tensile force that pulls them apart, then the energy expended in so doing should be the sum of the surface energies of each material. But, "because the two dissimilar materials were in contact, there were intermolecular forces present that are now missing since the materials were separated. That is, an interfacial energy may have been present before the materials were split apart. As this energy is missing after the two surfaces are separated, we must subtract it from the energy used to create the two new surfaces" [5]. This phenomenon is described by the concept of the work of adhesion (Figure 5.12) and is modeled by the Dupré Equation (Equation 5.8).

The Dupré Equation (Equation 5.8), dating to 1869 [30], provides a mathematical expression of this phenomenon:

$$W_A = \gamma_1 + \gamma_2 + \gamma_{12} \tag{5.8}$$

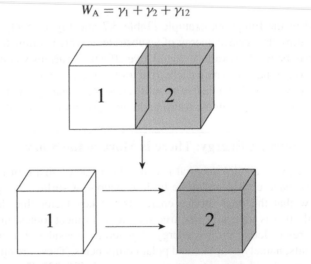

Figure 5.12 *Work of adhesion. The separation of a unit area of two, initially joined dissimilar materials (1 and 2 in upper half of diagram), requires that work is expended to create the two new surfaces, one on each of the materials (lower half of diagram). The work expended consists of the surface energies of the two materials, less an "interfacial energy" component as modeled by the Dupré Equation. Compare to Figure 5.7.*

where

W_A = work of adhesion;
γ_1 = surface energy of material 1;
γ_2 = surface energy of material 2; and
γ_{12} = interfacial energy between materials 1 and 2.

If we think of the work of adhesion in terms of a solid–liquid system, we could re-formulate Equation 5.8 as shown in Equation 5.9:

$$W_{SL} = \gamma_S + \gamma_{LV} - \gamma_{SL} \tag{5.9}$$

We know from Young's Equation, as expressed in Equation 5.3, that if we neglect the spreading pressure, π_e, then $\gamma_{LV} \cos \theta = \gamma_S - \gamma_{SL}$. If we then substitute $\gamma_{LV} \cos \theta$ into the Dupré Equation as expressed in Equation 5.9, we have:

$$W_{SL} = \gamma_{LV} + \gamma_{LV} \cos \theta = \gamma_{LV}(1 + \cos \theta) \tag{5.10}$$

Since the liquid surface tension, γ_{LV}, and the contact angle, θ, are easily obtained or measured parameters, we have here a method to calculate the work of adhesion or "adhesion energy" in solid–liquid systems.

5.6 Lignocellulosic Adherends

5.6.1 Adhesive Resin–Substrate Interactions

Our consideration of surface tension and surface energy has been driven by the premise that wetting of a substrate by an adhesive is a necessary condition toward the development of an adhesive bond. We have gone so far as to say that wetting is the first prerequisite of adhesion. When it comes to bonding of lignocellulosic substrates, the goal is to promote effective resin–furnish interactions. Ideally, even distribution of resin within adhesive joint will promote intimate contact between resin molecules and the furnish surface. In a practical sense, this involves liquid adhesive application, droplet formation, and fluid motion as materials are consolidated. An example would be the gluing of two solid pieces of wood together by first spreading adhesive onto the wood surface, followed by clamping the assembly together to allow for effective contact of the adhesive and substrate as the adhesive cures. This process is represented in Figure 5.13.

The surface interactions that dictate wetting behavior are in action in part (a) of Figure 5.13. These mechanisms continue to operate as the two substrates are brought together in part (b), with the wetting phenomenon now aided by the application of external force. Molecular motion, driven by clamping or consolidation pressure, accompanies the thermodynamic spreading/wetting process. Flow of adhesive into surface cavities of the substrate follows, driven by pressure, wetting dynamics, and capillary tension.

5.6.2 Adhesion as a Surface Phenomenon

Good adhesion requires transfer of resin to substrate, wetting of the substrate surface, controlled penetration into the substrate, flow of resin, gelling of the resin in a continuous film (applicable primarily to bonding of solid wood and veneer), cure of resin in place (i.e., little shrinkage

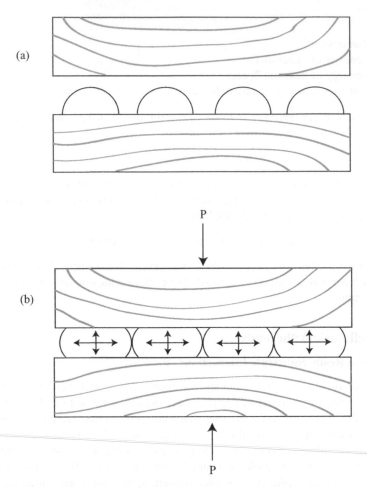

Figure 5.13 *Consolidation of two solid wood substrates with a liquid adhesive. (a) Liquid adhesive resin on the surface of the lower substrate wets the surface via thermodynamic interactions; (b) Wetting and adhesive penetration into both substrates is aided by application of clamping or consolidation pressure (P). Reproduced from [9] with permission from Springer Science + Business Media B.V © 1992.*

as it cures), and cure durability of the adhesive. Regardless of the adhesives or materials in question, all of these processes are essentially surface phenomena, occurring over a scale rarely exceeding a few millimeters in depth. When bonding very smooth, polished surfaces, this fact is readily acknowledged. The same is true with lignocellulosic, natural fiber surfaces, which are relatively rough and most definitely porous. Even so, adhesive penetration into substrates such as wood is typically limited to a depth of several cell diameters, amounting to no more than a few hundred micrometers (Figure 5.14).

5.6.3 Lignocellulosic Adherends Present Challenges to Adhesion

As the product of biological processes, lignocellulosic materials have characteristics that may aid or impede adhesion. For instance, we know that although the fundamental chemical

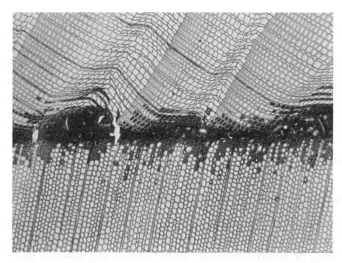

Figure 5.14 *Surface nature of adhesion: adhesive (dark material filling cell lumens) penetration is limited to a few cell diameters. Light micrograph of a fingerjoint between two solid wood substrates. Reproduced from [21] with permisson of the Forest Products Society © 2007.*

composition of plant cell walls is largely predictable for any plant species of interest, there are nonetheless variations with species and within individual plants that alter surface characteristics. These characteristics include the surface chemistry, which is a function of cell wall polymer composition in combination with the nonstructural chemical extractives. The latter may actually dominate the surface chemistry, with particular influence on the surface energy which is so important to wetting of substrate by adhesive. Surface chemistry is also influenced by mechanical, chemical, or thermal treatments introduced when plant materials are subjected to manufacturing processes.

Mechanical processing, in particular, has a significant effect on surface roughness and porosity. Whether it is a sawing or planing process on solid wood, or a chipping or fiberizing process implemented on plant fiber, the resulting surface topography of the material serves as an important variable affecting wetting and penetration of liquid adhesives. However, despite these challenges to adhesion, there are certain variables that are recognized as of primary importance in determining overall adhesion performance. These variables are highlighted in the following section, using wood as our teaching exemplar. As you study these variables, bear in mind that the fundamental principles illustrated with wood will also apply to other lignocellulosic, natural fiber adherends.

5.6.4 Wood Adherend Variables

5.6.4.1 *Wood Characteristics Affecting Adhesive Bond Performance*

A number of wood characteristics can affect the performance of an adhesive. These characteristics may be called "wood adherend variables." Wood adherend variables have been long recognized in a variety of ways. Marian [39] discussed surface properties of wood relative to adhesive bonding, citing surface geometry or topography, physicochemical characteristics, and "degree of intactness" of the wood substrate (extent of mechanical damage) as of importance. Collet [40], in a review of wood adhesion literature, cited surface wettability, specific gravity, "species factors," earlywood–latewood differences (in wettability), surface aging, surface roughness, and surface porosity as important factors in determining the performance of wood adhesive bonds. These examples provide an indication of the various material factors to be considered for bonding of natural fiber materials. Given the complex nature of adhesion in general, and of adhesion to lignocellulosic substrates in particular, it may be difficult to reduce the list of substrate adhesion variables to a manageable size. However, we would like to provide in the following section, a short list of what we have called the "top five wood adherend variables."

5.6.4.2   The "Top Five" Wood Adherend Variables

Perhaps we should not be so presumptuous as to venture a "top five" or even a "top ten" list of substrate variables, unless our purpose is understood. The purpose of this list is to summarize what may be nearly universally understood as key material characteristics that affect the adhesion of wood. As a result, this list is neither comprehensive nor prioritized. It is intended as a pedogical tool from which further discussion and understanding of the complexities of adhesion may be developed. So, with these caveats, here are the "top five" wood adherend variables: surface chemistry, surface roughness, permeability (and porosity), specific gravity, and moisture content.

5.6.4.2.1   Surface Chemistry   Although our top five list is not necessarily prioritized according to importance of each variable, surface chemistry is given the distinction as number one, due to its influence on surface energy. As we now know, the surface energy of a material is contributory to wetting by an adhesive. If the adhesive cannot wet the surface, the substrate cannot be adhesively bonded.

The inherent chemical composition of wood and other lignocellulosic substrates is obviously important to surface chemistry. However, these substrates undergo a variety of processes as they are converted from raw material into final products. Contamination by airborne particles, dirt, grease, and other foreign substances generally has a negative effect on surface properties. Exposure of wood to the air results in oxidation. Further oxidation of surfaces due to heating results in poor wetting. Wood surfaces may be heated by friction during machining (sawing, planing, chipping) or by drying. During drying processes, natural chemical extractives may be driven to surface. This is especially true of kiln drying of wood at elevated temperatures, jet drying of veneer, and of the drying processes used for particles and fibers. Many wood extractives are hydrophobic, with an adverse affect on surface energy. Further, they may inhibit curing reactions of adhesives, forming a weak boundary layer. Extractives that have migrated to the surface may also form a physical barrier, blocking micropores in the wood, thus impeding wetting and surface penetration of adhesives. In extreme cases, woods with high extractives content (>20%) are rendered non-bondable. A number of tropical woods fall into this category.

Table 5.8 provides some insight into the variation of surface energy between selected types of plant material, with or without chemical treatment. This table shows the dispersive component of surface energy for several types of lignocellulosic materials, with and without selected chemical treatments. As organic polymer systems, these are all materials with low surface energy ($\gamma_S < 100$ mJ/m^2).

Table 5.8 *Dispersive surface energy component, γ_S^D, for several lignocellulosic materials.*

Material	Treatment	γ_S^D at (T, °C), mJ/m^2
Cellulose	Untreated	48.5 (25°C)
Birch wood	None	43.8 (50°C)
Softwood Kraft pulp, bleached	Untreated	37.9 (40°C)
Lignin	Kraft	46.6 (50°C)
Eastern white pinewood	Unextracted	37 (40°C)
Kenaf powder	Unextracted	40 (40°C)
Eastern white pinewood	Extracted with toluene/ethanol (2:1 v:v)	49 (40°C)
Kenaf powder	Extracted with toluene/ethanol (2:1 v:v)	42 (40°C)

Source: Reproduced from [32] with permission of Taylor and Francis Group LLC-Books © 2005.

A comparison of interest here is that represented by the unextracted and extracted Eastern white pinewood and kenaf. The surface energy of the unextracted (i.e., not chemically treated) materials are relatively close, at 37 and 40 mJ/m^2 for pine and kenaf, respectively. Extraction with organic solvents increases surface energy of both materials, but more so for the pinewood. The extracted pine has a surface energy of 49 mJ/m^2 and the kenaf 42 mJ/m^2. Pine heartwood contains a rich variety of extractives, consisting primarily of terpenoid resins and resin acids. In contrast, kenaf has only a small amount of organic-soluble extractives. It is evident that removal of non-polar extractives from pinewood significantly increases surface energy. Similar results would be expected for other lignocellulosic substrates. This suggests that it is possible to use processing treatments to exercise a degree of control over the surface energy of natural fibers.

In industrial application, chemical extraction with organic solvents of lignocellulosic substrates solely for the purpose of altering surface energy is not ordinarily practical from an economic standpoint. However, there is contemporary interest in extraction of hemicelluloses from wood strands as a means to reduce hygroscopicity, increase dimensional stability and improve resistance to biodeterioration of wood strand products. Although surface energy modification is not the primary objective of such efforts, it is an important result. For example, hot water extraction of Southern yellow pine strands increased water drop contact angle and reduced surface energy [41]. Results from this study, summarized in Table 5.9, show a substantial reduction of the surface energy of the wood strands subjected to hot water extraction producing wood mass loss of 6.8–26.8%. We could expect that wettability of the extracted materials by aqueous (water-borne) adhesive resins would be reduced by this treatment, recalling that the surface tension of water at 20°C is 73.05 mJ/m^2 (refer to Table 5.5). This result

Table 5.9 *Surface energy of hot-water extracted Southern pine strands.*

Mass loss due to hot-water extraction, %	Surface energy, γ_S, mJ/m^2 (SD shown in parentheses)
0	41.3 (7.8)
6.8	26.1 (6.1)
9.5	21.4 (3.8)
14.1	23.7 (6.5)
20.0	22.3 (6.3)
26.8	25.1 (4.6)

Source: Reproduced from [41] with permission of the Society of Wood Science and Technology © 2011.

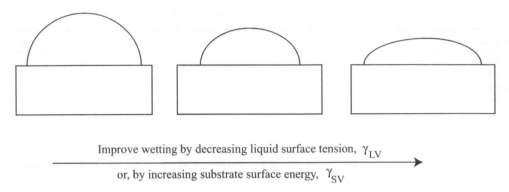

Improve wetting by decreasing liquid surface tension, γ_{LV}

or, by increasing substrate surface energy, γ_{SV}

Figure 5.15   *Wetting of wood substrates by liquid adhesive resin may be improved by (a) decreasing the surface tension, γ_{LV} of the adhesive or (b) by increasing the surface energy, γ_{SV}, of the wood substrate. The contact angle will decrease as γ_{LV} decreases or as γ_{SV} increases.*

is the opposite of that expected from extraction with organic solvents, which preferentially removes lower surface energy, hydrophobic extractives, whereas hot-water extraction primarily removes higher surface energy, hydrophilic extractives.

Within the wood-using industries, there are two primary approaches to improving wetting in practice: Alter the liquid resin properties, for example, reduce γ_{LV}, which results in a decrease in contact angle, or alter the surface properties of the substrate/furnish by increasing γ_{SV}, thereby decreasing θ (Figure 5.15).

Reduction of γ_{LV} may be accomplished by changing the resin's solvent system, or, more commonly, by adding a surface-active agent (surfactant) to the resin. The surface energy of wood, γ_{SV}, is increased by machining the wood surface. This may be accomplished by planing or sanding to expose "fresh" wood surfaces. It has long been recognized that wood surfaces become "deactivated" or "inactivated" with respect to adhesive bond performance as the surface ages. Surface aging occurs with exposure to air and other ambient conditions that result in oxidation and other changes to surface chemistry. These changes reduce surface energy. Removal of deactivated surface material by machining, followed by prompt application of adhesive resin, improves adhesive bond performance. This is something you can easily demonstrate for yourself. Find an old piece of lumber or veneer. Leave part of the wood untouched, but sand a couple of adjacent areas with a fine-grit sandpaper. Try using just two passes of the sandpaper in one area and perhaps four in another. Place a small drop of water on the unsanded and sanded areas and observe. You will likely see a much larger bead or contact angle formed by the drop on the unsanded area of the wood. Do you think you will see a difference between the two sanded areas?

Wood surfaces are also deactivated by heat treatment in both solid wood [37, 38] and wood fibers [42]. Heating of some kind is generally unavoidable, given the need to remove moisture by drying. Other processes, including machining, also heat wood. In addition, most wood-based composites are bonded with thermosetting adhesives that require elevated temperature to initiate cure. Nevertheless, minimization of both heat treatment and time from machining to application of adhesive will help to reduce the deleterious effects of surface deactivation, thereby improving wetting behavior and bond performance.

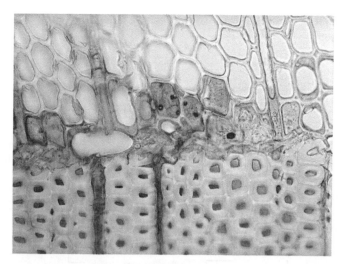

Figure 5.16   *Douglas fir laminated veneer lumber bonded with phenol formaldehyde adhesive resin. The veneer at the top is cut through an earlywood zone (large diameter, thin-walled tracheid cells). It is bonded to the veneer at the bottom, which is cut through a latewood zone (smaller diameter, thicker-walled cells). Light micrograph. Reproduced from [21] with permisson of the Forest Products Society © 2007.*

*5.6.4.2.2   **Surface Roughness***   Surface roughness is related to porosity and to surface preparation (machining). Smoother surfaces are generally better for bonding. For example, knife planed wood offers a better bonding surface than does sawn wood. A smooth surface allows for more intimate contact of adjacent substrates, thus promoting wetting, spread and penetration of adhesive into the substrate. Again, we must remember that effective wetting is the first prerequisite of good adhesion. "The roughness and porosity of substrates are generally suitable factors only insofar as the wettability by the adhesive is sufficient" [4]. The micrograph of a section of laminated veneer lumber in Figure 5.16 shows two smooth surfaces of Douglas fir bonded by phenol formaldehyde resin. The quality of the adhesive bond is partially attributable to the smooth surfaces of the substrates.

*5.6.4.2.3   **Permeability and Porosity***   Wood is a porous and permeable cellular solid. Flow of resin into cell cavities promotes the mechanical interlocking mechanism of adhesion upon subsequent resin cure. The permeable nature of wood permits the flow of adhesive into minute clefts within the cell wall, producing a mechanical interlocking effect on a miniature scale. Adhesive penetration occurs over a scale ranging from the anatomic level of cell lumens and pits (micrometer scale) to the ultrastructural cavities within the cell wall (nanometer scale) [43]. In addition, the anatomic structure of wood varies in three principal directions: longitudinal, tangential, and radial. This structural anisotropy has a profound effect wood's liquid permeability.

Permeability is defined as "a measure of the ease with which fluids move through a porous solid under the influence of a pressure gradient" [44]. Liquid permeability in the longitudinal direction is substantially greater than that in either the tangential or radial directions. In softwoods, the ratio of longitudinal to transverse permeability (i.e., either radial or tangential)

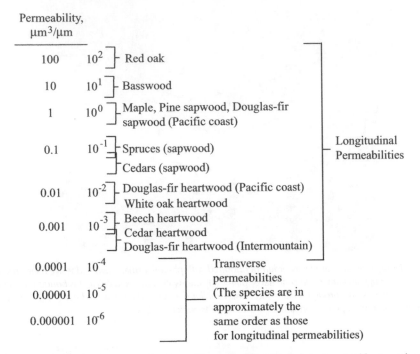

Figure 5.17   *Longitudinal and transverse liquid specific permeability of wood, summarized for several species. For most glued assemblies, the transverse permeability is that which is relevant to adhesive interactions. Reproduced from [44] with permission of Virginia Polytechnic Institute and State University © 1995.*

is on the order of 10,000 to 40,000:1. In hardwoods, the longitudinal to transverse permeability ratio is 30,000 to 4×10^8:1 [44].

The wide range of permeability of several species of softwoods and hardwoods is demonstrated in Figure 5.17. The values reported in this figure are for "specific permeability," which is the product of permeability and viscosity of liquid. The advantage of comparing the specific permeability of various media is that this metric is only a function of the porous structure of the medium (here, wood).

Due in large measure to the extreme longitudinal permeability of wood in general, adhesive bonding of end-grain (i.e., materials bonded on the cross-sectional surface) is unsuccessful. Glued lumber, veneer, particles and flakes are bonded together along their longitudinal surfaces, thus, it is the transverse permeability in the radial and tangential directions that govern adhesive interactions (Figure 5.18). Since this is the case, the range of permeability of interest with regard to adhesive bonding is between 10^{-4} and 10^{-6} $\mu m^3/\mu m$ (see lower part of Figure 5.17).

Kamke and Lee [43] have noted that "the permeability and surface energy are the two wood-related factors controlling adhesive penetration." The permeability of wood is directly related to porosity, but they are not necessarily the same. Generally, wood porosity is defined as shown in Equation 5.11:

$$\text{Porosity, \%} = \left(1 - \frac{G_M}{1.53}\right) 100 \qquad (5.11)$$

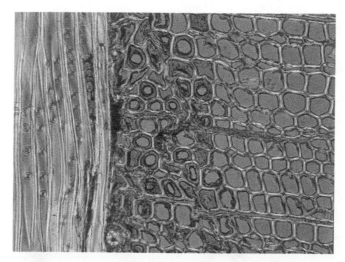

Figure 5.18 Adhesive joint in spruce plywood. Veneer at left is exposed on the radial-longitudinal surface, whereas the veneer at right is cut in cross-section. The difference in porosity of these surfaces is evident on visual inspection. Thinking three-dimensionally, however, it is evident that the two pieces of veneer are joined on the tangential-longitudinal plane of each. The individual plies of veneer are oriented at 90° to one another, as is typical of plywood cross-banded construction. Polarized light micrograph. Reproduced from [21] with permisson of the Forest Products Society © 2007.

where

G_M = specific gravity of wood based on volume at moisture content, M; and
1.53 = specific gravity of solid wood cell wall substance.

Porosity in this sense cannot account for microstructural changes in wood that directly affect overall liquid permeability. For example, it is well known that the bordered pits in the longitudinal tracheids of softwood may undergo "aspiration" during drying of the wood. Aspirated pits effectively block mass flow of liquids through the wood structure (Figure 5.19). Although liquids may be able to diffuse through the cell wall, the permeability of wood is nevertheless reduced by pit aspiration, with no effect on porosity as calculated by Equation 5.11. Micro- and nanoscale cavities in wood may likewise by obstructed by extractives or contaminants, also reducing permeability while leaving porosity unaffected. However, if one were to use some metric of porosity that is able to accurately quantify the volume of cavities at all structural levels, then porosity and permeability would presumably be equated.

The importance of adhesive penetration to adhesion performance is difficult to quantify. Kamke and Lee [43] stated: "Adhesive bond performance must be influenced by penetration. However, variability of penetration makes the development of correlations to bond performance difficult." Saiki [45] concluded from studies of wood bonding with epoxy and phenolic adhesives that "penetration of adhesives into the wall seems to have an effect on wood bonding only in the portion of the cell wall close to the lumen." In a review of literature concerning penetration of adhesive into wood cell walls, Frihart [46] found only one study that claimed that wall penetration improved bond strength. Our conclusion at this point is, therefore, that although wood porosity is recognized as an important wood adhesion variable, the absolute

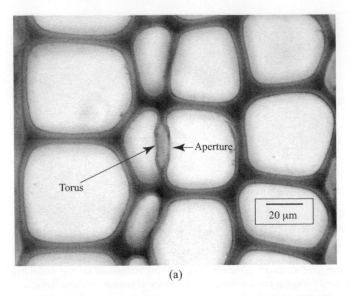

(a)

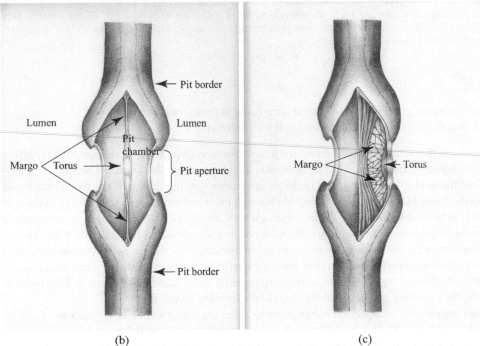

(b)   (c)

Figure 5.19 *(a) Transverse section of Ponderosa pine (Pinus ponderosa) showing an aspirated bordered pit pair. The torus is pressed against the pit aperture at the left of the pit pair. Light micrograph by D. D. Stokke. (b) Non-aspirated bordered pit pair, showing the solid torus suspended in the pit chamber by the margo. (c) Aspirated bordered pit pair in the secondary xylem of a softwood. Artist's renderings by Alexa Dostart.*

importance of adhesive penetration as affected by porosity and permeability remains something of a mystery.

5.6.4.2.4 *Specific Gravity*

Despite the elusive relationship of permeability and porosity to bond performance, the fact remains specific gravity is an important determinant of bond performance, likely due to the relationship of specific gravity to porosity. The strength of wood–adhesive joints generally increases with wood specific gravity up to $G_{12} = 0.70$ to 0.80 [47]. Above this approximate level of wood specific gravity, the percent wood failure (a measure of adhesive bond quality) decreases, meaning that high specific gravity wood has higher strength than the adhesive bond. This is not a desirable situation for bonded materials used in structural applications, as design values need to be based on the more predictable wood strength, rather than on the potentially more-variable adhesive bond. In other words, a quality adhesive bond is one in which failure is observed primarily in the wood substrate, not in the glue line.

If we accept that wood specific gravity in the range of 0.7–0.8 represents an upper limit for acceptable bond performance, we may also infer from Equation 5.11 that the minimum acceptable wood porosity is in the range of approximately 54–48%. Perhaps anything less than this does not allow for adequate wetting and penetration of extant adhesive formulations. An alternative interpretation is that porosity values lower than this range simply represent a higher proportion of solid cell wall substance, resulting in wood strength that exceeds that achievable by known adhesives.

5.6.4.2.5 *Wood Moisture Content*

Our list of the top five wood adherend variables ends with moisture content. As a hygroscopic material, wood (and other lignocellulosic substances) interacts not only with liquid water, but also with water vapor in the environment. For optimal adhesive bonding, the moisture content of the wood adherends must be monitored and controlled. Moisture content that is either too low or too high will impede adhesion. Since most industrial and consumer-use adhesives are water-borne, adhesive spread and penetration are affected by the moisture content of the substrate. Penetration and diffusion of water-borne adhesive will be greater into substrates with low moisture content, driven by the water concentration gradient. This can result in over-penetration of adhesive, producing a "starved joint" in which there is too little adhesive left at the interface to produce adequate joint strength. Conversely, a high moisture content in the wood substrate will retard adhesive penetration, also resulting in poor adhesion performance.

In studies on penetration of urea formaldehyde resin into beech (*Fagus sylvatica*) wood, Sernek et al. [48] found maximum adhesive penetration of six to eight μm at a wood moisture content of 9% (Figure 5.20). Brady and Kamke [49] found greater penetration of phenol formaldehyde adhesive into aspen wood at 15% moisture content than at 4% moisture content. Although the absolute influence of adhesive penetration on bond performance is difficult to quantify, results such as these indicate the effects of substrate moisture content on the movement of adhesive between the bondline and adherends.

Regardless of the overall effect of adhesive penetration, it is known that for many wood-based systems, a moisture content range of 6 to 14% will produce optimal adhesive bond performance [47].

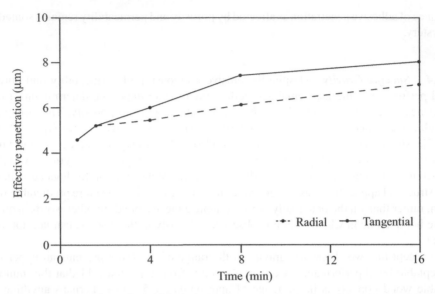

Figure 5.20   *Penetration of urea formaldehyde (UF) resin versus time in the radial and tangential directions within beech wood. Reproduced from [48] with permission of the Society of Wood Science and Technology © 1999.*

5.6.5   Mechanisms of Wood Bonding

As porous materials, wood adherends have an ample supply of void spaces which provide a capillary effect that augments surface wetting by adhesives. This leads to adhesive penetration into the substrate. For wood adherends, adhesive penetration of approximately 6–10 cell diameters (few 100 μm, maximum) is regarded as necessary for optimal adhesive bonding. However, actual resin penetration is may be substantially less than this, as we have seen in Figure 5.20. Resin flow and into the cell lumens and other interstices, followed by resin cure, reinforces the surface or interfacial layers of wood cells. Hardening of the resin results in interlocking over a surface scale of mm to nm.

There is no single, global theory of adhesion applicable to wood or any other bondable substrate. However, it is clear that thermodynamic adsorption (wettability) is a fundamental prerequisite for other mechanisms to operate. Thermodynamic adsorption may also function as an adhesion mechanism per se. This is accompanied by diffusion of polymeric adhesive into the polymeric substrate. It is unclear as to the degree to which adhesion develops due to diffusion, although this may contribute to the development of interpenetrating polymer networks. Covalent bonding, though theoretically plausible, is not well supported by experimental evidence. Mechanical interlocking is perhaps the predominant means of adhesion, with the caveat that one must be careful to specify the spatial scale over which this mechanism is deemed to occur. At the nanometer scale, interlocking may be conceived as the formation of an interpenetrating polymer network. Ultimately, secondary forces, including van der Waals forces, dipole interactions, hydrogen bonds and acid–base interactions, are likely the underlying, fundamental means by which adhesion develops in lignocellulosic substrates. It is apparent that spatial scale is highly significant when it comes to framing any discussion of adhesion.

5.6.6   Durable Wood Adhesive Bonds

Regardless of the mechanisms at work, the goal of wood bonding is to produce a bond that performs under the anticipated service conditions of the final product. In many instances, durability of the bond under exterior exposure is a major consideration. The adhesive bond of structural materials, especially, must maintain integrity, not only under mechanical loads, but must also resist degradation due to exposure to moisture and other environmental factors. Gardner [12] has provided a summary of criteria for durable, exterior-exposure wood adhesive bonds (Table 5.10). These may be regarded as water-proof bonds that withstand the combined environmental and load-imposed stresses over the engineered service life of the products. It is presumed that a bond meeting the majority of criteria listed in this table will result in satisfactory performance. Durable bonds minimally meet five of the first six criteria listed in Table 5.10, with the likely exception being the development of covalent bonds between adhesive and adherend. The final two criteria in the column at the right are crucial for durable bonds with rigid adhesives, a prime example of which is epoxy, used to bond carbon or glass fiber reinforcements to wood in fiber reinforced polymer (FRP) composites.

Table 5.10   *Characteristics of a durable wood adhesive bond.*

Satisfactory adhesion criteria, per Pocius [5]	Criteria met by durable wood adhesive bonds, per Gardner [12]
Choose an adhesive that is soluble or diffuses into the adherends.	Durable wood adhesives such as PF and pMDI resins have similar solubilities to the lignin in the wood cell wall.
Choose an adhesive with a critical wetting tension less than the surface energy of the adherend.	Most wood adhesives exhibit adequate wetting on properly prepared wood substrates.
Choose an adhesive with a viscosity low enough so the equilibrium contact angle can be attained during the assembly time.	The dynamic behavior of wood adhesive wetting ensures proper contact angles will be obtained during assembly.
Provided a microscopic morphology on the adherends.	Inherently, wood has microscopic and nanoscopic morphology.
Choose an adhesive compatible with the weak boundary layer or remove the weak boundary layer.	In producing a fresh surface for adhesive bonding, the chemical weak boundary layer is removed in wood, and mechanical damage inherent to the machining process may facilitate adhesive bonding in many types of wood composite elements. In addition, adhesives can be formulated to handle extractive contamination of the wood surface.
For exterior exposure, choose an adhesive which can provide covalent bonding between the adherend and the adhesive	Although speculated, covalent bonding between wood and adhesives has not been demonstrated.
	Similarity in dimensional behavior of the wood and adhesive under moisture stress.
–	Formation of a dimensionally stable wood surface through the use of a coupling agent (this is especially important in formation of durable wood–epoxy bonds).

Source: Developed from concepts presented by Gardner [12].

References

1. McBain JW, Hopkins DG. On adhesives and adhesion. *Journal of Physical Chemistry*, 1925;29(2):188–204.
2. Browne FL, Brouse D. Nature of adhesion between glue and wood. A criticism of the hypothesis that the strength of glued wood joints is due cheifly to mechanical adhesion. *Industrial and Engineering Chemistry*, 1929;21(1):80–84.
3. Truax TR, Browne FL, Brouse D. Significance of mechanical wood-joint tests for the selection of wood-working glues. *Industrial and Engineering Chemistry*, 1929;21(1):74–79.
4. Fourche G. An overview of the basic aspects of polymer adhesion. Part I: Fundamentals. *Polymer Science and Engineering*, 1995;35(12):957–966.
5. Pocius AV. *Introduction. Adhesion and Adhesives Technology, An Introduction*, 2nd ed. Munich: Carl Hanser Verlag; 2002. pp 1–14.
6. Packham DE. Adhesion. In: Packham DE, editor. *Handbook of Adhesion*, 2nd ed. West Sussex UK: John Wiley & Sons, Ltd; 2005. pp 14–16.
7. Bikerman JJ. Causes of poor adhesion. Weak boundary layers. *Industrial and Engineering Chemistry*, 1967;59(9):40–44.
8. Marra AA. Applications of wood bonding. In: Bloomquist RF, Chrisiansen AW, Gillespie RH, Myers GE, editors. *Adhesive Bonding of Wood and Other Structural Materials*. University Park, PA: EMMSE Project, Materials Research Laboratory, The Pennsylvania State University; 1981. pp 365–418.
9. Marra AA. *Technology of Wood Bonding: Principles in Practice*. New York: Van Nostrand Reinhold; 1992.
10. Frihart CR. Adhesive groups and how they relate to the durability of bonded wood. *Journal of Adhesion Science and Technology*, 2009;23:601–617.
11. Stehr M, Johansson I. Weak boundary layers on wood surfaces. *Journal of Adhesion Science and Technology*, 2000;14(10):1211–1224.
12. Gardner DJ. Adhesion mechanisms of durable wood adhesive bonds. In: Stokke DD, Groom LH, editors. *Characterization of the Cellulosic Cell Wall*. Iowa: Blackwell Publishing Professional; 2006. pp 254–265.
13. ASTM 2012. D2559-10a: Standard specification for adhesives for bonded structural wood products for use under exterior exposure conditions. West Conshohocken, PA: ASTM International.
14. ASTM 2010. D906-98: Standard test method for strength properties of adhesives in plywood-type construction in shear by tension loading. West Conshohocken, PA: ASTM International.
15. Frazier CE. The interphase in bio-based composites: What is it, what should it be? In: Humphrey PE, editor. *Proceedings of the 6th Pacific Rim Bio-Based Composites Symposium*. Portland: Oregon Department of Wood Science and Engineering, Oregon State University; 2002. pp 206–212.
16. Vick CB, Richter K, River BH, Fried AR Jr. Hydroxymethylated resorcinol coupling agent for enhanced durability of bisphenol-A epoxy bonds to Sitka spruce. *Wood and Fiber Science*, 1995;27(1):2–12.
17. Johns WE. The chemical bonding of wood. In: Pizzi A, editor. *Wood Adhesives Chemistry and Technology*. vol. 2 New York: Marcel Dekker, Inc.; 1989. pp 75–96.
18. Pizzi A. *Brief Nonmathematical Review of Adhesion Theories as Applicable to Wood. Advanced Wood Adhesives*. New York: Marcel Dekker, Inc.; 1994. pp 1–18.
19. Schultz J, Nardin M. Theories and mechanisms of adhesion. In: Pizzi A, Mittal KL, editor. *Handbook of Adhesive Technology*, second, revised and expanded edition. New York: Marcel Dekker, Inc; 2003. pp 53–67.
20. Packham DE. The mechanical theory of adhesion. In: Pizzi A, Mittal KL, editors. *Handbook of Adhesive Technology*, second, revised and expanded edition. New York: Marcell Dekker, Inc.; 2003. pp 69–93.
21. Kutscha NP. *A Compilation of Micrographs on Wood and Wood Products*. Madison, WI: Forest Products Society; 2007.
22. Hansen CM, Björkman A. The ultrastructure of wood from a solubility parameter point of view. *Holzforschung*, 1998;52(4):335–344.

23. Wellons JD. Adhesion to wood substrates. In: *Wood Technology: Chemical Aspects ACS Symposium Series 43*. Washington, DC: American Chemical Society; 1977. pp 150–168.

24. Zhou X, Frazier CE. Double labeled isocyanate resins for the solid-state NMR detection of urethane linkages to wood. *International Journal of Adhesion and Adhesives*, 2001;21(3):259–264.

25. Yelle DJ, Ralph J, Frihart CR. Delineating pMDI model reactions with loblolly pine via solution-state NRM spectroscopy. Part I. Catalyzed reactions with wood models and wood polymers. *Holzforschung*, 2011;65:131–143.

26. Yelle DJ, Ralph J, Frihart CR. Delineating pMDI model reactions with loblolly pine via solution-state NRM spectroscopy. Part 2. Non-catalyzed reactions with the wood cell wall. *Holzforschung*, 2011;65: 145–154.

27. Bikerman JJ. *The Science of Adhesive Joints*. New York: Academic Press; 1961.

28. Gardner DJ, Shi SQ, Tze WT. Comparison of acid-base characterization techniques on lignocellulosic surfaces. In: Mittal KL, editor. *Acid–Base Interactions: Relevance to Adhesion Science and Technology*. Zeist, The Netherlands: VSP; 2000. pp 363–383.

29. Lee L-H. *Fundamentals of Adhesion*. New York: Plenum Press; 1991.

30. Petrie EM. *Theories of Adhesion. Handbook of Adhesives and Sealants*, 2nd ed. New York: McGraw-Hill; 2007. pp 39–58.

31. Young T. An essay on the cohesion of fluids. *Philosophical Transactions of the Royal Society of London*, 1805;95:65–87.

32. Tshabalala MA. Surface characterization. In: Rowell RM, editor. *Handbook of Wood Chemistry and Wood Composites*. Boca Raton, FA: CRC Press, Taylor & Francis; 2005. pp 187–211.

33. Fox HW, Zisman WA. The spreading of liquids on low energy surfaces I. Polytetrafluoroethylene. *Journal of Colloid Science*, 1950;5(6):514–531.

34. Fox HW, Zisman WA. The spreading of liquids on low-energy surfaces. III. Hydrocarbon surfaces. *Journal of Colloid Science*, 1952;7(4):428–442.

35. Zisman WA. Influence of constitution on adhesion. *Industrial and Engineering Chemistry*, 1963;55(10):19–38.

36. Šernek M. Comparative analysis of inactivated wood surfaces [Dissertation]. Blacksburg, VA: Virginia Polytechnic Institute and State University; 2002.

37. Šernek M, Kamke FA, Glasser WG. Comparative analysis of inactivated wood surface. *Holzforschung*, 2004;58(1):22–31.

38. Gérardin P, Petric M, Petrissans M, Lambert J, Ehrhrardt JJ. Evolution of wood surface free energy after heat treatment. *Polymer Degradation and Stability*, 2007;92(4):653–657.

39. Marian JE. Wood, reconsituted wood and glued laminated structures. In: Houwink R, Salomon G, editors. *Adhesion and Adhesives: Volume 2, Applications*. 2nd ed. Amsterdam: Elsevier Publishing Company; 1967.

40. Collett BM. A review of surface and interfacital adhesion in wood science and related fields. *Wood Science and Technology*, 1972;6:1–42.

41. Zhang Y, Hosseinaei O, Wang S, and Zhou Z. Influence of hemicellulose extraction on water uptake behavior of wood strands. *Wood and Fiber Science*, 2011;43(3): 1–7.

42. Garcia RA, Riedl B, Cloutier A. Chemical modification and wetting of medium density fibreboard produced from heat-treated fibres. *Journal of Materials Science*, 2008;43(15):5037–5044,

43. Kamke FA, Lee JN. Adhesive penetration in wood—a review. *Wood and Fiber Science*, 2007;39(2):205–220.

44. Siau JF. *Wood: Influence of Moisture on Physical Properties*. Blacksburg, VA. Department of Wood Science and Forest Products, Virginia Polytechnic Institute and State University, 1995.

45. Saiki H. The effect of the penetration of adhesives into cell walls on the failure of wood bonding. *Mokuzai Gakkaishi*, 1984;30(1):88–92.

46. Frihart CR. Wood adhesion and adhesives. In: Rowell R.M, editor. *Handbook of Wood Chemistry and Wood Composites*. Boca Raton, FA: CRC Press, Taylor & Francis; 2005. pp 215–278.

47. Frihart CR, Hunt CG. Adhesives with wood materials: bond formation and performance. *Wood Handbook: Wood as an Engineering Material*. General Technical Report FPL-GTR-190. Madison, WI: USDA Forest Products Laboratory. 2010. p 10-1–24.
48. Šernek M, Resnik J, Kamke FA. Penetration of liquid urea-formaldehyde adhesive into beech wood. *Wood and Fiber Science*, 1999;31(1):41–48.
49. Brady DE, Kamke FA. Effect of hot-pressing parameters on resin penetration. *Forest Products Journal*, 1989;38(11/12):63–68.

6

Adhesives Used to Bond Wood and Lignocellulosic Composites

6.1 Introduction

This chapter will provide information on the major types of adhesives utilized to bond wood and natural fiber composites. We will begin with a discussion of important descriptors of adhesive characteristics, including molecular weight, viscosity, gel time, and tack. The primary sources and uses of important industrial adhesives will be presented. For a number of the adhesive resins described, a discussion of the main reactions involved in their chemical synthesis will precede a description of cure chemistry, as cure is often an extension of the synthetic reactions. In these cases, our approach will be to present rudimentary information on resin synthesis and cure of adhesive resins, with an aim to provide an accurate and understandable view of how these resins work. We will deliberately omit a number of details or fine points of these reactions, as these are topics for more advanced study. Even so, we think it is useful for students to consider the basic concepts of synthesis, as this will lend a greater understanding of how these materials cure and then function as adhesives. Readers desiring more detail on the complexities of the chemical syntheses and cure reactions are encouraged to dig into the references listed at the end of the chapter. We encourage particular attention to articles in the excellent handbook edited by Pizzi and Mittal [1], which we have cited extensively. Our goal for this chapter is to provide a fundamental working knowledge of the basic chemical nature of adhesives that will serve the student or practitioner well in their effort to use adhesives efficiently and appropriately.

6.2 The Nature of Wood Adhesives

Although the mechanisms of wood adhesion fail to conform to a singular overarching theory of adhesion, we nonetheless have a basic understanding of how many adhesives "work." In similar fashion, though the chemistry of adhesives may be complex and not completely understood,

Introduction to Wood and Natural Fiber Composites, First Edition. Douglas D. Stokke, Qinglin Wu and Guangping Han.
© 2014 John Wiley & Sons, Ltd. Published 2014 by John Wiley & Sons, Ltd.

many have underlying synthesis and cure reactions that are well characterized. Indeed, the adhesives of major importance—as determined by the total industrial amounts used—have basic similarities in the chemical reactions involved in their synthesis and curing mechanisms. Cure of the adhesives, particularly those characterized as thermosetting, is generally the result of chemical reactions that are similar, if not identical, to the reaction(s) occurring in the synthesis of the resin. Thus, understanding of chemical synthesis leads to understanding of adhesive cure mechanisms.

6.2.1   Most Wood Adhesives Are Organic Polymers

Most wood adhesives may be classified as organic polymers. These adhesive organic polymers may be obtained either from "natural" sources (e.g., plants or animals) or by synthesis from petrochemical and related resources (e.g., natural gas). In either case, the inherent characteristics of the organic polymers determine the manner in which they may be used and the level of their performance as adhesives.

Organic polymers are carbon-based molecules composed of repeating chemical units. We have seen that wood and other plant materials are, fundamentally, composed of natural organic polymers. Although the characterization of the chemical nature of organic polymers may be quite complex, there are some important descriptors of adhesive resins that are relatively simple to understand, measure, and use. These include molecular weight, viscosity, gel time, and tack.

Sidebar 6.1   Adhesive Resins

Among the terminology encountered in adhesives science and practice is the word "resin." Broadly, a resin is a natural organic substance, such as that exuded by plants. Although natural resins may be used as adhesives, adhesive resins include organic molecules produced by industrial chemical synthesis. Often, the word resin is used to denote a polymeric material, generally supplied in liquid form, which will produce adhesive characteristics as it cures or further polymerizes. An adhesive resin is therefore a mixture of organic polymers that confer adhesion, typically upon loss of solvent (solvent loss curing) or as a result of chemical- or heat-induced polymerization or cure. In the case of a resin that cures by polymerization, that is, growth in size of the adhesive molecules, the resin may be called a prepolymer.

6.2.2   Molecular Weight, Viscosity, Gel Time, and Tack Are Important Attributes of Polymeric Adhesive Resins

6.2.2.1   Molecular Weight

Polymeric adhesive resins are composed primarily of carbon, hydrogen, and oxygen. Some also contain nitrogen. These elements have atomic weights of 12, 1, 16, and 14, respectively. If the chemical structure of a polymer is known, the molecular weight is simply the sum of the atomic weights of all of the atoms comprising the molecule.

While a given type of adhesive resin will have a known fundamental structure, the resin will invariably contain molecules of differing molecular weights; that is, there will be a statistical distribution of molecular weight within the resin. This results from the varying degrees of polymerization of molecules arising during biosynthesis of natural polymers or in the process of chemical synthesis of polymers in the laboratory or by industrial manufacturing.

Molecular weight distribution is generally characterized by the number average molecular weight or the weight average molecular weight. The number average molecular weight, $\bar{M}_n$, is the arithmetic mean or average of the molecular weights of the resin polymers:

$$\bar{M}_n = \frac{\sum_i N_i M_i^2}{\sum_i N_i} \tag{6.1}$$

The weight average molecular weight, $\bar{M}_w$, is defined as:

$$\bar{M}_w = \frac{\sum_i N_i M_i^2}{\sum_i N_i M_i} \tag{6.2}$$

For Equations 6.1 and 6.2, N_i is the number of moles of polymer of a given size, wherein size is represented by the degree of polymerization or chain length for the ith polymer size class, and M_i is the molar mass or weight of the ith polymer size class (polymer species).

The ratio of M_w to M_n is called the polydispersity index, a measure of the width of the molecular weight distribution. For typical polymers, the polydispersity index ranges from 1.5–2.0 to 15–20 [2].

In addition to the number average and weight average molecular weights, polymers may also be characterized by metrics known as the Z average molar weight and the viscosity average molar weight. Molecular weight is measured by a variety of methods, including osmometry, light scattering, or size exclusion chromatography.

Obviously, molecular weight and degree of polymerization, d.p., are related. If the molecular weight of both the polymer and the fundamental repeat unit comprising the polymer are known, the d.p. may be calculated by division. For example, polythethylene is synthesized by the polymerization of the monomer ethene, with the chemical formula C_2H_4, molecular weight $(12 \times 2) + (1 \times 4) = 28$. Let us say that $\bar{M}_n$ of a particular sample of polyethylene is found to be 300,000. The average degree of polymerization is therefore shown in Equation 6.3 as:

$$\text{d.p.} = \frac{\bar{M}_n}{\text{m.w.}} = \frac{300,000}{28} = 10,714 \tag{6.3}$$

Molecular weight is an important indicator of the characteristics of adhesive resin prepolymers. Chemical syntheses are carried out until the prepolymer reaches a desired molecular weight, at which point the synthesis is quenched or ceased. The resin is then held in this condition (to the degree possible) until it is applied and used as an adhesive. We say, "to the degree possible," in that the resin will typically undergo further, albeit slow, polymerization during storage. As a result, liquid adhesive resins have a limited storage life. Storage life is often monitored by determination of molecular weight or viscosity of the liquid.

6.2.2.2   *Viscosity*

While the molecular weight of an adhesive prepolymer is important, it is not always readily measured. Though the chemical manufacturer may routinely perform this measurement, the end-user may not have the necessary equipment or expertise to do so. Therefore, an easily measured adjunct is useful. Viscosity is such a parameter. Viscosity is a measure of the "thickness" of a liquid or its resistance to flow, and is related to molecular weight. Fluidity is the ability of a liquid to flow, that is, it is the reciprocal of viscosity. In general, as the molecular weight of a mixture of liquid polymers increases, so does its resistance to flow, or viscosity. Viscosity is therefore often used as a practical adjunct to or surrogate for the molecular weight. In the case of chemical synthesis of adhesive resin prepolymers, the progress of the reaction and its desired end point are commonly monitored by viscosity. Similarly, viscosity may be monitored as an indicator of resin shelf life. Shelf life, or the length of time over which resin remains usable, is limited due to the ongoing, slow polymerization and build of molecular weight of resin over time. That is, it is not possible in practical terms to completely terminate the polymerization reactions in a mixture of resin prepolymers. Some adhesives are supplied to the end-user as a two-part system, for example, a base resin plus catalyst or hardener. The two components are mixed just prior to use. In a case such as this, the term pot life is used to refer to the amount of time over which the adhesive remains usable after mixing of the two components. Eventually, the resin will increase in viscosity and form a gel. The gelled resin is then unusable, signaling the end of the pot life.

Viscosity is represented by the lowercase Greek letter eta, η. Strictly, "viscosity of a fluid can be defined as the ratio of shear stress to shear rate during flow, where shear stress is the frictional force exerted by the fluid per unit area (τ), and shear rate is the velocity gradient

---

Sidebar 6.2   Viscosity Units

Fundamentally, viscosity is understood as the force required to shear a given volume of fluid through a specified distance in a standard unit of time. A variety of units are used to express viscosity. Commonly, the standard unit of time is fixed at 1 second. Force may be expressed in Newtons (one Newton, N, is equal to 100 grams force), pounds, kg, or dynes (one dyne is 1 g·cm/s^2 or 10 μN), with distances correspondingly expressed in centimeters, inches, or meters. Note also that within the International System of Units (SI), one Pascal (Pa) is a force of one Newton exerted on one square meter of surface area, or 1 Pa = 1 N/m^2. The centimeter–gram–second (CGS) system, a nineteenth century precursor to the SI system is important here, as the poise (P), or more commonly, centipoise (cP or cp or cps) is a well-known and oft-used unit of viscosity. Recall that viscosity is the ratio of shear stress to shear rate, $\eta = \tau/\dot{\gamma}$. The units typically used to express all three are shown in Table 6.1.

Table 6.1   *Units used to express shear stress, shear rate, and viscosity in the SI, US Customary, and Centimeter–Gram–Second (CGS) systems.*

System	Shear stress, τ	Shear rate, $\dot{\gamma}$	Viscosity, η	Note
SI	N/m^2 = Pascal, Pa	1/s where s = second	N·s/m^2 = Pa·s	See note below for the CGS system
US Customary	lb/in.^{2} = psi	1/s	lb·s/in.^{2}	6,895 Pa·s = 1 lb·s/in.^{2}
CGS	dyne/cm^2	1/s	dyne·s/cm^2 = 1 poise	Centipoise (cP, cp, or cps) is the more common unit of viscosity in the CGS system, where 1 cP = 10^{-2} Poise = 10^{-3} Pa·s = 1 mPa·s

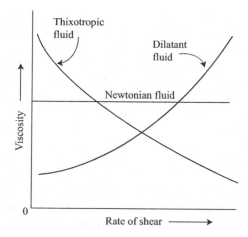

Figure 6.1   *Viscosity versus shear rate for Newtonian, dilatant, and thixotropic fluids.*

perpendicular to the flow direction $(\dot{\gamma})$" [3]. Instruments designed to measure viscosity have rotating plates or spindles. When immersed in a fluid, the rotation of the plate or spindle is resisted by the fluid. The force required to rotate the plate or spindle in the liquid is converted to a measure of viscosity.

6.2.2.3   *Newtonian, Dilatant, and Thixotropic Fluids*

When measuring viscosity, it is important to know that fluids may exhibit differing responses to stirring or shearing motions. Three possible responses are shown in Figure 6.1, representative of Newtonian, dilatant, and thixotropic fluids.

A Newtonian fluid is one that has a constant viscosity, regardless of shear rate. Water is a prime example of a Newtonian fluid, with a viscosity of 0.001 Pa·s (1 mPa·s or 1 cP) at 20°C. Although the viscosity of water decreases with temperature, it does not vary with shear rate. Think of what might happen if water did change viscosity with shear. If it were to thicken as shear rate increases, that is, behaving as a dilatant fluid, it might be impossible to swallow a glass of water without choking!

Thankfully, for our sake as humans, water is a Newtonian fluid. In contrast, non-Newtonian, dilatant fluids are rather interesting from a phenomenological perspective. One well-known example of such a system is cornstarch, a natural polymer, mixed with water. It is actually possible for a person to run across a large tank full of cornstarch–water mixture without sinking. As a person runs across the fluid, the impact of their feet induces a dilatant response that increases the fluid's viscosity such that their weight is supported. However, if that same person stands still on the same cornstarch–water mixture, they will promptly sink into the fluid because of its lower viscosity at low shear.

Thixotropic liquids are also non-Newtonian fluids, with a shear response that is opposite that of dilatant fluids. Thixotropic fluids decrease in viscosity as shear is increased (Figure 6.1). Most wood adhesives are thixotropic. As organic polymers often with linear polymer structure, wood adhesives in liquid form may be thought of as existing at rest as long, tangled chains dispersed within a carrier liquid, the latter often water. If the mixture is stirred, the induced shear helps to "untangle" and straighten the polymers, allowing them to flow more readily. This is observed as a decrease in viscosity as shear, or stirring rate, is increased.

The thixotropic behavior of adhesive resins is advantageous in practice, allowing for more efficient dispersion of adhesives onto wood furnish. When bonding solid wood, application of clamping pressure to pieces to be joined causes shear in the glueline, resulting in better spread of the adhesive and improved wetting of the substrate. Once the pieces are firmly clamped, the adhesive viscosity increases, decreasing the propensity for dripping of adhesive from the glueline as it cures. Application of resin to fibers, particles, or wood strands is typically accomplished by generation of droplets with pneumatic pressure through an orifice or by high-speed centrifugal application (spinning disk application). In either case, shear induced by flow through an orifice or across the surface of a high-speed disk reduces resin viscosity, allowing it to be more readily dispersed. But, once the resin droplet contacts a piece of furnish, the shear force is reduced and the resin viscosity increases, allowing it to more readily stay in place where it is needed.

6.2.2.4   *Gel Time*

Another indicator of the nature of adhesive resins is their gel time, defined as the time interval during which a resin changes from a liquid to a nonflowing gel as a result of the initiation of cure. Gel time is both an indicator of the working life of an adhesive and its pot life, the former representing the time available to apply an adhesive and form the final product before the adhesive has set, and the latter the time that the adhesive remains a viable resin in a processing or storage container. Gel time is also used, along with viscosity, to monitor the reaction stage during adhesive resin synthesis. As the prepolymer molecular weight builds, viscosity tends to increase and gel time decreases as the polymer becomes more "advanced."

Gel time is typically measured simply and directly, particularly in the case of its use in monitoring the reaction stage during polymer synthesis. One method is to place a drop of adhesive on a hot plate at a specified temperature (e.g., 150°C for a phenolic resin) and then to observe how long it takes for the droplet to become a gel-like substance. This provides a rapid assessment of the level of advancement of the resin prepolymer.

6.2.2.5   *Tack*

Tack is "the property of an adhesive that enables it to form a bond of measureable strength immediately after adhesive and adherend are brought into contact under low pressure, that is, stickiness" [4]. This dry or cold tack, as it is known, is important in composites manufacture as it is a property that allows a mattress of formed wood furnish to remain intact as it moves through the manufacturing system prior to final consolidation. For example, fibers used in the manufacture of medium density fiberboard have liquid adhesive resin applied as they flow at high speed through a pneumatic "blow line." The fibers then enter a forming machine that deposits the fiber into a thick mattress onto a moving conveyer. Just enough pressure is applied to the mattress to compress it into a layer that may be 30 or more centimeters (one foot or more) in height. This is known as a prepress operation. The mat then travels to the main press for final consolidation into a panel that may be 1.3–3.8 cm (about 0.5–1.5 in.) in thickness.

Adhesive tack allows the mat to retain its integrity between the prepress and the final pressing operation. Tack is similarly important in the manufacture of other wood and fiber composites.

Although methods are known to quantify tack of pressure-sensitive adhesives [5], it is difficult to do so for wood furnish to which adhesive has been applied. In practice, machine operators or production workers in wood composites operations may conduct a qualitative test of tack by squeezing a handful of resin-blended furnish in their hand and then observing the relative integrity of the ball of material so formed. With experience, the degree of tack or tackiness may be surmised with a qualitative test such as this. Often, tack is not explicitly considered unless problems arise with mat integrity during production. Tack then is one consideration in troubleshooting the issue at hand.

6.3 Adhesives Used to Bond Wood and Other Natural Fibers

There are many ways that organic polymers may be characterized or classified. One classification we have used is that of "natural" (i.e., occurring in nature, such as the complex molecules produced by plants) or "synthetic" (i.e., manufactured by humans using methods of chemical synthesis). Adhesives or adhesive raw materials of natural origin include blood protein, milk protein, starch, tannins, and lignins. Examples of synthetic adhesives are urea and phenol formaldehyde resins, isocyanates, and polyvinyl acetate. Classification of organic polymers as natural or synthetic is a classification according to origin. In the case of adhesive polymers, it is also advantageous to characterize them according to their intended end-use category, which includes the adhesive's mechanical strength and moisture resistance, and according to their behavior in terms of their response to heat.

6.3.1 Classification of Adhesives by Origin

If we think in terms of natural versus synthetic origin of adhesives or their precursors, in some sense we are also classifying adhesives by the era in which one source was predominant. Prior to World War II, most wood adhesives were of natural origin. Development of technology to convert petrochemicals into plastics and other materials used in the war effort launched the era of synthetic adhesives. Although adhesives of natural origin have a number of positive attributes, in general, they cannot match synthetics in terms of cure reaction rate, consistency of performance, and water resistance (the latter of selected synthetics intended for use in exterior environments). Cost of adhesives is also typically in favor of the synthetics. Nevertheless, adhesives of natural origin are still used in selected applications, and are the topic of ongoing research and development. Table 6.2 presents a list of selected natural and synthetic adhesives used to bond wood and natural fibers.

6.3.2 Classification of Adhesives by Structural Integrity and Service Environment

In addition to categorizing adhesives according to origin or source, another useful classification is to place them in groups suggestive of appropriate use, both in terms of structural integrity and suitability for different service environments (Table 6.3). Structural integrity refers to the expected level of strength performance of the adhesive. Products intended for use in structures (e.g., framing or sheathing of buildings intended for human habitation) have greater strength requirements than those used for esthetic purposes (e.g., decorative paneling, art objects).

Table 6.2   *Classification of wood adhesives by origin.*

Natural origin	Synthetic origin

Carbohydrate
  Starch
  Cellulose
	Amino Resins
	  Urea formaldehyde (UF)
	  Melamine formaldehyde (MF)
	  Melamine-urea formaldehyde (MUF)

Protein
  Plant-based (e.g., soybeans)
  Animal-based
    Casein (milk protein)
    Blood
    Protein from hides, bones, sinew
      Liquid ("hide" glue)
      Solid (hot animal glues)
	Phenolic Resins
	  Phenol formaldehyde (PF)
	  Resorcinol formaldehyde (RF)
	  Phenol-resorcinol formaldehyde (PRF)

Lignocellulosic extracts
  Tannins
  Lignins
	Isocyanate-based resins
	  Polymeric isocyanate (pMDI, MDI, or isocyanate(s))
	  Polyurethane(s)

–	Other
	  Polyvinyl acetate (PVA)
	  Epoxies
	  Elastomeric
	    Contact adhesives
	    Mastics
	    Hot melts

Source: Adapted from [6].

Table 6.3   *Classification of wood adhesives by structural integrity and service environment.*

Structural integrity	Service environment	Adhesive type
Structural	Fully exterior (withstands long-term water soaking and drying)	Phenol formaldehyde Resorcinol formaldehyde Phenol-resorcinol formaldehyde Emulsion polymer/isocyanate Melamine formaldehyde
	Limited exterior (withstands short-term water soaking)	Melamine-urea formaldehyde Isocyanate/pMDI Epoxy
	Interior (withstands short-term high humidity)	Urea formaldehyde Casein
Semistructural	Limited exterior	Cross-linked polyvinyl acetate Polyurethane
Nonstructural	Interior	Polyvinyl acetate (PVAc) Animal Soybean Elastomeric construction Elastomeric contact Hot-melt Starch

Source: Reproduced from [6].

Accordingly, classification of adhesives by strength or structural integrity is important. A separate, but often closely related classification is according to service environment, which includes likelihood of exposure to rain, extreme temperatures, or other environmental factors. Usually, service environment is used to describe the degree of exposure to moisture expected when the adhesively bonded product is in use or service. Typically, those products intended for structural uses also need to have a high degree of water resistance, given the likely exposure to moisture during construction. Products used in structural applications may also need to withstand periodic rewetting by rain, or the effects of prolonged elevated humidity if used in subtropical or tropical environments. As a result, a "fully exterior" adhesive may be required, that is, one that is insoluble in water when cured and which resists degradative effects of water over time. In some cases, a "limited exterior" structural adhesive is sufficient, meaning that the adhesive is insoluble in water when cured, but the adhesive is likely to chemically break down with continuing exposure to water over time. Those products designed mainly for esthetic purposes are often intended for indoor environments where the likelihood of contact with liquid water is minimal, thus a moisture-resistant (not water proof) bond is sufficient.

6.3.3 Classification of Adhesives by Response to Heat

It is useful and practical to characterize adhesives according to their origin or source and suitability for given uses and/or service environments. Another meaningful characterization of adhesive resins is according to their response to heat. By observing what happens to an adhesive polymer when it is heated, it may be classified as having a thermoplastic or thermosetting nature.

6.3.3.1   *Thermoplastic and Thermosetting Polymers*

A thermoplastic polymer is one that is solid within a certain temperature range, but as temperature is increased, the polymer softens into a rubbery and then liquid state, with a return to the solid state upon subsequent cooling. A thermosetting polymer (thermoset) is one that undergoes a chemical reaction upon heating, resulting in polymerization or cure into a solid that does not soften upon subsequent heating. Heating of a thermoset that has chemically cured may result in thermal degradation, but not softening in the sense observed with thermoplastics. The majority of wood and natural fiber composites is bonded with adhesives that are thermosetting.

There is actually a continuum of polymer behavior with respect to temperature. That is, polymers may exhibit some aspects of either thermoplastic or thermosetting behavior, depending on conditions. Nevertheless, the thermoplastic versus thermoset categorization is useful with respect to adhesives, as it provides insight into the expected curing and subsequent performance of polymeric adhesives.

6.3.3.2   *Most Wood-Based Composites Are Bonded With Thermosetting Adhesives*

As discussed in Chapter 4, most wood-based composites are consolidated under heat and pressure. Heating of the lignocellulosic substrate results in softening of the natural polymers comprising the material, aiding in the effective compaction or consolidation of the product

Table 6.4   *Major raw material components required for the synthesis of the main resins used for composite panels.*

Resin type	Main synthetic components
Amino resins (UF and melamine formaldehyde)	Urea (derived from natural gas) Formaldehyde (derived from methanol, which is derived from natural gas)
Phenolic resins	Phenol (derived from petroleum) Formaldehyde
Isocyanate resins	Aniline (derived from petroleum) Formaldehyde Phosgene (derived from carbon monoxide and chlorine)

under pressure. Another important function of the heat applied to the material is to initiate and carry forward the chemical polymerization or cure reaction of thermosetting adhesive resin used to bond the material. Thermosetting adhesive resins are particularly useful for making composites, in that they are relatively stable at room temperature or normal ambient conditions. Only when sufficient heat is applied will the resins polymerize to a solid. This statement must be qualified, however, by the fact that even "stable" adhesive prepolymers or base resins, particularly in the liquid phase, undergo slow polymerization such that they will eventually gel and then harden, given sufficient time at normal ambient conditions. Nevertheless, in an ordinary production situation, it is heat that provides the primary catalyst for the cure of thermosetting adhesives.

Within North America, wood and natural fiber composites panel products are bonded with three primary types of thermosetting resins, namely urea formaldehyde (UF), phenol formaldehyde (PF), or polymeric diisocyanate (pMDI), also known as isocyanate. As summarized in Table 6.3, UF forms a water-*resistant* bond and is used exclusively for interior applications, whereas PF, with a fully exterior, *waterproof* bond, is used mainly for exterior application and/or structural products. pMDI, with its water-resistant to waterproof bond, is used mainly as a core resin in oriented strand board (OSB). When pMDI is used for the OSB core (middle "layer" of the panel), PF is used as a face resin in the surface layers.

UF, PF, and pMDI (isocyanate) resins are all synthetics dependent upon raw material extracted from the earth [7]. The main components required for the chemical synthesis of these resins are summarized in Table 6.4. Notice that UF is an amino resin, sharing this category with melamine formaldehyde. Given the need for petrochemical and natural gas resources to make these resins, their price volatility can be considerable.

6.4   Amino Resins

Amino resins include the important, widely used thermosetting adhesives urea formaldehyde, melamine formaldehyde, and melamine-fortified urea formaldehyde. Amino resins derive their name from the fact that one of the fundamental starting materials for their synthesis, urea, contains an amino ($-NH_2$) group. These resins are all examples of thermosetting, organic adhesive polymers.

6.4.1   Urea Formaldehyde Resins

Urea formaldehyde, or UF, resins represent the main type of thermosetting adhesive used for wood-based panels and other consolidated products intended for use in interior environments. UF is quick setting, and when cured, it is durable (in reasonably dry environments), hard, and colorless. Although UF remains highly significant in terms of use, concerns over the release of formaldehyde from UF-bonded materials have prompted changes in their formulation and in usage patterns. This is an area of ongoing research, development, regulation, and changing industrial practice that warrants continuing scrutiny by students and practitioners alike.

UF is an amino resin or amine-formaldehyde class of adhesive synthesized by the reaction of urea and formaldehyde to obtain stable, but reactive intermediates. UF in liquid form has good storage stability due to reduced reactivity of intermediates in alkaline conditions (pH >7).

6.4.1.1   Synthesis: Methylolation

The synthesis of UF involves two types of reactions. The first step is an addition reaction between urea and formaldehyde under alkaline conditions to form mono-, di- and trimethylol ureas. The second step involves acid condensation reactions of methylol ureas that build polymer size. Once the resin prepolymer reaches the desired stage, the reaction is quenched, and the reactive intermediates are held in alkaline conditions until application of the resin to bond the desired substrates.

The reaction of the starting materials, urea and formaldehyde, to form monomethylolurea is shown in Figure 6.2. This first step is very significant in that it illustrates a reaction of major importance in the synthesis of both amino and phenolic resins, the methylolation reaction. Methylolation, also known as hydroxymethylation, refers to the addition of a specific alcohol group, the methylol group ($-CH_2OH$). In Figure 6.2, note the methylolation of urea residue, here depicted at the right end of the monomethylolurea molecule. The methylol group is a reactive site that is key in the build of molecular weight of the resin prepolymer, and also in the final cure reaction of UF. Other intermediates, such are dimethylol urea, are also formed. In the case of the dimethylol intermediate, polymer build from both ends is possible. Other chemical species that may form are unstable and quickly decompose to the mono- or dimethylol forms.

The reaction of urea and formaldehyde is carried out in aqueous solution at moderate temperatures (e.g., up to 100°C) at neutral or slightly alkaline conditions (pH $\geq$7). In order for the reaction to drive toward the formation of methylolated intermediates, an excess of formaldehyde to urea is required. This excess is ordinarily expressed as the mole ratio, that is, the ratio of formaldehyde to urea, expressed on a molar basis (i.e., number of molecules of

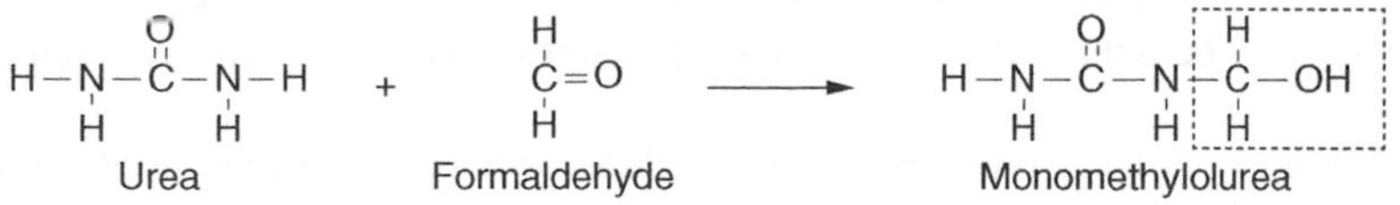

Figure 6.2   *Reaction of urea and formaldehyde to form monomethylolurea, the primary step in the synthesis of urea formaldehyde resin. The reaction is carried out under moderate temperature and mild alkaline conditions. The methylolation reaction, also known as hydroxymethylation, is of major importance in the synthesis of both amino and phenolic resins.*

each reactant). In North America, mole ratio of UF resins is generally understood as the ratio of formaldehyde to urea. In other parts of the world, the ratio may be expressed as urea to formaldehyde. In this text, we will adopt the convention of mole ratio = formaldehyde:urea. It is also important to recognize that general reference to "mole ratio" means the final mole ratio of a resin at the completion of synthesis. This is often heard in the context of the effect of mole ratio on formaldehyde emissions from products bonded with UF. This usage of the term, mole ratio, is to be distinguished from mole ratio at various stages of synthesis.

In order to achieve desired resin characteristics, the mole ratio is controlled at various stages of synthesis, rather than simply placing the required amounts of urea and formaldehyde in the reaction vessel all at once [8]. During the first stage, that is, methylolation, a mole ratio of formaldehyde to urea of 2.0–2.2 is typical [9]. The pH is generally slightly alkaline, perhaps in the range of 8.0. The addition reaction of urea and formaldehyde is exothermic. When the exotherm subsides, perhaps after 10–30 minutes of reaction time, the methylols have formed [9] and the mixture is then ready for the next stage of synthesis.

6.4.1.2   *Condensation*

Methylol ureas formed in the first stage of synthesis are condensed into larger molecules in the second stage. The second-stage condensation reaction occurs very rapidly under acidic conditions. At this stage, pH is adjusted to 5.0–5.3 by addition of acid, accompanied by a second addition of urea to consume the excess formaldehyde remaining from the first stage, so that the final mole ratio is reduced to <1.1–1.5 or more. Most contemporary resins will have a final mole ratio of 1.1 or less.

The products of the second-stage condensation reaction are called reactive intermediates, as these molecules are intermediate in size relative to the first-stage methylol ureas and the final, cured and hardened adhesive. The reactive intermediates generally consist of molecules with a fairly wide range of molecular weights (m.w.), from a few hundred to a few thousand m.w.

Production of the reactive intermediates is monitored and controlled by viscosity, that is, viscosity is used as a practical adjunct to or surrogate for the molecular weight (recall discussion in Section 6.2.2.2). Increasing viscosity is correlated to polymer growth; this indicates how "advanced" the resin is. In addition, solids content of the resin is important. As an example, a simple resin made from starting materials, urea and formaldehyde, having 50% solids, will have a theoretical solids content of 61% upon completion of the synthesis reaction [8]. Such a resin may have a viscosity of approximately 0.1 Pa·s (equal to 100 mPa·s or 100 centipoise, cP or cps). Vacuum distillation is used to remove water, so that a resin ready for shipment to a panel manufacturing facility will have a solids content on the order of 65%, increasing the viscosity to 0.2 Pa·s (200 cP). The pH is adjusted to an alkaline condition, typically 7.5–8 for storage. Under these conditions, the reactive intermediates are relatively stable, that is, their reaction, though ongoing, is slowed such that and acceptable storage life of a few weeks may be achieved. This assumes that storage is also at moderate temperatures (Figure 6.3).

In industrial practice, large pressure-rated vessels known as reactors are used for synthesis of UF resin in batches, typically in the range of 37,500–75,000 L (10,000–20,000 gallons) capacity. Reactors are fitted with a condenser to remove water liberated by the condensation reaction characterizing polymerization. The condenser also captures formaldehyde, which is refluxed into the reactor.

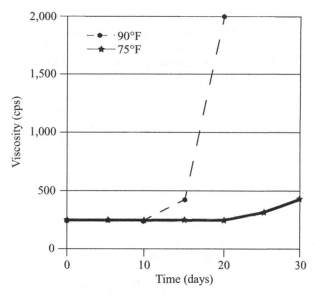

Figure 6.3   *Effect of temperature on storage life of UF resin prepolymer (reactive intermediates). Example assumes alkaline storage conditions. Reproduced from [8] with permission of the Forest Products Society © 1999.*

At one time, the final mole ratio of formaldehyde to urea was typically well in excess of 1.5, whereas it is typically below 1.1 today. This, along with other measures, has significantly reduced the release of formaldehyde from products bonded with UF.

6.4.1.3   Cure Chemistry: Condensation Reactions

Cure of UF resin is simply a continuation of the condensation reactions characterizing the second stage of synthesis. Cure is initiated by reduction of pH and/or the addition of heat. Since the reactive intermediate prepolymer is held in alkaline conditions, an acid is typically added concurrent with blending or applying the resin to the wood or fiber furnish. In some cases, the wood is sufficiently acidic that little to no extra acid is needed. Examples of acidic furnish include Southern pines (*Pinus* sp.), Douglas fir (*Pseudotsuga menziesii*), and oaks (*Quercus* sp.). Other species commonly used in composites, for example, aspen (*Populus tremuloides*), birches (*Betula* sp.), or true firs (*Abies* sp.) are not sufficiently acidic to promote UF resin cure. Typically, acidic salts are used, for example, ammonium chloride, ammonium sulfate, or ammonium nitrate, added at rates of 0.5–4% catalyst solids as a percentage of adhesive resin solids. In laboratory practice, a mineral acid, such as hydrochloric acid, may be alternatively used as a curing agent. Though commonly referred to as catalysts in North America, the European term, hardener (or curing agent), is more accurate, given that these acid salts are not catalysts in the true chemical sense [10]. A true catalyst accelerates a chemical reaction without being consumed. The hardeners for UF are, in fact, consumed in the curing reaction. Although many other chemicals may be used as curing agents, the most common industrial hardeners are ammonium chloride or ammonium sulfate [9].

With or without addition of an acid hardener, the condensation of UF is promoted by heat. Process temperatures of 150–160°C in particleboard manufacturing, for example, coupled with the affect of acidic hardener and/or acidic furnish, results in rapid polymerization or cure of the adhesive. In industrial practice, press cycle times of ≤120 to perhaps 200 seconds are representative, depending on thickness of the final product, press technology, and other manufacturing parameters.

Polymerization of the reactive intermediates builds polymer chain length and molecular weight. This is a condensation reaction, resulting in the release of water as a by-product concurrent with the buildup of the resin polymer. Polymer condensation may be linear or cross-linked (Figure 6.4). The resin changes from a low-to-high viscosity liquid, to a gel, and ultimately to a solid. Resin hardening corresponds to development of mechanical strength of adhesive bond [11].

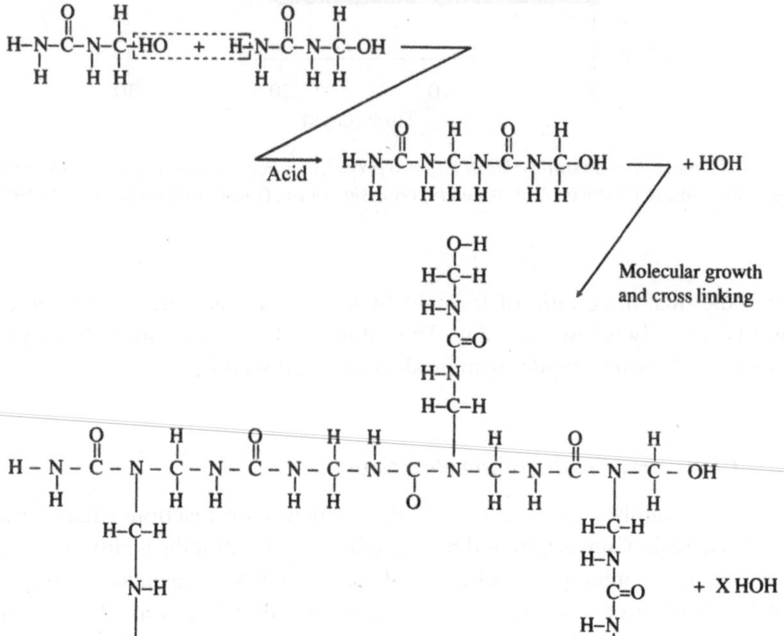

Figure 6.4 Condensation of methylolureas is the curing reaction of urea formaldehyde resin. The condensation of two monomethylolureas is shown in the upper left. A portion of a cured, cross-linked resin is depicted at the bottom of the figure. Reproduced from [4] with permission from Springer Science+Business Media B.V © 1992.

Resin cure continues after the product is discharged from the hot pressing operation. UF-bonded panels are usually cooled quickly after pressing to counteract heat-mediated decomposition of the adhesive bonds. However, cooling of panels is not always necessary with some types of UF developed in recent years.

Lower final molar ratios (formaldehyde to urea) are favored in contemporary formulations to reduce formaldehyde emissions. Most are <1.1. Formaldehyde scavengers may be used as an additional strategy.

6.4.2   Melamine Formaldehyde Resins

Like UF, melamine formaldehyde (MF) is an amine formaldehyde or amino resin. Unlike UF, MF forms fully waterproof bonds. Thus, MF may be used in a wider range of applications than UF. It is suitable as a replacement of UF, albeit considerably more costly, to bond interior products. MF is used to impregnate paper to form waterproof materials used in tables and kitchen countertop laminates, for example. It may also be used to bond materials for exterior environments, such as siding, or for products used in high-humidity interior locations, such as gymnasium locker rooms or indoor swimming pool areas.

6.4.2.1   *Synthesis*

The synthesis of MF has striking similarities to the process for UF. The first step is the methylolation (hydroxymethylation), of melamine via its reaction with formaldehyde (Figure 6.5). Melamine is very reactive toward formaldehyde, resulting in nearly complete methylolation of the melamine, with up to six molecules of formaldehyde reacted with each melamine molecule [12]. The second stage of synthesis is similar to that for UF, in that the methylolated melamines react, one with another, via condensation to form the reactive intermediates.

6.4.2.2   *Cure Chemistry*

As for UF, an acidic hardener is added to aid polymerization of MF, again depending to some extent on the inherent acidity of the furnish. However, MF polymerization can also occur under neutral and slightly alkaline conditions [12], and as a result, MF has a very limited storage life in liquid form. It is therefore often supplied in power form. MF, as a thermosetting resin, may

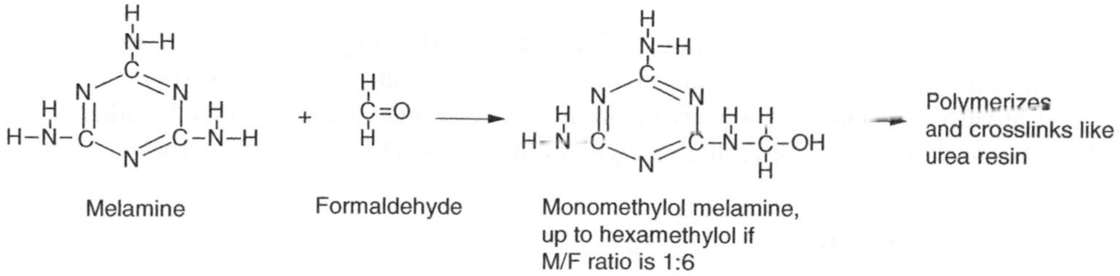

Figure 6.5 Reaction of melamine with formaldehyde to form monomethylol melamine (hydroxymethyl melamine). Note the methylol group (CH₂OH) at the lower right of the reaction product, and compare this to the hydroxymethylation of urea as shown in Figure 6.2.

be cured in a temperature range as low as 115–127°C. Polymerization of the methylolated melamine results in a highly branched, cross-linked material that is hard, clear (i.e., colorless and glassy), waterproof and durable. Despite its relatively high cost, melamine is widely used for surface coatings and laminates. An example of the latter is Formica brand materials, popular for kitchen counter surfaces. Preparation of impregnated paper laminates with MF requires use of hardeners other than the ammonium salts ordinarily used in bonding particleboard or other wood-based materials [12].

Melamine has seen increased use in recent years in the manufacture of composite flooring due to its hardness, abrasion resistance, and moisture durability. MF solids content is upward of 30% in these applications. Europeans make wide use of MF for exterior products (e.g., hardboard siding).

6.4.2.3   *Melamine-Fortified UF Resins*

MF and UF may be blended to form "fortified UF" or "melamine-urea formaldehyde" (MUF) resins to take advantage of MF's superior qualities and the low cost advantage of UF. These MUF resins may also have substantially reduced off-gassing of formaldehyde as compared to UF-only. This stems from the waterproof nature of MF, owing to its tenacious resistance to hydrolysis of the formaldehyde–melamine bond.

The preferred route to MUF resins is to carry out copolymerization of urea and melamine with formaldehyde during synthesis, as opposed to mixing of UF and MF preparations. Melamine:urea proportions of 50:50 to 30:70 are typical for MUF resins [12].

6.5   Phenolic Resins

Along with amino resins, phenolics are the most important adhesive resins for the wood-based industries. Phenolic resins form structural, fully exterior (waterproof bonds), making this the resin of choice for large volumes of construction products, such as plywood and OSB sheathing. Cured phenolic resin is typically dark brownish-red in color, limiting it in certain applications where the color may not be desired from an esthetic standpoint. There are two major types or classes of phenolic resins, namely, resoles and novolacs. Resoles are cross-linking, thermosetting resins, whereas novolacs are resins with some thermoplastic properties of the so-called B-stage intermediates.

6.5.1   Resoles (Resols)

Resole resins are also called resols (note difference in spelling) by some authors. Regardless of spelling, this type is the most important of the phenolic resins for use as a wood adhesive. Novolacs are also used as wood adhesives, but in a limited range of applications. In either case, the first step of synthesis is similar: the methylolation of phenol by reaction with formaldehyde.

6.5.1.1   *Resole Synthesis*

The synthesis of phenolic resoles consists of two steps or stages, sometimes referred to as the "A" and "B" reaction stages. The A stage in resole synthesis is the alkaline-catalyzed reaction of phenol with formaldehyde, forming monomethylol phenol (Figure 6.6). Does this look

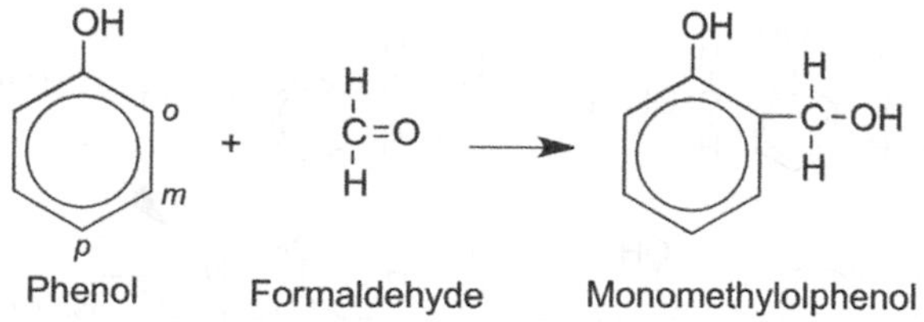

Figure 6.6 *A-stage resole synthetic reaction of phenol with formaldehyde to form monomethylol phenol. Methylolation may occur at the ortho, meta, or para (o, m, p) positions under alkaline conditions. Ortho methylolation is illustrated here.*

familiar? It should, assuming you have been reading this chapter in sequence. Methylolation, or hydroxymethylation, was also seen as the first stage in the synthesis of UF and MF resins.

For the preparation of resoles, the A stage is carried out under alkaline conditions with a molar excess of formaldehyde. The molar excess is typically on the order of P:F ratio = 1:1.1 to 1:3. As a side note, please observe that the mole ratio convention here is the reverse of that commonly used in reference to UF in North America. This highlights the point that one needs to always be aware of varying conventions as one reads scientific and technical literature. Always be cognizant of definitions. Returning to the main topic of this paragraph, these conditions promote a high degree of methylolation of the phenol molecules in the reaction solution. These methylolated phenols are highly reactive.

The B stage is the condensation of the methylolated intermediates to produce water insoluble, but fusible products (Figure 6.7). The synthetic reaction is monitored by viscosity and terminated prior to complete polymerization. Resin characteristics are tailored, in part, by the degree of polymerization, or advancement, of the resin. Resin solids content at the termination of the B-stage may be in the 40, 55, or 60% range, depending on the degree of resin advancement with viscosity in a range of 0.1–0.2 mPa·s (100–200 cP).

Resins supplied as adhesives are invariably B-stage prepolymers. A-stage molecules are small and as such, easily penetrate the cell wall of wood or other natural fiber furnish. This can be advantageous if the goal is to dimensionally stabilize the cell wall, as in a process known as Impreg or Compreg, but is a disadvantage for fiber-to-fiber or particle-to-particle adhesion. The larger B-stage molecules are more suitable as adhesives, partly because the molecules in the B-stage are large enough to resist over-penetration into the cell wall.

Because of their reactivity, B-stage resoles supplied by resin manufacturers in liquid form have limited storage life. Spray drying to form a powdered resin may extend shelf life to as much as a year, with the additional benefit of lower shipping cost due to the elimination of water weight. However, spray drying is a relatively expensive process, adding to the cost of the resin itself.

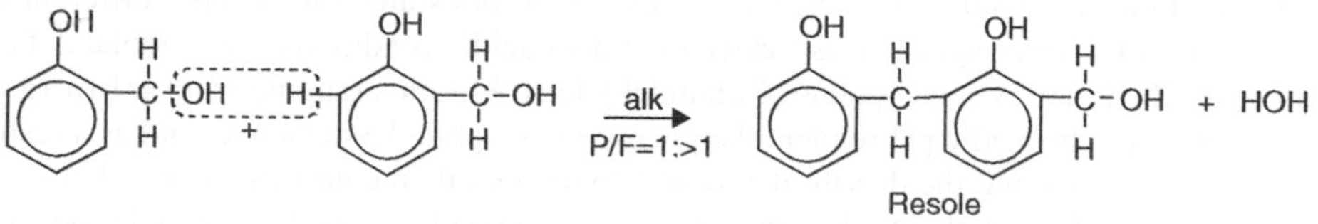

Figure 6.7 *B-stage condensation of a phenolic resole resin.*

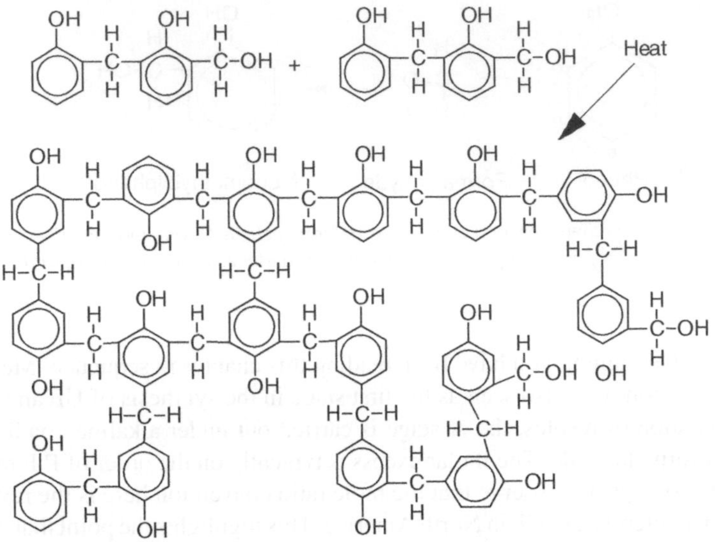

Figure 6.8 *C-stage (cure) of a resole. Only heat is required to initiate polymerization. Formation of methylene bridges (–CH$_2$–) creates a cross-linked, three-dimensional network that is stable and waterproof.*

6.5.1.2   Cure Chemistry

Curing of resole is sometimes referred to as the C stage. Given the high degree of methylolation of the B-stage phenol moieties, this reaction requires only heat as a hardening agent or "catalyst." The linkages formed between the phenolic groups may be called methylene bridges. Although resoles will cure at lower temperatures, process (press) temperatures are typically 190–200°C. A three-dimensional, cross-linked, water insoluble structure is formed (Figure 6.8).

6.5.2   Novolacs (Novolaks)

Novolacs, or novolac resins, represent a second type of phenolic adhesive. An alternative spelling that also appears in the literature is novolak. Their similarities and differences with the resoles are highlighted by considering their synthesis and cure chemistry.

6.5.2.1   Synthesis

The first step in the synthesis of novolac resins is similar to that of resoles in that it involves the methylolation of phenol. However, the reaction conditions are considerably different in that the A stage of novolac synthesis is carried out under acidic conditions with a molar deficit of formaldehyde. In the A stage, phenol:formaldehyde (P:F) ratios ranging from 1:1 to 1:0.6 are representative. The acidic pH renders the phenol nucleophile less reactive, and as a result of this lower reactivity and the dearth of formaldehyde, less than complete methylolation of the phenol occurs. Monomethylol phenols then undergo a condensation reaction with phenol, forming linear, rather than cross-linked intermediates (Figure 6.9). These intermediates are

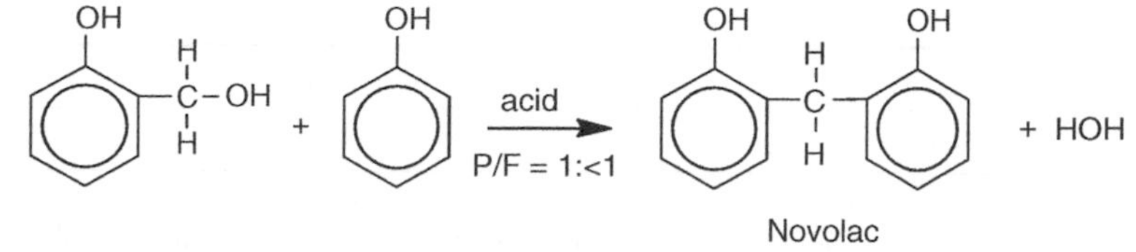

Figure 6.9 *Reaction of A-stage monomethylol phenol with phenol to form a B-stage novolac phenolic resin. Note that the B-stage phenol nuclei at the right are not methylolated.*

linked by methylene ($-CH_2-$) bridges, and lack the hydroxymethyl substituents present in B-stage resoles. Since the novolac intermediates lack methylol groups, they are incapable of further polymerization without a source of additional formaldehyde. Accordingly, B-stage novolacs are very stable, with an indefinite shelf life. The resin at this stage is soluble and partially thermoplastic. It may be supplied as dry sheets or flakes.

6.5.2.2   Cure Chemistry: Formaldehyde Donors

In order for a novolac to cure, both heat and a hardener in the form of additional formaldehyde are required. Addition of formaldehyde will enable the formation of reactive methylol groups on the phenolic nuclei of the B-stage intermediates, thus permitting polymerization via condensation reactions into a linear structure (Figure 6.10). Formaldehyde donors for cure, known as hardening agents, include formalin (formaldehyde in aqueous solution), paraformaldehyde (formaldehyde in dry powder or pellet form) and hexamethylenetetramine ("hexa" hardener). The latter, hexa hardener, is a compound that decomposes to release formaldehyde upon heating. It is a preferred formaldehyde donor in industrial practice in that it may be added to the furnish in the resin blender, but will not release formaldehyde until the material is in the pressing operation where the formaldehyde will be captured by reaction with the novolac resin.

6.5.3   Comparison of Key Attributes and Uses of Resoles and Novolacs

At this point, you should have some appreciation for the usefulness of formaldehyde as a starting material for the chemical synthesis of the two most important classes of wood and

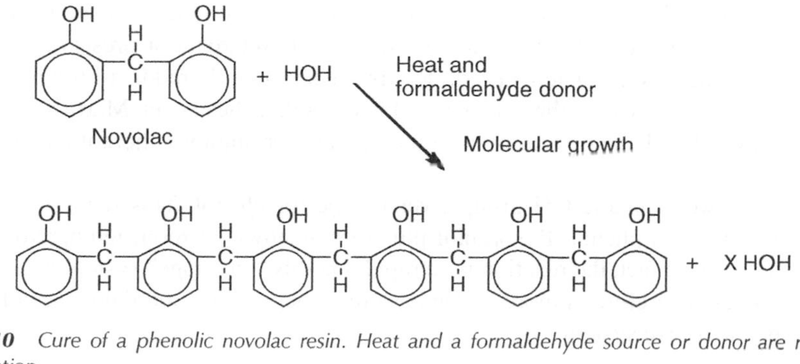

Figure 6.10 *Cure of a phenolic novolac resin. Heat and a formaldehyde source or donor are required for polymerization.*

Table 6.5   *Comparison of resole and novolac phenolic resins.*

Attribute	Type of phenolic resin	
	Resole	Novolac
A-stage conditions	alkaline; P:F = 1:>1–1:3	acidic; P:F = 1:1–1:0.6
A-stage characteristics	Low molecular weight, may be used as an impregnating resin for laminating materials	Limited methylolation of phenol makes this less reactive than resole type
B-stage characteristics	Useful as a bonding (adhesive) resin; limited shelf life due to reactivity, that is, continues to polymerize in storage; thermosetting	B-stage molecules lack methylol groups; stable with indefinite shelf life; soluble and thermoplastic
C-stage (cure)	Only heat is required for cure; cross-linked, branched structure is formed	Heat plus formaldehyde donor required for cure; cured molecules are linear
Form of adhesive	May be supplied as a liquid (aqueous solution) or as spray-dried powder	Supplied as liquid or as dry sheets or flakes

fiber adhesives, the amino and phenolic resins. For both of these classes of adhesives, the first step in synthesis is the reaction of formaldehyde with another material (urea or phenol) to form reactive methylolated intermediates. Resoles and novolacs are interesting in that the A-stage methylolation occurs under very different reaction conditions, leading to the unique attributes of these two types of phenolics. It is instructive to consider this and other contrasts between the two. Table 6.5 is provided for this purpose.

6.6   Resorcinol Resins

The thermosetting UF and PF resins are the most important wood adhesives, as measured by total usage. Related adhesive systems occupy special niches. An example is melamine, a chemical cousin of UF. Likewise, resorcinol is closely related to phenol in terms of chemical structure, but its uses are more limited due to its highly reactive nature and significantly higher cost as compared to phenolic resins.

Resorcinol formaldehyde (RF) adhesives have a short pot life (when mixed in a two-part system), generally 2–4 hours. Because it forms a fully structural, waterproof bond, and it cures at room temperature, RF is very well suited for bonding of large glue-laminated beams. RF also works well for difficult-to-bond substrates, such as fire retardant or preservative-treated wood. Like phenolic resins, when cured, RF produces a dark reddish-brown glueline. The glueline is also known as the adhesive film or bulk adhesive in Marra's chain-link analogy of adhesive bond anatomy and the contemporary terminology thereof (refer back to Table 5.1).

Resorcinol has two aromatic OH groups, versus one for phenol. This renders resorcinol much more reactive than phenol. Resorcinol plus formaldehyde at room temperature reacts quickly to form a gel. In fact, the reaction is so rapid that this "one stage" system is unsuitable for wood adhesives. Therefore, a more stable base resin is prepared and then stored for later use in a two-part adhesive system.

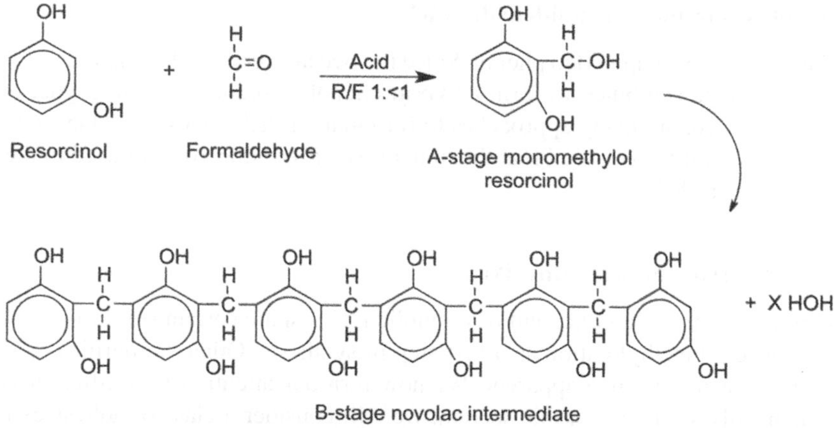

Figure 6.11 *Formation of B-stage RF base resin. The A-stage produces monomethylol resorcinol, which condenses to form a stable novolac B-stage intermediate. Addition of a formaldehyde hardener to the B-stage resin results in full cure into a three-dimensional, cross-linked network.*

6.6.1 Synthesis

Resorcinol will react with formaldehyde under either acidic or alkaline pH, and is 10–15 times more reactive than phenol under similar conditions. In order to control the exothermic reaction of resorcinol, the A-stage base resin synthesis is carried out under alkaline conditions with a molar deficit of formaldehyde, that is, R:F = 1: <1. Notice that this is the same approach taken in the synthesis of novolac phenolics, that is, a molar deficit of formaldehyde is specified. These conditions create monomethylol resorcinol, which then condenses to form a linear, stable B-stage intermediate base resin (Figure 6.11). The fact that the intermediate reaction products are terminated with resorcinol groups, rather than methylols, means that the B-stage product has an indefinite shelf life, like that of the novolak phenolics. Addition of alcohol to the mixture also extends storage life of the liquid base resin [11].

6.6.2 Cure Chemistry

Cure of a B-stage RF resin requires only the addition of a hardener (formaldehyde). In this case, solid paraformaldehyde is preferred over formalin, as the system is so reactive that it is better to use a formaldehyde source that is of a more "slow release" variety. Hexa hardener is not used, as RF is generally cured at room temperature, with full cure achieved at or above 21°C (70°F). RF base resin is mixed with the hardener in a pH range of 7–9, with higher pH promoting a faster reaction. The reaction is exothermic, that is, heat is released as the reaction occurs. Following initial mix, the adhesive is allowed to react for a short time (perhaps 15 minutes), and then it is ready to use, with a pot life of 2–4 hours [11]. The fact that this resin forms a fully structural and exterior-grade bond at room temperature makes this an ideal system for glue-laminated beam manufacturing. Two-part RF systems are also used for finger jointing of lumber.

6.6.3   Phenol-Resorcinol-Formaldehyde (PRF)

Resorcinol is expensive compared to phenol. Since the two are very similar chemically, phenol and resorcinol may be combined to form a hybrid phenol-resorcinol-formaldehyde, or PRF adhesive system. There are many approaches to formulating PRF adhesives. One example is to add RF as a "hardener" to resole PF at the time of use. PRF adhesives are used in the same type of applications as RF.

6.7   Polymeric Isocyanate Adhesives

We have seen that amino, phenolic, and resorcinolic resins share certain similarities, particularly their reliance on methylolation as a first step in synthesis. Other similarities of various sub-sets of these resins are also apparent. We now turn our attention to a different type of adhesive, chemically speaking. In this section, we will consider a class of adhesives known as polymeric isocyanates. These adhesives are rapidly growing in importance, particularly for the manufacture of OSB and other wood strand composites. From a chemical standpoint, isocyanates are of the general structure, R–N=C=O, where R is an aliphatic or aromatic component and –N=C=O is the isocyanate group. Aromatic isocyanates are more important as wood adhesives. They are thermosetting and form structural bonds with limited exterior to fully exterior moisture durability, depending on actual formulation of the adhesive. Polyurethane adhesives, which are also based on isocyanate chemistry, form semistructural bonds for limited exterior environments. Polyurethanes are increasingly used as consumer adhesives, that is, adhesives for general home usage, as well as in a wide variety of industrial or commercial applications aside from the forest products industry.

6.7.1   Isocyanate Synthesis

The first step in the synthesis of isocyanates is the HCl-catalyzed reaction of formaldehyde with aniline. This produces a mixture of reactive intermediates, known as isomeric polyamines [13]. Under typical industrial conditions, the main monomeric product of this step is the 4-4′ methylenedianiline isomer (Figure 6.12).

The mixture of intermediates is then neutralized and dried, after which it is reacted with phosgene, $COCl_2$. This second stage of the synthesis process, that is, phosgenation, converts the amino groups ($-NH_2$) to isocyanate groups ($-N=C=O$). The products of this reaction are many and varied, reflective of the mixture of polyamines generated in the first step. However, the primary monomeric product is 4,4′ diphenylmethane diisocyanate, recognized as the most

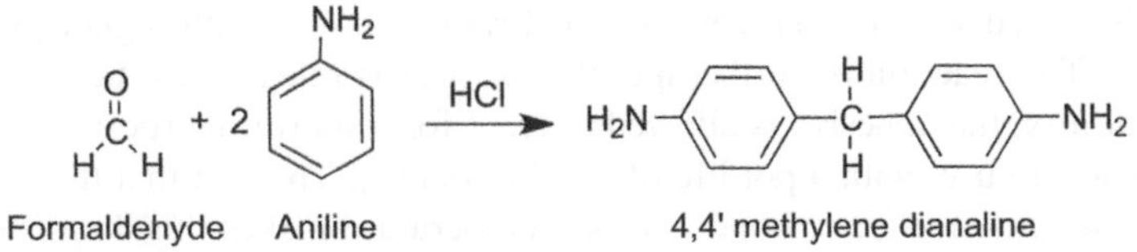

Formaldehyde   Aniline   4,4' methylene dianaline

Figure 6.12   *Reaction of formaldehyde with aniline produces a mixture of polyamides, with 4,4' methylene-dianiline the predominant monomer. This is the precursor of the most important isocyanate used as a wood binder.*

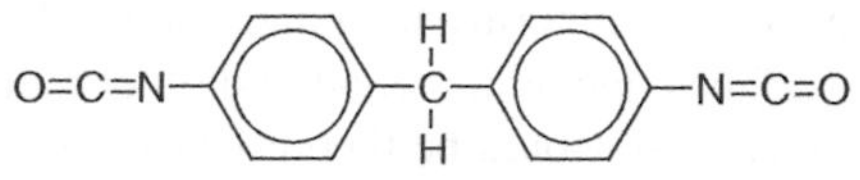

$$O=C=N-\underset{}{\bigcirc}-\overset{\overset{\text{H}}{|}}{\underset{\underset{\text{H}}{|}}{C}}-\bigcirc-N=C=O$$

Figure 6.13 *4, 4' diphenylmethane diisocyanate or 4,4' methylene diisocyanate (MDI), the most important polyisocyanate isomer used as a wood adhesive. MDI is the product of phosgenation of 4,4' methylenedianiline.*

important polyisocyanate for use as a wood adhesive (Figure 6.13). Readers interested in the particulars of the phosgenation process are referred to the detailed review by Twitchett [14].

6.7.2 Isocyanates Used as Wood Adhesives

Diphenylmethane diisocyanate, also known as methylene diisocyanate (MDI) is the non-volatile polyisocyanate that is useful as an adhesive. The output of industrial synthesis is not a single type of molecule, but a mixture of related structural isomers, wherein structural isomers are molecules with the same molecular formula but arranged in different orders. As indicated in Figure 6.13, the one structural isomer most commonly recognized as a wood adhesive is of the 4,4' variety. But, the 2,4' and 2,2' methylene diisocyanates are also present, along with other polyisocyanates. Accordingly, wood adhesives are a mixture of these molecules, generally referred to in the forest products industry as polymeric isocyanates or pMDI(s) or MDI(s) or simply isocyanate(s). Frazier [13] noted the imprecise nature of this terminology, including the fact that the molecules in question are not polymers at all, but dimers, with a variety of oligomers included in adhesive preparations. Despite the inaccuracy of these terms from a chemical nomenclature standpoint, the terms polymeric isocyanate and pMDI are widely used in general communication concerning isocyanate adhesives.

pMDI wood adhesives have low molecular weight (number average 255–280 g/mol; weight average 470–550 g/mol), low viscosities on the order of 175–250 mPa·s (175–250 cP), and surface tension in the range of 41–46 mN/m, making them well suited for effective wetting of wood adhesion substrates [13]. They also have low volatility, making them relatively safe to handle in liquid form, with the caveat that skin contact should be avoided. Aerosoled pMDI is a health hazard, in that if inhaled, the pMDI may react with moisture within the respiratory system. Over time, this may lead to respiratory ailments. It should further go without saying that pMDIs, nor any adhesive resins, are not to be ingested.

6.7.2.1 Chemical Reaction of Isocyanates and Polyols

Isocyanate adhesives are said to "stick to anything." This includes metal, which means that isocyanates will bond wood substrates to press platens! This generally restricts their use to bonding the core (interior portion) of products such as OSB, which typically have a phenolic face (surface layer) resin. The affinity of isocyanates to a variety of substrates may be illustrated by their reaction with hydroxyl (–OH) chemical groups. As illustrated in Figure 6.14,

$$\text{R-N=C=O + HO-R'} \rightarrow \text{R-NH}\underset{\underset{\text{OR'}}{|}}{C}=O$$

Figure 6.14 *Reaction of an isocyanate with an alcohol. The reaction product contains a urethane bridge or linkage, –COOR'.*

isocyanate groups react with an alcohol to form a urethane bridge. Hydroxyl groups are present on a wide variety of organic surfaces, including wood and other natural fiber. Although reaction such as that shown in Figure 6.14 should be theoretically possible with wood, chemical evidence for the same is generally regarded as debatable. As we have considered in Chapter 5, effective adhesion in wood substrates does not seem to be inherently dependent upon the formation of covalent bonds. Rather, mechanical interlocking and other underlying specific adhesion mechanisms seem to predominate.

6.7.2.2   *Reaction with Moisture*

Isocyanates react readily with water, including water on the surface of substrates and moisture in the air. Reaction with water vapor and/or that found in the wood or fiber furnish therefore speeds the resin cure process. However, carbon dioxide is released when isocyanate reacts with water. This is known as a "blowing reaction." This may be used to advantage in the manufacture of foamed polyurethanes (polyurethanes are based on isocyanate chemistry). However, excessive blowing of isocyanate foams the adhesive, which can have an adverse affect on bond formation and performance. Therefore, careful control of furnish moisture content, as well as ambient humidity, is needed to effectively utilize isocyanate binders.

6.7.2.3   *General Attributes of pMDI Adhesives*

Some of the general attributes of pMDI adhesives are summarized in Table 6.6.

6.7.3   **Polyurethane Adhesives**

Polyurethanes represent a diverse group of materials, including hard plastics, elastomers, rigid foams, soft foams, insulating materials, and adhesives. Polyurethanes are based on isocyanate chemistry, the same chemistry as that of pMDI wood binders. However, when we speak of polyurethane adhesives, the implications are that the general uses and a number of characteristics of the materials differ from the specific uses and formulations of pMDI. Within the forest products industry, polyurethane adhesives are used for laminating thermal sandwich

Table 6.6   *Advantages and disadvantages of pMDI wood binders as generally compared to PF or UF.*

Advantage/favorable attribute	Disadvantage/unfavorable attribute
Dry strength comparable to other resins	High cost
Good water resistance; >UF, but < PF long term	Bonds to metal; must use platen release agent or use only as core resin
It is a fully organic resin (no water in adhesive formulation) thus it is possible to use higher furnish moisture content	Has low vapor pressure (this is an advantage), but vapors are irritating to skin and eyes
Low resin application rates (typically <4%)	Airborne pMDI aerosols/blended particulates can cause respiratory problems for factory workers
Cure/setup of adhesive is generally rapid, translating into lower process times/higher production rate	Sticks in resin blenders. May have to resort to jackhammers to periodically clean out blenders
Can effectively wet and bond recalcitrant materials, such as Ag particles (e.g., straw particleboard)	–

panels, bonding gypsum board to joists in modular homes, bonding of veneers to composite core stock, and for gluing plywood flooring in home construction. Urethanes are also used as patching materials for face veneers in plywood manufacturing [15]. Polyurethane adhesives are also significant in the consumer market for bonding wood and other household items.

Generally speaking, polyurethanes are of two components systems, consisting of "A side" and "B side" ingredients. The A side is the isocyanate, commonly toluene diisocyanate (TDI) or MDI. The B side is the polyol, or hydroxyl-containing molecule(s). The A and B components are kept separate physically or chemically (e.g., by emulsion) until use. Mixing of the isocyanate and polyol results in chemical reaction, producing polyurethane, that is, urethane-linked molecules or materials. The isocyanate may be 100% solids, solvent-borne, or emulsified in water. In order to keep isocyanate from reacting until it is used, a number of chemical blockers have been developed [15].

Polyurethane glues for consumer use usually look like a one-part system, that is, they are supplied in a single bottle, but they may contain both the isocyanate and polyol components in the same liquid mixture. Emulsification or other chemical blocking methods are employed to prevent the adhesive from curing until it is used. Moisture from the air or substrate helps to initiate the cure reaction. Some consumer polyurethane adhesives may be a one-part system, again using moisture to induce the cure reaction. Some suppliers claim the foaming resulting from the blowing reaction of isocyanate and water is advantageous in that it imparts gap-filling characteristics to the adhesive. Polyurethanes effectively bond many household substrates, given the propensity of isocyanate to react with hydroxyl-containing surfaces.

6.8 Epoxy Adhesives

Epoxies are well-known as another class of adhesives with a reputation of being able to "stick to anything." Although this is an exaggeration, epoxies nevertheless exhibit excellent adhesive characteristics, suitable for applications requiring high strength, durability, and chemical resistance. Epoxy resins are 100% solids, that is, they are not carried in a solvent. As a result, there is no solvent loss upon cure, resulting in very low shrinkage of the adhesive, a beneficial aspect for gap-filling between bonded surfaces. Epoxies consist of two primary components, the epoxy and a catalyst or reactive hardener. Epoxy adhesive formulations also contain other additives, mainly fillers. Fillers are used to increase gap-filling ability, modify other adhesive characteristics, or reduce cost. Examples of filler materials include titanium dioxide, ferric oxide, silica, and calcium carbonate. Fillers may be added at rates of 50–300 parts by weight of resin [16]. Epoxies have high viscosity, with values in the range of 6–12 Pa·s (6,000–12,000 cP) representative.

Epoxy adhesives are expensive and therefore of limited use in the forest products industry. However, there is one niche that is an active area of research and development, and that is in the bonding of fiber-reinforced polymer (FRP) composite-wood laminates. In particular, bond durability in these systems is a focal point for research.

6.8.1 Synthesis

An epoxide is a cyclic ether with three atoms in the ring, two carbons and one oxygen. Ethylene oxide, or oxirane, is the simplest epoxide, representing the fundamental reactive group characterizing epoxy adhesives (Figure 6.15). The three-membered ring is very nearly

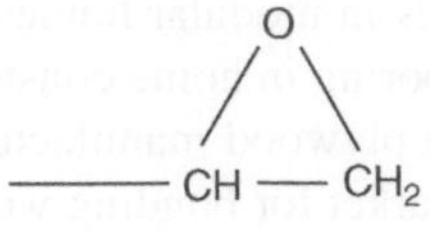

Figure 6.15 *Oxirane group, the reactive chemical group characterizing epoxy adhesives.*

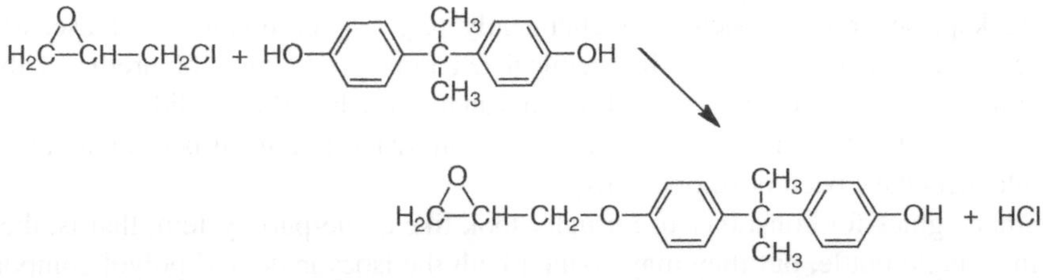

Figure 6.16 *Reaction of epichlorohydrin with bisphenol A, the first step in the synthesis of a typical epoxy.*

an equilateral triangle, with bond angles of approximately 60° at each corner. The triangle is said to be highly strained, making the ring highly reactive, especially with OH groups. The oxirane group reacts with hardeners to form an insoluble, infusible (not easily melted), thermoset resin.

Although a huge variety of epoxy adhesive systems exist, most are based on reaction of epichlorohydrin with bisphenol A (Figure 6.16). The initial condensation product reacts with additional epichlorohydrin to yield diglycidyl ether of bisphenol A, DGEBA (Figure 6.17). The resulting resins are known as DGEBA epoxies.

6.8.2   Cure Chemistry

In order to cure, the epoxy (such as that represented in Figure 6.17) is mixed with a separate hardener component. Many classes of compounds are used as hardeners, depending on application, including polyamines, polyamides, polysulfides, urea resins, phenolic resins, and acid anhydrides. Some of these compounds are toxic; therefore, it is important to always read and follow the manufacturer's label directions when using epoxy adhesives. The classic epoxy curing mechanism is illustrated by the reaction of epoxy with a primary amine (Figure 6.18). The result is a hard, chemically resistant cross-linked polymer network.

In addition to the DGEBA epoxies, of interest here are the novolac phenolic epoxies, produced by the reaction of epichlorohydrin with phenolic novolac resins (Figure 6.19). These

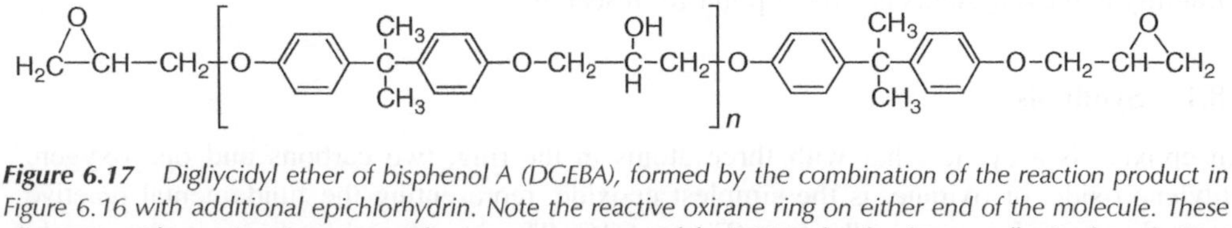

Figure 6.17 *Digliycidyl ether of bisphenol A (DGEBA), formed by the combination of the reaction product in Figure 6.16 with additional epichlorhydrin. Note the reactive oxirane ring on either end of the molecule. These oligomers, with n = 1 to 4, represent the epoxy component of the DGEBA family of epoxy adhesive formulations.*

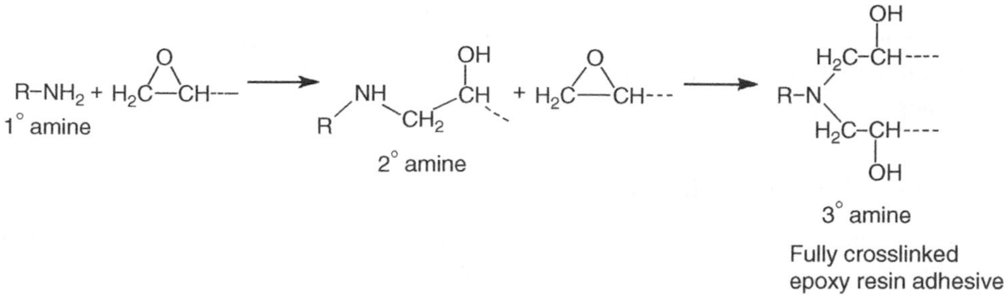

Figure 6.18 *Curing reaction of an epoxy with a primary (1°) amine to form a cross-linked polymer network. Reproduced from [17] with permission of Taylor and Francis Group LLC-Books © 2005.*

resins produce a more highly cross-linked cured network, and are therefore more brittle, but have higher temperature resistance than, for example, typical DGEBA epoxies. As such, the novolac epoxies may be used as modifiers of the DGEBA type [16].

6.8.3   Durability of Wood-Epoxy Bonds

Despite the excellent chemical affinity of epoxies for wood surfaces, wood-epoxy bonds are not durable in long-term exterior exposure due to significant differences in the dimensional stability of substrate and adhesive. Movement of wood, mainly in response to fluctuation in wood moisture content, places strain on the wood-epoxy interface, in that cured epoxy is dimensionally stable upon exposure to moisture. As a result, failure of the gluebond occurs at the wood-epoxy interface. The desire to bond fiber-reinforced-polymer (FRP) materials to wood glulam beams has provided impetus for research and development concerning durability of wood-epoxy bonds. The approach is to bond high strength and stiffness FRP materials containing, for example, carbon fiber, to the upper and lower surfaces of wood glue-laminated beams. This improves the engineering properties of the beams significantly with little increase in size or bulk of the beams. However, lack of bond durability in exterior environments has prevented wider adaptation of this technology [17].

One approach that has proven successful to improve bond durability is through the use of hydroxymethyl resorcinol (HMR) primer or coupling agent in conjunction with wood-epoxy systems [18, 19]. It is hypothesized that the HMR dimensionally stabilizes the surface

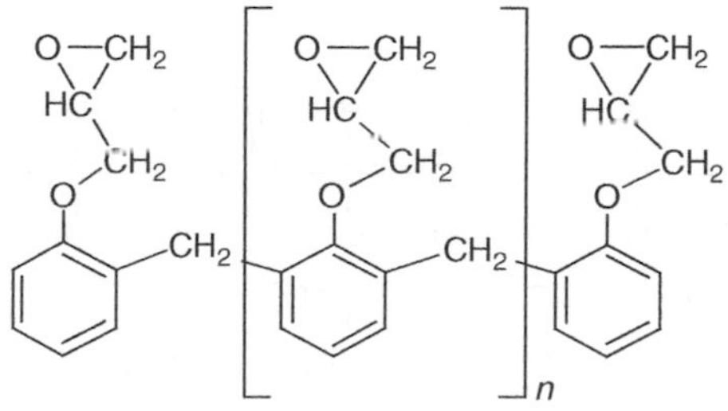

Figure 6.19 *A phenolic novolac epoxy, produced by the reaction of epichlorohydrin with a novolac.*

layer of wood, leading to greater bond performance with stable adhesives such as epoxies or other polymers [20]. The HMR coupling agent has also proven effective in bonding other polymer-reinforced materials to wood, including those containing polyurethanes, polyesters, or vinylesters.

6.9   Polyvinyl Acetate Adhesives

Polyvinyl acetates (PVAc) are thermoplastic polymer adhesives, best known as common "white glue" for wood joinery and household use. Elmer's[TM] is a well-known consumer brand of PVAc. Though long used in wood joinery in the furniture construction industry, PVAc has historically seen little use in wood-based composites. Some shift in usage has occurred in recent years, however, concurrent with the need to reduce formaldehyde emissions and/or the use of formaldehyde-containing or formaldehyde-based adhesives, in response to regulatory considerations. For example, cross-linked PVAc adhesives are increasingly used for bonding hardboard door skins to wood-based door cores, supplanting UF resin in such applications.

In general, PVAc adhesives are more flexible and more susceptible to thermal softening and creep (time-dependent deformation or flow under mechanical load) than UF resins. Like UF, PVAc bonds are also susceptible to moisture-induced degradation. Nevertheless, these adhesives are effective for bonding a variety of porous substrates including wood, paper, and other natural fiber materials.

6.9.1   Synthesis

Polyvinyl acetate is synthesized by the free-radical polymerization of vinyl acetate to form the homopolymer of PVAc (Figure 6.20). The acetate group, CH_3COO (abbreviated OAc), imparts polarity to the vinyl backbone. The PVAc polymers are emulsified as small, spherical droplets or globules, typically in a mixture of water and polyvinyl alcohol, the latter described as the "main protective colloid and thickener" [21]. The emulsions are prepared such that they are relatively stable at room temperature, provided that evaporation of water is prevented, generally by storing the mixture in a closed container. Freezing will disrupt the emulsion, making the resin unusable.

6.9.2   Solvent Loss Cure Mechanism

PVAc emulsions or suspensions are stabilized within the solvent such that loss of a small amount of water results in coalescence of the PVAc molecules, producing an adhesive film.

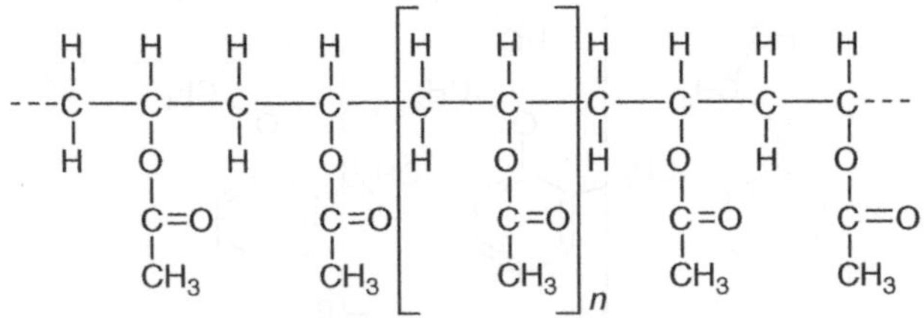

Figure 6.20   *Structure of polyvinyl acetate (PVAc).*

Water loss occurs via sorption into substrate and/or by evaporation. Strength of the glueline is developed by molecular entanglement of PVAc molecules with one another. On porous substrates such as wood, the adhesive is able to form hydrogen bonds with the adherend due to the polarity of PVAc. Bond formation may occur in as little as 10 minutes, with full strength development in 24 hours.

6.9.3   Modified PVAc

PVAcs are modified in a number of ways, for example, by copolymerization with ethylene to form ethylene vinyl acetate (EVA), or by addition of external plasticizers, such as dibutyl phthalate. Yellow carpenter's glue is one such modified form, sometimes known as aliphatic PVAc [11]. In actuality, the term, "alphatic" is not particularly helpful, in that PVAc in general has an aliphatic backbone. Nevertheless, compared to the unmodified forms, modified PVAcs may be more thixotropic, faster setting, more tolerant of low temperatures, have greater creep resistance, or less sensitivity to moisture and heat.

Other modifications may be introduced to impart thermoset properties to PVAc, resulting in improved thermal stability, moisture resistance, and better creep performance. These include covalent cross-linking of PVAc with glyoxal, formaldehyde, or isocyanates, or ionic bond formation with organic titanates, chromium nitrates, aluminum chloride, or aluminum nitrates [17]. The crosslinker(s) must be added just prior to application of the adhesive to the substrate, making these two-part resin systems. It is developments such as these that have allowed increased use of PVAc in applications such as door manufacturing.

6.10   Hot Melts and Mastics

Though not used directly in the manufacture of wood and fiber composites, hot-melt and mastic adhesives are important for bonding materials to composite substrates, and other related applications. The general characteristics of hot melts and mastics will be described in this section.

6.10.1   Hot Melts

Hot-melts are 100% solid thermoplastics, supplied in the form of pellets, chips, blocks, or slugs. As thermoplastics, they soften and become semi-molten upon heating and set (harden) within seconds upon removal of the heat source. The "cure" mechanism is therefore described as a "heat-loss adhesive system." Within the forest products industry, hot-melts are used extensively for edge-banding, veneer splicing, and similar applications. For these uses, the two most common types of hot melts are ethylene-vinyl acetates (EVA) and polyamides. The advantages of hot melts for applications such as these have been enumerated as follows [22,23]:

1. Ease of application via high speed equipment;
2. Formation of strong, permanent, durable bonds within a few seconds of application;
3. No environmental hazard and minimal wastage because of 100% solid systems;
4. Ease of handling;
5. Absence of highly volatile or flammable ingredients;
6. Excellent adhesion to both wood and plastics;

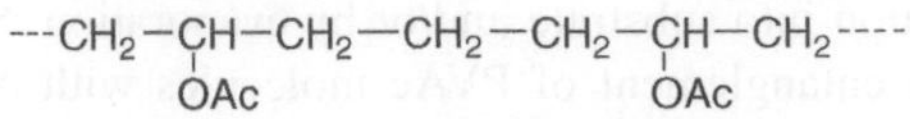

$$---CH_2-CH-CH_2-CH_2-CH_2-CH-CH_2----$$
$$\quad\quad\quad OAc \quad\quad\quad\quad\quad\quad\quad\quad OAc$$

Figure 6.21   Ethylene-vinyl acetate (EVA), a typical hot-melt adhesive. Note similarity of chemical structure with that of PVAc (see Figure 6.20).

7. Wide formulation possibilities to suit individual requirements (e.g., color, viscosity, application temperature, and performance characteristics); and
8. Cost effectiveness.

6.10.1.1   Ethylene-Vinyl Acetate Adhesives

Ethylene-vinyl acetate (EVA) is closely related to PVAc. It is synthesized by the copolymerization of vinyl acetate and ethylene. The structure of EVA is therefore very similar to that of PVAc, with wider spacing of the acetate groups due to interposition of ethylene into the aliphatic backbone of the molecule (Figure 6.21). Although the molecule is more flexible than PVAc, EVA is generally prepared to high molecular weight, rendering a material that is solid at room temperature. Upon heating, the material softens for application as an adhesive. These materials are very high in viscosity, on the order of 50–60 Pa·s (50,000–60,000 cP) at 200°C [23]. Removal of the heat source results in very rapid setting of the adhesive. Because of the extremely high viscosity, bond strength is relatively poor due to the limited wetting and penetration into the substrate. Therefore, hot melts are not suitable for structural applications.

A typical use of EVA hot melts in the forest products industry is for bonding edge banding (veneers, laminates, or other decorative materials) to composite cores, as in tabletops or kitchen and bath countertops. EVA glues are also popular for hobby use, industrial assembly, and packaging, supplied for stick, slug, and granular applicators.

6.10.1.2   Polyamide Hot Melts

Polyamides represent a second type of hot melt of importance in the wood industries. A typical application is for veneer splicing. Polyamides have a low melt viscosity, which aids rapid glue spread and wetting of the substrate. These attributes, coupled with high tack and rapid setup upon cooling, make polyamides ideal for splicing. Polyamides are also used for edge banding and other related applications, but they are higher cost than EVA, and are thus limited to specific uses that justify the cost differential.

6.10.2   Mastics

Mastics are very high viscosity and tack adhesives with a paste-like consistency. Mastics are used for general-purpose home, shop, and construction adhesive bonding. An example of construction use is for bonding of flooring to joists. Mastics are also used to bond ceramics, plastics, metals, asphalt, or acoustical tile. Other construction applications are for bonding hardboard and wood paneling, wallboard, furring strips, and countertop laminates. Mastics may be used for industrial production of wooden I-beams.

The composition of these adhesives varies widely. They are high solids, low solvent materials, blended with a variety of additives, including tackifiers (rosin, resin, phenolics, resorcinolics), fillers (carbon black, chalk, zinc oxide, clays), and plasticizers, antioxidants, curing agents, and other materials [11]. Polyurethane adhesives for construction use are commonly supplied mastic form. Some are rubber based, either natural or synthetic, either as latex in water or a dispersion in solvent. Mastics are suitable for construction use, in that they require only nailing pressure to effect adequate bonding, with excellent gap-filling ability. These adhesives generally have good dry and wet strength. However, they do not stand up well to repeated cyclic moisture exposure and are likely to harden and become brittle over time due to reaction with oxygen and ozone [17].

6.11   Adhesives from Renewable Natural Resources

This chapter may well have begun with the topic of adhesives from renewable natural resources, as the first adhesives used to bond wood and other materials were derived from animals [24], extending back in human history four millenia from the present [25]. However, in today's world, adhesives from natural sources have been largely supplanted by synthetics. This is not to say that there are not some important uses of natural adhesives today, but these are limited as compared to the amount of synthetics used by the natural fiber composites industries. Nevertheless, a resurgence of interest in green or environment friendly adhesives and chemicals has spawned vigorous contemporary research and development in this area. It is therefore important to be cognizant and observant of this class of adhesives.

The research literature concerning adhesives from natural materials is formidable and fascinating. By providing this brief introduction, it is our hope that readers will be motivated to learn more about this subject. Our approach will be to provide an overview of the main classes of natural adhesives for wood and related fiber, with further explanation regarding adhesives from lignins, tannins, and soybean protein. These three sources are highlighted as they are currently used in specific niches of composite manufacture and may hold promise for greater use in the future.

6.11.1   Classes of Natural Materials for Adhesives

Natural materials for adhesives may be categorized by their origin from either plants or animals. Those from plants include carbohydrates, lignocellulosic extracts (typically, phenolics), and proteins. Carbohydrate materials used as adhesives include starches, gums, and cellulose derivatives, wherein a derivative represents cellulose modified by chemical reaction to add various substituents to the cellulose backbone. Starch is the food storage form of glucose in plants, widely available from potatoes, tubers, fruits, grains such as corn and wheat, and other plant parts. Gums represent a wide range of hydrophobic and hydrophilic non-starch carbohydrates isolated from plants.

Lignocellulosic extracts useful as adhesives include lignins, typically by-products of chemical pulping operations, and condensed tannins extracted from wood and bark of some species of trees. Plant proteins are generally obtained from grains, with the proteins obtained from soybeans of primary interest with respect to adhesives. Natural adhesive materials from animals include the protein, collagen, that is obtained from hides, bones, and sinew; casein (milk protein); and blood protein. The general attributes and some of the use properties and uses of these adhesive materials are summarized in Table 6.7.

Table 6.7 reveals that the carbohydrate- and animal protein (hide, bone, sinew)-based adhesives, though useful in various aspects of the paper and furniture industries, have no

Table 6.7 General attributes and uses of selected adhesives from natural sources.

Source	Form and color	Preparation and application	Strength properties	Typical uses
Carbohydrate, including cellulose derivatives, starch and gums	Films, powders, hot melts, white to yellow to dark brown	Methods of application vary widely; cellulose derivatives may be solvent-borne or solid hot-melts; starches are prepared variously and mixed with borax, plasticizers, water-resistance additives (e.g., UF, MF, or RF polymers, polyvinyl acetates, etc.), viscosity stabilizers, fillers and other additives and are generally applied as liquid formulations ranging from watery to paste-like viscosities; gums are dispersed in hot or cold water to form gel-like materials	Low strength relative to other adhesive classes, but adequate for the intended purposes; water or moisture resistance varies with type of derivative; none of these materials are intended for use in the wood and fiber composites that are covered in this book	Cellulose derivatives are used as paper sizings and coatings, wallpaper adhesives, leather processing aids, additives to paints, solvent- and hot-melt adhesives, and so on.; starch is a common adhesive for book binding, corrugated box manufacturing, wettable adhesives for envelopes and stamps, and as a paper and textile sizing agent; non-food, industrial uses of gums include laundry products, textile processing, paints additives, pressure-sensitive tape, denture adhesives, pharmaceutical tablet binders, and so on
Lignocellulosic residues and extracts, primarily lignins and condensed tannins	Powder or liquid; may be blended with phenolic adhesive; dark brown bondline	Blended with extender and filler by user; adhesive cured in hot-press ranging from 130°C to 205°C (266°F to 401°F), depending on type of lignocellulosic extract used in adhesive formulation	Good dry strength; moderate to good wet strength; durability improved by blending with phenolic adhesive	Partial replacement for phenolic adhesive in composite and plywood panel products; condensed tannins may be formulated as standalone adhesives for plywood and composites

Source/Type	Form and appearance	Application	Properties	Uses
Soybean, protein; soy protein (and a carbohydrate fraction) is generally obtained as a by-product of soy oil extraction	Powder with added chemicals; white to tan, similar color in bondline	Mixed with cold water, lime, caustic soda, and other chemicals; applied and pressed at room temperature, but more frequently blended with blood adhesive; contemporary research has introduced chemical cross-linkers and alternative applications	Moderate to low dry strength; moderate to low resistance to water and damp atmospheres; moderate resistance to intermediate temperatures	Softwood plywood for interior use, now replaced by phenolic adhesive. New fast-setting resorcinol-soybean adhesives for finger-jointing of lumber has seen limited use; new formaldehyde-free adhesive has likewise seen limited use for particleboard and interior plywood
Animal, protein; from hides, bones, sinew; protein from fish skin also has adhesive uses	Solid and liquid, brown to white bondline	Solid form added to water, soaked, and melted; adhesive kept warm during application; liquid form applied directly; both pressed at room temperature; bonding process must be adjusted for small changes in temperature	High dry strength; low resistance to water and damp atmosphere	Assembly of furniture and stringed musical instruments; repairs of antique furniture; "hide glues" are also preferred for high-end laminated table tennis paddles
Casein, protein	Powder with added chemicals; white to tan bondline	Mixed with water; applied and pressed at room temperature	High dry strength; moderate resistance to water, damp atmospheres, and intermediate temperatures; not suitable for exterior uses	Was the original adhesive for structural glue-laminated timbers; now replaced by synthetics in this application; interior doors
Blood, protein	Solid and partially dried whole blood; dark red to black bondline	Mixed with cold water, lime, caustic soda, and other chemicals; applied at room temperature; pressed either at room temperature or 120°C (250°F) or greater	High dry strength; moderate resistance to water and damp atmosphere and to microorganisms	Interior-type softwood plywood, sometimes in combination with soybean adhesive; mostly replaced by phenolic adhesive

Source: Adapted from Reference 6. Information on carbohydrate adhesives from Reference 26.

relevance to the composites represented in this book. With respect to contemporary use of natural adhesives for wood and natural fiber composites (including plywood and glue-laminated timbers), there are only two categories of significance, lignocellulosic residues and extracts, and plant proteins. The former refers to lignins and condensed tannins, and the latter to soybean meal or soy protein. Also of importance for bonding solid wood, wood laminates and related composites, mainly from a historic perspective, are animal proteins from milk and blood.

6.11.2   Lignins in Adhesive Formulation

Interest in using lignin for industrial purposes has a long history, with patents from the late 1800s describing the use of spent sulfite pulping-process lignin as a wood and paper adhesive. However, despite much research, including a virtual explosion of patents on lignin adhesives in recent years, the use of "technical lignin" worldwide remains relatively low. Only about 20% of these materials are used, mainly for dispersants, oil-drilling mud, pelletizing materials, and adhesive additives. The remainder of the spent lignins from pulping operations are burned as a fuel in the pulp factories in which they are generated.

Spent lignin is the term for the lignin residue left after isolation of the cellulosic fibers that are the desired products of pulping. The term, technical lignin, is generally synonymous with spent lignin, and represents the starting material for potential modification into adhesives. In actual fact, only a fraction of technical lignins are suitable for use in adhesives, specifically those obtained from the sulfite pulping process, in a slurry known as spent sulfite liquor (SSL). Unfortunately, lignins from the worldwide dominant Kraft (sulfate) pulping process are not generally found to be amenable for use as an adhesive, mainly due to difficulties in separating lignin from the pulping chemicals in the so-called black liquor generated in the process.

Lignins are polyphenolic; therefore, their use in manners similar to petrochemically derived phenol are anticipated. Unfortunately, the condensation reactions of lignins are less effective than those of phenol due to fewer free reaction sites on the aromatic nuclei of lignin. Furthermore, technical lignins are also reduced in reactivity by their lack of purity. For example, it is almost a certainty that some carbohydrate fraction of the cell wall remains in spent lignins due to the existence of complex and somewhat recalcitrant *in situ* cell wall lignin-carbohydrate complexes. The practical outcome of the lower reactivity of technical lignins is twofold: First, difficulty in chemically modifying them for use as adhesives, and second, the general necessity of higher press temperatures and/or longer press times when used as adhesives. These issues have, despite much research, generally limited lignins to use as extenders for synthetic phenolic adhesives. An extender is a material added to a base resin to reduce the amount of base resin required in an adhesive formulation, generally driven by cost-reduction goals. Extenders are also materials considered to impart some adhesive functionality. Extenders are contrasted with fillers, which reduce the amount of base resin in a formulation, but have no appreciable adhesion functionality.

A variety of approaches to lignin adhesive formulation and use have been attempted, including cross-linking by condensation reactions or oxidative coupling, employment of long press times and post-heating treatment, curing with sulfuric acid or hydrogen peroxide, methylolation of lignin, or combining lignin with PF or UF resins [27]. In a sense, use of lignin as a common, reliable adhesive or value-added chemical feedstock remains an elusive "holy grail" of renewable resource research.

6.11.3 Plant-Derived Tannins as Adhesives

Like lignins, tannins are phenolic plant-derived extracts. These polyphenolic compounds were historically used for processing or "tanning" leather, and are obtained from the bark or wood of certain tree species. Two classes of these natural compounds are recognized, namely, hydrolyzable tannins and condensed tannins, with the latter proven much more useful for wood adhesive formulations. Condensed tannins are also more widely available, constituting more than 90% of world production of commercial tannins [28].

Tannins from plant sources are produced primarily in the Southern Hemisphere, and have been used been used industrially as adhesives since the early 1970s, particularly in South Africa and Australia. Commercial use of tannin adhesives is also know in Zimbabwe, Chile, Argentina, Brazil, and New Zealand [17]. At present, applications for condensed tannin adhesives include the manufacture of exterior plywood, glued-laminated timbers, particleboard, and in finger-jointing of lumber products.

6.11.3.1 Chemical Structure of Condensed Tannins

Condensed tannins are natural polymeric flavonoids obtained by extraction primarily from the following trees: *Acacia* sp. (wattle or mimosa) bark, *Schinopsis* sp. (quebracho) wood, *Tsuga* sp. (hemlock) bark, *Rhus* sp. (sumac) and some *Pinus* sp. (pine) bark [28, 29]. Flavonoids are polyphenols with the $C_6C_3C_6$ carbon skeleton, the polymers of which are the condensed tannins [2]. The "structure of the flavonoid constituting the main monomer of condensed tannins" [29] is shown in Figure 6.22. In this structure, the "B" ring may have either two or three phenolic hydroxyl groups, the former designated as a catechol type, and the latter as a pyrogallol type. The catechol type is represented in extracts of quebracho and chestnut wood and wattle, pine, *Eucalyptus* and *Betula* bark [2, 28].

6.11.3.2 Condensed Tannin Wood Adhesives

Unlike lignins, condensed tannins are very reactive, much more so than synthetic phenol under similar conditions. As a result of their extreme chemical reactivity, it is not possible to make useful tannin resoles. Tannin molecules, methylolated by reaction with formaldehyde, will condense very rapidly with other tannin nuclei, resulting in poor stability and shelf life. Accordingly, tannin adhesives are made solely as novolacs. Perhaps you have noted that this is

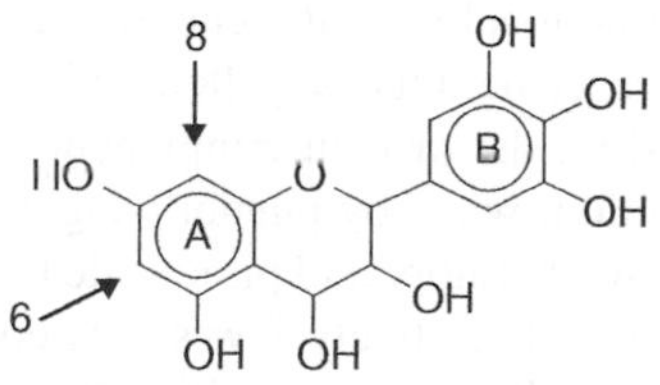

Figure 6.22 *Condensed tannins useful as wood adhesives are natural polymers based on the flavonoid structure shown here. Methylolation of the A ring by reaction with formaldehyde occurs at the 6 and 8 positions (indicated by arrows) almost as readily as the methylolation of resorcinol by formaldehyde. Hydroxymethylated condensed tannins (not shown) cross-link via condensation reactions.*

reminiscent of the comparison of the chemical properties of resorcinol versus phenol. In fact, the following comparison has been observed: "Assuming the reactivity of phenol to be 1 and that of resorcinol to be 10, the A rings" (of a flavonoid of the type shown in Figure 6.22) "have a reactivity of 8 or 9" [28]. Tannin-formaldehyde adhesives are thus obtained by hardening of natural polymeric flavonoids via polycondensation with formaldehyde, with the typical formaldehyde hardeners, paraformaldehyde or hexamethylenetetramine (hexa hardener) added at the point of use in the same manner as for resorcinolic novolacs. Other condensed tannin systems, for example, in combination with pMDI, are also known [28].

6.11.4   Soy Protein Adhesives

Soybeans (*Glycine max*) are the major source of oilseeds worldwide. Whole beans are composed of approximately 40% protein, 34% carbohydrates, 21% fat (oil) and 5% ash on a dry weight basis [30]. Most of the oil extracted from soybeans is for food use, although there are some contemporary industrial (non-food) uses. Soybean meal is the protein and carbohydrate fraction remaining after oil extraction, and soybean flours are produced by grinding and size-fractionating the defatted soy meals. Meals and flours are used in a variety of human and animal food products. Soy protein isolates are purified protein products, that is, the carbohydrate fraction of the meal or flour is removed. Historically, industrial (non-food product) use of both the oil and protein fractions of soybeans was more significant than at present, with only 0.5% of soy protein and 2.6% of oil used industrially today [31].

One of the significant, largely historic uses of soy flours is represented by soy adhesives, at one time the major adhesive used to bond interior plywood. Effective formulations of soybean, soy- and blood protein mixtures, and casein (milk protein) adhesives for wood products are well known, as enumerated by Lambuth [24]. However, these adhesives are not competitive in today's markets, due to the general superiority of synthetics with respect to speed of cure, strength of gluebond, and moisture, temperature and microorganism resistance.

In spite of the general limitations of soy adhesives, there is a small but vigorous research community working to overcome the technical impediments to greater use of this resource. These efforts are driven, in large measure, by environmental and regulatory pressures aimed at reducing formaldehyde emissions from products, and the reduction or complete elimination of formaldehyde in manufacturing processes. Although soy isolates (purified protein) may produce better overall adhesive performance, in most cases, defatted soy flours are preferred as feedstock for new adhesive formulations primarily due to economic considerations.

Soy flour or meals require pretreatment, typically with alkali, in order to disperse them in water. Under certain conditions, alkaline treatment may also partially denature the protein, allowing for improved adhesive functionality and/or chemical reactivity. One approach toward soy adhesive formulation is to react defatted soy flour with phenol and formaldehyde at neutral pH in an effort to cross-link the protein component of the flour with phenol [32]. Adhesives formulated in this fashion, with soy:phenol weight ratios of up to 7:3, generally produce acceptable performance in laboratory and pilot-scale fiberboard or particleboards, but tend to suffer from high viscosities at relatively low solids contents, and long press times. A better use of such systems may well be as spray-dried powders for compression molding applications. However, Westcott et al. [33] reported improvements in PF-cross-linked-soy adhesive characteristics and performance via a process consisting of soy protein denaturation, modification, and co-polymerization with phenol and formaldehyde.

Li [34] and Li et al. [35] employed polyamidoamine-epichlorohydrin (PAE) resins, well-known as wet strength agents in currency paper, as the cross-linking agent with alkaline soy protein isolate. This resulted in a completely formaldehyde-free adhesive, that is, one with no formaldehyde used at any stage of its manufacture. This type of adhesive has been used on a limited commercial scale for the manufacture of hardwood plywood and particleboard. Innovations such as this may become of even greater importance in the future, driven in large measure by environmental considerations.

6.11.5   Animal Protein Adhesives

As shown in Table 6.7, a number of animal proteins may be used to bond wood and other porous substrates. These proteins include collagen (protein obtained from animal hides, bones, connective tissue), casein (from milk), and albumin and other blood proteins. Of these, casein and blood proteins have been used historically as adhesives in composite materials as defined in this text. Sellers, Jr. [36] presented data that showed peak US consumption of casein and blood protein adhesives in the years 1973 and 1960, respectively.

6.11.5.1   Casein

Casein is the name given for various phosphoproteins found in mammalian milk. Cow's milk consists of approximately 88% water, 5% carbohydrate, 4% fat, and 3% protein, with 80% of the proteins consisting of casein. Casein is isolated from skim (defatted) milk by forming curds, accelerated by the addition of acid. It is washed and then alkali treated with hydrated lime or sodium hydroxide. Other ingredients may be added, for example, copper salts as a fungicide. It is then dried and supplied in dry power form. Casein adhesive is mixed with water at the point of use, as it has a limited pot life in liquid form. When used as an adhesive for solid wood, casein forms a bond with strength properties sufficient for use in structural members, such as glued-laminated beams. Despite its excellent strength, the gluebond formed has only moderate moisture resistance, limiting its overall usefulness in structural applications. As a result, casein has been supplanted from its former use in glulam by synthetics. In addition, shifts from natural adhesives in general to synthetics have been driven not only by limited moisture durability and other performance issues, but also economics, which are generally skewed heavily toward the side of synthetics. It should also be noted that it is highly questionable whether natural or renewable materials alone would be available in sufficient quantities to meet the current and future demands for bonding wood and fiber materials.

6.11.5.2   Blood Protein

Albumin and other blood serum proteins from cattle, swine, and sheep have a long history as wood adhesives. With respect to wood-based composites, blood was a historically important adhesive for plywood. Spray drying technology is used to prepare soluble adhesive-grade blood powder. Other drying technologies, such as ring- and flash drying, are used to produce blood meal used mainly in animal food products. Due to the higher temperatures used in these processes, blood meal is insoluble and is not suitable for use as an adhesive per se, but may be used as an extender, wherein an extender is an additive that imparts some adhesive function in the adhesive formulation. Blood is thus used contemporarily as an extender in phenol formaldehyde plywood adhesive formulations. Supplies of adhesive-grade blood have declined in the United States since its peak use in 1960. Blood protein adhesives have lower structural strength but better moisture resistance than casein. As with casein, the use of blood as an adhesive has been practically eliminated by synthetics having superior performance and price attributes.

6.11.6   Adhesives Future

The field of bio-based adhesives is a fertile area of research, having potential for significant advancements in the use of renewable resources. Underlying factors for this research include interest in reducing dependency on petrochemical resources and the reduction of formaldehyde use in industrial processes. Current trends in synthetic adhesives use within the wood industries are likewise shifting a number of processes away from formaldehyde-containing adhesives. Recent research on frictional "wood welding" may lead to new processes to bond wood without any adhesive at all [37]. Until such time, adhesives from synthetic and natural sources will remain vital to the manufacture of wood and natural fiber composites.

References

1. Pizzi A, Mittal KL, editors. *Handbook of Adhesive Technology*, 2nd ed. revised and expanded. New York: Marcel Dekker, Inc.; 2003.
2. Sjöström E. *Wood Chemistry: Fundamentals and Applications*. New York: Academic Press; 1981.
3. Groover MP. *Fundamentals of Modern Manufacturing: Materials, Processes, and Systems*. Hoboken, NJ: John Wiley & Sons, Inc.; 2007
4. Marra AA. *Technology of Wood Bonding: Principles in Practice*. New York: Van Nostrand Reinhold; 1992.
5. ASTM 2010 Standard test method for tack of pressure-sensitve adhesives by rolling ball, D3121-06. Annual Book of ASTM Standards. West Conshohocken, PA: ASTM International.
6. Vick CB. Adhesive bonding of wood materials. *Wood Handbook—Wood as an Engineering Material*. Gen Tech Report FPL-GTR-113. Madison, WI: USDA Forest Service, Forest Products Laboratory; 1999. p. 9-1–9-24.
7. White JT. Wood adhesives and binders. What's the outlook? *Forest Products Journal*, 1995;45(3):21–28.
8. Graves G. Urea-formaldehyde resins: yesterday, today, and tomorrow. In: Bradfield J, editor. *Resin & Blending Seminar*. Portland, Oregon and Charlotte, North Carolina: Composite Panel Association; 1998. pp 3–10.
9. Pizzi A. Urea-formaldehyde adhesives. In: Pizzi A, Mittal KL, editors. *Handbook of Adhesive Technology*, 2nd ed, revised and expanded. New York: Marcel Dekker, Inc; 2003. pp 635–652.
10. Gunnells D, Griffin K. Catalyst systems. In: Bradfield J, editor. *Resin & Blending Seminar*. Portland, Oregon and Charlotte, North Carolina: Composite Panel Association, 1998.
11. Marra AA. Characteristics and composition of adhesives. In: *Technology of Wood Bonding: Principles in Practice*. New York: Van Nostrand Reinhold; 1992. pp 61–103.
12. Pizzi A. Melamine-formaldehyde adhesives. In: Pizzi A, Mittal KL, editors. *Handbook of Adhesive Technology*, 2nd ed, revised and expanded. New York: Marcel Dekker, Inc; 2003. pp 653–680.
13. Frazier CE. Isocyanate wood binders. In: Pizzi A, Mittal KL, editors. *Handbook of Adhesive Technology*, 2nd ed, revised and expanded. New York: Marcel Dekker, Inc; 2003. pp 681–694.
14. Twitchett HJ. Chemistry of the production of organic isocyanates. *Chemical Society Reviews*, 1974;3(2):209–230.
15. Lay DG, Cranley P. Polyrethane adhesives. In: Pizzi A, Mittal KL, editors. *Handbook of Adhesive Technology*, 2nd ed, revised and expanded. New York: Marcel Dekker, Inc.; 2003. pp 695–718.

16. Goulding TM. Epoxy resin adhesives. In: Pizzi A, Mittal KL, editors. *Handbook of Adhesive Technology*, 2nd ed, revised and expanded. New York: Marcel Dekker, Inc.; 2003. pp 823–838.

17. Frihart CR. Wood adhesion and adhesives. In: Rowell RM, editor. *Handbook of Wood Chemistry and Wood Composites*. Boca Raton, FL: Taylor & Francis; 2005. pp 215–278.

18. Vick CB, Richter KH, River BH, inventors. The United States of America as represented by the Secretary of Agriculture, assignee. Hydroxymethylated resorcinol coupling agent and method for bonding wood. US Patent 5,543,487; 1996.

19. Vick CB, Richter KH, River BH, Fried AR. Hydroxymethylated resorcinol coupling agent for enhanced durability of bisphenol-A epoxy bonds to Sitka spruce. *Wood and Fiber Science*, 1995;27(1):2–12.

20. Gardner DJ. Adhesion mechanisms of durable wood adhesive bonds. In: *Characterization of the Cellulosic Cell Wall*. Stokke DD, Groom LH, editors. Ames, IA: Blackwell Publishing; 2006. pp 254–265.

21. Geddes K. Polyvinyl and ethylene-vinyl acetates. In: Pizzi A, Mittal KL, editors. *Handbook of Adhesive Technology*, 2nd ed. revised and expanded. Marcel Dekker, Inc; New York, 2003. pp 719–729.

22. Quixley NE. Hotmelts for wood products. In: *Wood Adhesives*. New York: Marcel Dekker, Inc.; 1989. pp 211–215.

23. Pizzi A. Hot-melt adhesives. In: Pizzi A, Mittal KL, editors. *Handbook of Adhesive Technology*, 2nd ed. revised and expanded. New York: Marcel Dekker, Inc.; 2003. pp 739–745.

24. Lambuth AL. Protein adhesives for wood. In: Pizzi A, Mittal KL, editors. *Handbook of Adhesive Technology*, 2nd ed. revised and expanded. New York: Marcel Dekker, Inc.; 2003. pp 457–477.

25. Pearson CL. Animal glues and adhesives. In: Pizzi A, Mittal KL, editors. *Handbook of Adhesive Technology*, 2nd ed. revised and expanded. New York: Marcel Dekker, Inc.; 2003. pp 479–494.

26. Baumann MGD, Conner AH. Carbohydrate polymers as adhesives. In: Pizzi A, Mittal KL, editors. *Handbook of Adhesive Technology*, 2nd ed. revised and expanded. New York: Marcel Dekker, Inc.; 2003. pp 495–510.

27. Pizzi A. Natural phenolic adhesives II: Lignin. In: A Pizzi, Mittal KL, editors. *Handbook of Adhesive Technology*, 2nd ed. revised and expanded. New York: Marcel Dekker, Inc; 2003. pp 589–598.

28. Pizzi A. Natural phenolic adhesives I: Tannin. In: Pizzi A, Mittal KL, editors. *Handbook of Adhesive Technology*, 2nd ed. revised and expanded. New York: Marcel Dekker, Inc; 2003. pp 573–587.

29. Pizzi A. Wood/bark extracts as adhesives and preservatives. In: A Bruce, Palfreyman JW, editors. *Forest Products Biotechnology*. London: Taylor & Francis; 1998. pp 167–182.

30. Perkins EG. Composition of soybeans and soybean products. In: *Practical Handbook of Soybean Processing and Utilization*. Erickson DR, editor. Champaign, IL: AOCS Press and the United Soybean Board; 1995. pp 9–28.

31. Johnson LA, Myers DJ. Industrial uses for soybeans. In: Erickson DR, editor. *Practical Handbook of Soybean Processing and Utilization*. Champaign, IL: AOCS Press and the United Soybean Board; 1995. pp 380–427.

32. Kuo Ml, Myers DJ, Heemstra H, Curry D, Adams DO, Stokke DD. inventors; (October 23). Soybean-based adhesive resins and composite products utilizing such adhesives. Iowa State University Research Foundation, Inc., assignee. US Patent 6,306,997; 2001.

33. Westcott JM, Frihart CR, Traska AE. High-soy-containing water-durable adhesives. *Journal of Adhesion Science and Technology*, 2006;20(8):859–873.

34. Li K, Peshkova S, Gen X. Investigation of soy protein-Kymene® adhesive systems for wood composites. *Journal of the American Oil Chemists' Society*, 2004;81(5):487–491.

35. Li K. inventor (August 7). Formaldehyde-free lignocellulosic adhesives and composites made from the adhesives. State of Oregon Acting by and through the Oregon State Board of Higher Education on Behalf of Oregon State University, assignee. US Patent 7,252,735; 2007.

36. Sellers T. Jr. *Plywood and Adhesive Technology*. New York: Marcel Dekker, Inc.; 1985.

37. Pizzi A. Recent developments in eco-efficient bio-based adhesives for wood bonding: opportunities and issues. *Journal of Adhesion Science and Technology*, 2006;20(8):829–846.

7

Technology of Major Wood- and Fiber-Based Composites: An Overview

7.1 Introduction

The primary learning objectives of this book are to provide the following:

- A basic understanding of the fundamental composition and properties of wood and other plants of industrial interest and the response of these materials to thermocompression processes, that is, consolidation behavior;
- Insight into the fundamentals of adhesion;
- Information on the synthesis and cure behavior of adhesive polymers important for bonding wood and natural-fiber composites.

It is our view that if a student or learner acquires a grasp on these concepts, then their application to a variety of manufacturing scenarios is possible. Indeed, although it may appear at first glance that the number of manufacturing processes for wood- and fiber-composites is relatively few, there are many variations on these themes that introduce a multitude of options. This realization, in combination with the severe limitations of space within a book such as this, dictates that the treatment in the following pages will be in the form of a broad overview. We suggest that a basic cognition of process flows may be combined with study of other resources devoted to specific process technologies (as cited herein) as a logical learning sequence for the application of the primary learning objectives of this book. Additionally, in this information age, it is undeniable that simple Internet searches will invariably produce a number of good-quality videos that document process flow and equipment for many of the products described herein. It is suggested that the student search out and view these resources with the fundamentals of material characteristics, consolidation, and adhesion firmly in mind, as a means to synthesize and integrate one's learning in these areas.

Introduction to Wood and Natural Fiber Composites, First Edition. Douglas D. Stokke, Qinglin Wu and Guangping Han.
© 2014 John Wiley & Sons, Ltd. Published 2014 by John Wiley & Sons, Ltd.

7.2   Wood and Natural Fiber Composites as a Material Class

An underlying assumption for this text is that we may classify adhesively bonded wood and natural fiber products, inorganic-wood fiber substances, and natural fiber-plastic materials as composites, wherein a composite is a combination of at least two or more materials which maintain their identity in the combination at some arbitrary scale of structure. As we explored to some extent in Chapter 1, wood-based composites are often ignored in many books or other works devoted to composites, perhaps largely due to the general conception of composites as reinforced matrix materials. It is also possible that natural materials are often overlooked as valid raw material for composites because they may be viewed either as unsophisticated or as materials with unpredictable properties relative to synthetics. Nevertheless, those in the wood- and natural fiber-industries recognize the validity of wood and fiber composites as a material class. For example, Stark et al. [1] use the word composite "to describe any wood material adhesively bonded together."

7.3   Taxonomy of Adhesive-Bonded Composites Technology

In Chapter 1, we considered one approach to classifying wood and fiber composites, based on the type of furnish and the form of the final manufactured material, that is, either lumber-like or panel products (refer to Table 1.3). If we adopt this approach, we find that lumber-like adhesive-bonded products may be made from solid-sawn lumber, veneer, and strands. That is useful, but we may just as easily segregate products into categories of structural (i.e., engineered wood products) versus nonstructural products, with materials in each category made not only with a variety of wood furnish types but also appearing in both lumber-like and panel product form (Sidebar 7.1).

Sidebar 7.1   Engineered Wood Products

Structural composites are those materials that are designed with specific engineering properties in mind for the primary purpose of using the product as a load-bearing member in a structure. In most cases, such products would be subject to manufacturing process qualification, inspection, grading procedures, and other such requirements to allow the specification of design properties for the material. The term, engineered wood product, is typically synonymous with structural wood composite, and in many cases, the engineered wood product terminology is preferred.

APA–The Engineered Wood Association is a nonprofit trade association representing manufacturers of engineered wood products in the United States and Canada. As an example of the types of materials included under the umbrella of "engineered wood products," here is a list of product categories that are produced by member companies of APA–The Engineered Wood Association:

- Glued-laminated timber (Glulam);
- Plywood;
- Oriented strand board (OSB);
- Structural composite lumber (SCL);
- I-joists;

- Rack-resistant panel siding;
- Specialty products, including radiant barrier sheathing, furniture frames, truck and recreational vehicle bodies, signs, and so on.

Engineered wood products as a material class are dependent upon effective adhesive bonding for their manufacture and superior performance as compared to solid wood. Structural composites may be made from smaller, lower-quality trees than solid-sawn materials, thus making use of renewable, short-rotation forests. Properties-reducing defects, such as knots, are either dispersed or completely removed, making engineered products more uniform, with greater predictability of mechanical properties as compared to their solid-sawn counterparts. Materials may be designed to take advantage of higher-strength or stiffness wood positioned within the composite to make more efficient use of high-quality wood resources, an example being the placement of high-strength wood laminates in the top and bottom layers of glued-laminated beams. These advantages are possible due to effective adhesive bonding technology.

A graphic that logically combines more than one approach to product classification and also introduces the approximate density range of the final product and the form of wood raw material input is shown in Figure 7.1. We have chosen to use this general taxonomy to structure the ensuing overview of manufacturing operations. In essence, the topics are thus outlined according to the form of the wood raw material input into the process.

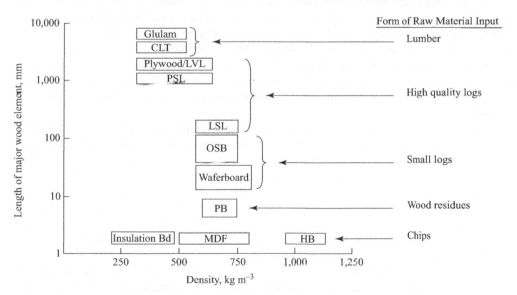

Figure 7.1 *Classification of adhesively bonded wood composites according to size of the fundamental unit of wood furnish (left vertical axis) and approximate density range of final product (horizontal axis). The form of the raw material input for these product categories is shown in the associated column at right. Glulam, glued-laminated timber; CLT, cross-laminated timber; LVL, laminated veneer lumber; PSL, parallel strand lumber; LSL, laminated strand lumber; OSB, oriented strand board; PB, particleboard; MDF, medium density fiberboard; HB, hardboard. Adapted from [2] with permission from Springer Science + Business Media B.V © 1988.*

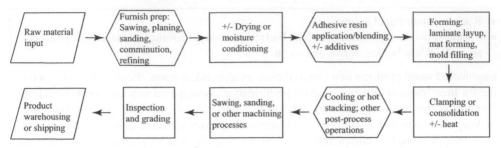

Figure 7.2   *Generic process flow diagram for adhesively bonded wood and natural fiber composites.*

7.4   A Generic Process Flow

Although the specific details of manufacturing the various types of wood- and natural-fiber composites vary, sometimes significantly, generic elements may be identified for most processes. Figure 7.2 shows a generalized process flow for adhesive-bonded materials.

Regardless of process, the first step is the intake of raw material. In some cases, the raw material may be in the form of logs, pulp bolts, or other relatively unmodified tree stems or limbs. In many cases, lumber, veneer, chips, or other forms of material are generated elsewhere prior to receipt at the composite manufacturing facility. These raw materials may be in suitable condition for immediate use, or they may require further processing prior to the actual composite manufacturing steps. One example of in-plant processing or preparation of raw material prior to composite fabrication is the surfacing of lumber before gluing of laminates in glulam beams. Another example is comminution (size reduction by chipping, hammermilling, or related means) of wood materials to prepare the furnish for particleboard manufacturing. If necessary, these intermediate operations are generally followed by drying or conditioning, that is, adjustment of furnish moisture content. Raw material quality control sampling for moisture content, density, and/or other attributes is routinely conducted at this stage.

Following sizing, drying, or other preparatory operations on the raw material, the next step is typically the application of adhesive. Adhesive resins, either in neat (unmodified) or blended, modified form (through addition of extenders, fillers, etc.) may be applied by spreaders or via formation of droplets, the former for veneers and other laminates, and the latter for strands, particles, or fibers. Other additives, such as wax, catalysts, or formaldehyde sources, may also be blended with the furnish at this point. With few exceptions, the application rate for adhesives is generally calculated as adhesive solids as a percentage of dry weight of wood or fiber furnish. Additives may be calculated in similar fashion, whereas catalyst solids application may be computed as a percentage of adhesive resin solids.

Once the adhesive is applied, the material moves to a forming stage. Large glued-laminated materials and plywood veneers are assembled, whereas flake, particle, and fiber-based materials are formed into thick mats of adhesive-blended furnish. Some materials are placed into closed molds. Invariably, the assembly or mat formation stage is followed by clamping, compression, or consolidation, involving an increase in pressure on the material. Generally speaking, as the size of the individual lignocellulosic furnish decreases, the consolidation pressure increases. For mat-formed and molded materials, this is almost universally accompanied by the application of heat, as discussed in Chapter 4. Here, the dynamics of heat, pressure, material chemistry and properties, and adhesion interact to produce the characteristics inherent in the manufactured composite.

At the completion of the clamping, pressing, or consolidation process, the formed material moves to postprocessing operations. In some cases, the hot-pressed material must be quickly cooled to prevent heat-induced deterioration of the adhesive bond, and in others, the product is hot-stacked to permit further cure of the adhesive. Other materials may be further treated with heat or chemicals to impart properties such as enhanced moisture resistance. Materials typically undergo sawing or shaping to final size, may be edge-coated, wrapped, or packaged, and are then inspected and graded before warehousing or shipment directly to the customer.

Viewed in this generic manner, the process of wood composite manufacturing is rather simple. However, the numerous variables introduced by the raw material as well as the particulars of processing increases the overall complexity of the system for many of the composites under consideration. Our goal here, as well as throughout the remainder of this chapter, is to provide conceptually concise descriptions of the manufacturing sequence for major adhesive-bonded wood- and natural-fiber composite materials.

7.5 Technology of Adhesive-Bonded Materials Based on Form of Raw Material Input

7.5.1 Glued-Laminated Timber and Cross-Laminated Timber

Glued-laminated timber (Glulam) and cross-laminated timber (CLT) are both structural products made from graded lumber as input into the manufacturing process. Glulam is a product used for structural beams and columns, and is thus "timber-like" in form, whereas CLT is a panel product that combines the functionality of both wood framing and structural sheathing into a single material. The development of CLT has permitted entry of wood into new markets, primarily high-rise construction.

7.5.1.1 Glulam

To laminate is to "make by building up in layers." Wood-based laminates include beams, columns, or timber-like materials made by gluing dimension lumber together in layers and plywood, constructed by the lamination of relatively thin wood veneers. Glued-laminated timber, better known as glulam, was one of the first structural, adhesive-bonded wood products, used to construct an auditorium in Switzerland in 1893. Engineering principles were applied to the design of glulam arches used in the construction of a building at the US Forest Products Laboratory in Madison, Wisconsin, in 1934 [3]. Initially, the adhesives used for glulam manufacturing possessed the necessary strength for these applications, but did not have sufficient moisture resistance capability for exterior use. The best-known example of such an adhesive is casein, the protein derived from cow's milk. The post–World War II development of waterproof, petroleum-derived synthetic adhesives such as phenol- and resorcinol-formaldehyde enabled the use of engineered gluelam arches, beams, and columns in both interior and exterior structural applications in the construction of churches, schools, gymnasiums, indoor arenas and swimming pools, bridges, and so on. Indoor stadiums with open spans exceeding 152 m (500 ft.) have been constructed with glulam arches supporting the roof. Glulam members are produced in depths of 152–1,828 mm and greater (6–72 in. and more) and in lengths exceeding 30 m (100 ft.). Glulam is used for projects in which the functional strength and esthetic beauty of wood may be featured.

The process for glulam manufacturing has been summarized as consisting of five basic steps: drying and grading lumber, end-jointing lumber to create long laminations (lams), face-gluing of lams, and finishing and fabrication [3]. Within the United States and Canada, one could also add the essential preliminary and ongoing "steps" of qualification and inspection. Manufacturers of glulam products must first undergo third-party inspection of their facilities, personnel, and processes before entering into manufacturing operations, and then must implement and maintain ongoing internal, daily quality control testing and submit to random third-party plant visits and inspections. These procedures are in place to assure that quality standards and specified engineering properties are maintained for purposes including public safety when glulam is used in structural applications. The American National Standards Institute (ANSI), the Canadian Standards Association (CSA) and the Japanese Agricultural Standard (JAS) are examples of certification standards to which manufacturers and their products must comply, under third-party inspections from member organizations such as APA–The Engineered Wood Association.

Dimension lumber is the basic input raw material. Within the United States, lumber thickness for this purpose are typically 35 mm ($1^3/_8$ in.) for Southern pine and 38 mm ($1^1/_2$ in.) for Western species [4]. For products manufactured under ANSI/AITC A190.1, the maximum moisture content of laminations is 16%, with the average moisture content specification of 12% designated, along with a maximum laminate moisture content variation of 5% permitted [5]. These moisture tolerances are required to minimize dimensional change between lams in the final product, thus reducing drying-induced defects. Moisture tolerances may be adjusted to match the expected in-service long-term equilibrium moisture content.

Lumber within moisture tolerances is graded by visual or mechanical means, the latter having the advantage of assigning more precise allowable bending strength and stiffness values to each piece of lumber. Lumber is sorted by strength and stiffness values. For glulams intended for use as beams, those lams with higher properties are designated for use in the critical tension side of the beam. This makes the most efficient use of the wood raw material and permits the design of glulams according to the engineering requirements of the structure to be built with the product. Lams may or may not be treated with preservative chemicals prior to assembly.

Lams are invariably required to be longer than the lumber from which they are made. The length of lams is tailored by end-jointing of graded lumber by either scarf, or more typically, higher-strength finger joints. Immediately prior to layup, the wide surfaces of the lams are surface planed or sanded to activate the material for adhesive bonding, that is, this process improves the surface energy of the substrates in order to achieve adequate wetting of the substrate by the adhesive. A cold-setting PRF is typically used, allowing for manufacture of large timbers without the capital investment of heated presses; however, some manufacturers accelerate the adhesive cure by use of radio-frequency (RF) curing. The assemblies cure under pressure in clamping beds which apply mechanical or hydraulic pressure for typically 6 to 24 hours [3]. The product then receives a final surfacing to remove excess adhesive, followed by any fabrication operations, for example, drilling of holes for fasteners, and so on. Some products are pressure-treated with an oil-borne or creosote preservative at this point. The glulams are then inspected and shipped.

7.5.1.2   Cross-Laminated Timber

CLT is a new panelized wall system that has allowed wood to enter high-rise housing and commercial structures markets. Using this technology, multistory building (10–12 stories) may be constructed in a matter of weeks. The material usage is classified similarly to heavy timber construction, which means it may attain sufficient fire rating for the markets for which it was developed. Unlike glulam, in which lumber pieces comprising the produce is assembled parallel to one lumber, layers are laid up in cross-banding manner, similar to plywood. That is, a first layer (lamella) of lumber is established, then a second lamella is placed on top at 90° to the first, with a third layer perpendicular to the second (core) and parallel to the first layer. In addition to sawn lumber, CLT may be fabricated with SCL. Panels may be 3, 5, 7, or more layers thick. Panels approximately 3 m wide by 16 m long and up to 400 mm thick (10 × 52 ft. by 15.7 in. thick) are bonded with structural adhesives. CLT has been used in Europe for over 15 years and is now being introduced to North America [6]. Specifications for lumber species grades and moisture content, lamination sizes, tolerances, adhesive requirements, and required performance characteristics for CLT products are given in American National Standard ANSI/APA PRG 320-2012 [7].

7.5.2   Plywood, Laminated Veneer Lumber and Parallel Strand Lumber

Plywood, laminated veneer lumber (LVL), and parallel strand lumber (PSL) are structural composites made from wood veneer. The veneer process starts with logs of sufficient size and quality to yield veneer in the desired grades, the standards for which have changed considerably over time. Logs are typically stored either in ponds or under water sprinklers to reduce degrade due to shrinkage-induced checking. Water storage also eliminates most forms of biological degradation by reducing the amount of oxygen available to wood-destroying organisms. Logs are then debarked and "cooked" in heated water to soften the wood in preparation for peeling, or in the case of nonstructural, decorative hardwood plywood, veneer slicing. Peeling temperatures vary by species, generally ranging from 10°C to 71°C (50–160°F) for softwoods [8]. Modern production facilities have automated lathe chargers and computer-controlled lathes which produce veneer at rates of up to 6 m/s. Veneer may be platen or jet impingement dried, the latter at temperatures up to 200°C with air velocities of 20 m/s. Contemporary final moisture content of veneer is typically 6–12% or greater, as compared to typical values of 3–6% in prior years [9]. Veneer is clipped, ultrasonically graded, and repaired (patched) prior to layup of the product.

7.5.2.1   Plywood

Structural plywood is typically made from coniferous wood, although hardwood may be used in some cases, for example, Finnish birch plywood. Plywood is made of by cross-banding of adjacent plies or layers, wherein a ply is a single sheet of veneer, and a layer may be either one ply or multiple plies, for example, three-ply, three layer versus three-ply, four layer construction. Face plies are oriented with the wood grain direction parallel to the long dimension of the plywood panel to maximize bending strength and stiffness, with adjacent plies oriented perpendicular to the faces to provide dimensional stability (refer to Figure 1.4). Dimensional stability is also achieved by balanced construction wherein face plies on both surfaces must match one another in terms of dimensional performance.

In the United States, structural plywood is typically bonded with PF. The base resin is mixed in water with extenders, fillers, and other additives, which may include blood meal, wheat flour, nut shell flour, corncob flour, lignin, or others [10]. The glue formulation may have PF resin solids content on the order of 30%, with total solids content of 65%. Glue is applied to veneer by rollers, curtain coaters, or with a foam adhesive extruder. Examples of application rates are given in Section 7.6.3. The veneer is then laid up with higher quality/higher strength veneer for the surface plies, cold pressed for 3–5 minutes to promote wetting of the substrate by adhesive, and then hot-pressed in multi-opening hydraulic presses. Consolidation pressure of 0.75–1.5 MPa is applied with process temperatures for PF-bonded plywood ranging from 140°C to 160°C for 3.5 minutes for 9.5 mm thick panels to 12 minutes for 25 mm plywood [9]. PF-bonded products are hot-stacked after removal from the press to promote further cure of the adhesive.

7.5.2.2   *Laminated Veneer Lumber*

LVL is the premier material in the product class known as structural composite lumber (SCL). It, like plywood, is manufactured from rotary-peeled veneer that is ultrasonically graded to allow use of the highest stiffness veneers on the surface layers of the LVL to enhance mechanical performance of the product and to improve the efficiency of raw material usage. Unlike plywood, the veneer in all layers are laid up parallel to the long dimension of the LVL. PF resin is typically used to bond the veneer into large "billets" 90 mm thick, 1.2 m wide, and 35 m long (3.5 in. × 4 ft. × 80 ft.). Billets may be bonded together in a second gluing operation to produce a thicker product. The LVL production process is shown in Figure 7.3. LVL provides a good example of the material properties improvements achieved in that the coefficient of variation (COV = standard deviation expressed as a percentage of the mean) for strength and stiffness of LVL is 10–15%, as compared to COV of 25–40% for solid-sawn lumber and timber [11].

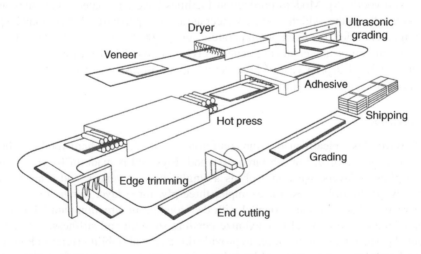

Figure 7.3   *LVL production process. Reproduced from [11] with permission of the Forest Products Society ©*
1997.

7.5.2.3   *Parallel Strand Lumber*

Like LVL, PSL is also an SCL made from veneer. However, unlike LVL, in which the veneer is the same width as the product, PSL is formed from long strips of veneer, each about 3.2 mm thick × 19 mm long (0.125 × 0.75 in.), enabling a higher usage rate of the log raw material input. Nelson [11] showed total log volume conversion into the final target product as 40%, 52%, 64%, and 76% for sawn lumber, LVL, PSL, and laminated strand lumber (LSL), respectively. The tradeoff is that LVL has the highest engineering properties, followed by PSL and LSL, all of which are superior to solid-sawn. PRF resin is cured by microwave technology (radio frequency bonding), allowing the production of thicker billets than possible for LVL in a one-step bonding process.

7.5.3   Strand Composites

7.5.3.1   *Oriented Strand Board*

The first adhesive-bonded structural panel product, other than plywood, was waferboard, developed in the early 1960s. Small logs were processed into thin wood wafers, wherein a wafer is square rather than elongated, as is a strand. The wafers were re-constituted into a panel that could be used for wall and roof sheathing. Subsequent development of technology to produce longer flakes or strands, along with the means to orient the strands, led to the superior OSB product. OSB strands are 0.3–0.8 mm (0.012–0.031 in.) thick and 50–150 mm (2–6 in.) long. Following strand drying and blending with resin, a mat is formed with surface strands parallel to the long dimension of the panel, with core strands oriented perpendicular to the faces, in much the same manner as plywood is cross-banded. Usually, different resins are used for the face and the core, with the core resin designed to cure either at lower temperature or more quickly than the face. This might be accomplished by use of two different PF resin formulations, or by use of PF face resin and MDI core resin. In Europe, PMDI is used for both face and core, at rates of 3–6% and 4–10%, respectively [12]. Resin application rates in North American plants are low, as shown in Table 7.1. Production capacities are enormous, upwards of 2,000 m³/day from a single factory.

7.5.3.2   *Laminated Strand Lumber and Oriented Strand Lumber*

LSL and oriented strand lumber (OSL) are SCL products manufactured with technology similar to OSB, with the difference that strands throughout are oriented parallel to long dimension

Table 7.1   Resin and wax application rates for OSB production in North America.

Resin type	Samples	Average consumption of weight percentage of oven-dried wood		
		Face	Core	
			PF	PMDI
Powdered PF	9	2.30	2.35	–
Liquid PF	11	3.82	3.66	2.28
Wax	20	1.14	1.14	–

Source: Reproduced from [13].

of the product. The only real difference between LSL and OSL is in the length of the wood strands used in their fabrication. LSL is made with strands 150–300 mm (6–12 in.) long, whereas OSL relies on 75–150 mm (3–6 in.) strands.

7.5.4   Particleboard

Particleboard (PB) is one of several nonstructural adhesive-bonded composites. Nonstructural composites are those materials that are not primarily intended as load-bearing members of a structure. Although materials in this classification may have certain mechanical properties requirements to serve their intended function, they are generally specified for their performance as substrates for flooring (as in particleboard flooring underlayment), as substrates for laminates (e.g., medium density fiberboard tabletop core), or as decorative products (e.g., interior paneling). These materials may rightly be considered as engineered in the sense that the process and product are designed to meet specified performance criteria, but they are not generally classified as engineered wood products, since that terminology is reserved for structural composites. It is here, in the nonstructural composites realm, that nonwood fiber substitutes or alternatives find their greatest application. In many instances, nonwood plant fiber may be partly or wholly substituted for wood in identical or nearly identical manufacturing processes and resulting products. For example, wheat-straw particleboard may be a direct substitute for wood-based particleboard in certain applications. The generic process for manufacturing these two types of particleboard is similar, but the specific implementation must necessarily vary due to the differing characteristics of the input raw material.

Particleboard was developed in Germany in 1941 as a means to use primary wood processing residue to make panel products. Even if residues are used (e.g., planer shavings), the material is typically further comminuted (broken down or reduced in size) and screened to separate fine and coarse particles. Fine particles are suitable for use in the faces of the particleboard to impart a smooth surface and to modify vertical density profile, with coarser particles reserved for use in the core. Particleboard is used as core stock for cabinets and furniture, as a laminating substrate, for shelving and other utility uses and for flooring underlayment.

Furnish is dried to 2–8% mc and blended with resin and wax, the latter to improve moisture resistance. Particleboard, intended for interior use, is usually bonded with UF resin, although melamine-fortified UF (MUF), MF or PF resin may be used for products requiring higher moisture resistance. Resin application rates of 6–10% solids are typical, with higher rates for specialty products. Consolidation in multi-opening or continuous presses is carried out at 2–4 MPa pressure and 150°C to upward of 220°C, with lower temperatures used in stationary platen presses and higher temperatures in high-speed continuous presses. Continuous presses up to 50 m long are with line speeds of 80–90 m/min for 2 mm board versus 3–5 m/min for 38 mm panels. These developments have enabled production output of 300 m³/day for particleboard [12]. Final product density is >800 kg/m³ (>50 pcf) for high-density PB, 640–800 kg/m³ for medium density (40–50 pcf) or <640 kg/m³ for low density PB (<40 pcf).

7.5.5   Medium Density Fiberboard and Hardboard

7.5.5.1   *Medium Density Fiberboard*

Medium density fiberboard (MDF) and hardboard (HB) are both produced from pulp chips. Pulp chips are wood materials that have specified dimensions suited to the pulping and refining processes to be employed. In the case of MDF or hardboard, nonchemical thermomechanical pulp (TMP) or chemical-thermal mechanical pulp (CTMP) is utilized. Full chemical pulping processes, such as those used to make paper pulp, are not necessary and are not justified on an economic basis for the production of MDF and HB.

MDF has captured a significant share of the market once occupied by particleboard, due to the uniformity of the product, its quality as a substrate for bonding of laminates, and the machinability of the edges for decorative effects. Wood chips are "cooked" in pressure vessels. Steam pressures of over 8 MPa were used in the original Masonite process, in which the chips were literally exploded into fibers by sudden release of the pressure [14]. High pressure processes yield chips with lignin-rich surfaces that may be consolidated into hardboard without adhesive as a result of the autoadhesion characteristics of the thermoplastic lignin. However, these processes are energy intensive and liberate large amounts of hemicellulose by-products, including organic acids and air pollutant hydrocarbons. Lower pressure cooks (0.6–1 MPa), coupled with subsequent mechanical refining is the approach used today. Still, with temperatures of 170–195°C at these pressures and the diluent effect of moisture, the chips are heated well above the glass transition temperature of lignin, allowing for the generation of lignin-rich fiber surfaces during refining[12].

Following refining, the wet fiber at approximately 50% moisture content is conveyed through a long pneumatic tube drier with air velocity of up to 15 m/s (3,000 fpm) with wet end temperatures of 260–343°C (500–650°F). Fiber exits in approximately 5 seconds at 8–12% moisture content with an exit temperature of 65–88°C (150–190°F). Alternatively, a two-stage drier may be used. For products bonded with UF resin, fiber is conveyed and resin is added in a lower-temperature blow line or in a resin blender in order to avoid pre-cure of the resin in the dryer. UF resin application rates of 8–10% solids is representative, but may vary up to 15% for some products [15,16]. The mat is formed, prepressed and then consolidated at 0.5–5 MPa at 180–210°C [12]. The final product has a density ranging from 500 to 900 kg/m³ (31–56 lb/ft.³), in thicknesses of 2–200 mm (0.08–7.9 in.).

7.5.5.2   Hardboard

Hardboard (HB) manufacturing is similar to that for MDF, except that the final product is thinner at 1.6–12.7 mm (0.06–0.5 in.) and has a higher density, usually around 1,000 kg/m³ (62.4 pcf). Resin application is low at 0–2% solids. Hardboard may be formed in either a wet or dry process, with the former less common today, resulting in smooth one side (S1S) product from the wet process or S2S from the dry. The panels are consolidated in the range of 8 MPa (>1,000 psi) at 190–235°C.

7.6   Laboratory Panel Calculations

The manufacture of modern wood- or lignocellulose-based composites is dependent upon the effective use of adhesive resins. Although the quality of the final product depends on a large number of variables (wood/lignocellulose furnish characteristics, heat, pressure, press time, etc.), few are as basic as the resin content and final board density. Calculation of board parameters such as resin and wax content is simply an application of equations for density and moisture content, and conversion to appropriate units, and is essential in both the laboratory and factory production environments.

7.6.1   Material Needs

In order to fabricate small panels in the laboratory, the material needs must be calculated with a few simple resources: Working knowledge of wood density, specific gravity, and the moisture content equation (recall Equations 3.1, 3.2, 3.13, and 3.14), paper and pencil, a little brain power, and this handy Equation 7.1:

$$W_{panel} = \rho_p \times V_p = W_o + W_{rs} + W_{sw} + W_a \qquad (7.1)$$

where

W_{panel} = mass of panel, oven-dry basis;
ρ_p = density of panel, OD mass and volume "out of press" basis;
V_p = volume of panel, "out of press";
W_o = oven-dry mass of wood furnish (particles, flakes, fibers, etc.);
W_{rs} = mass of resin solids;
W_{sw} = mass of slack wax solids; and
W_a = mass of solids of any other additives (e.g., preservatives).

Since the density and volume of the panel are quantities that you choose or specify, the equation can be solved for the various material needs as follows: As shown in Equation 7.1, the amount of resin is typically expressed as the amount of resin solids as a percentage of the oven-dry mass of wood furnish. For example, if you have a board that has 6% resin, this means that the *solid material* found in the resin amounts to 6% by weight of the oven dry mass of wood furnish. This solid material is typically called "resin solids" or "nonvolatiles content" of resin. In this example, $W_{rs} = 0.06 W_o$. Other additives (except for catalyst content) are calculated in similar fashion. Let us say we want to make a laboratory particleboard panel at 0.65 specific gravity, measuring 1.25 cm × 60 cm × 60 cm = 4,500 cm^3 out of press, with 6% resin solids and 1% wax solids. The mass of the panel, oven-dry basis, is simply the specific gravity that we have chosen multiplied by the volume of the panel (also something you choose) and density of water (ρ_w), thus (Equation 7.2):

$$W_p = G_{panel} \times V_{panel} \times \rho_w = 0.65 \times 4,500 \text{ cm}^3 \times 1.0 \text{ g/cm}^3 = 2,925 \text{ g} \qquad (7.2)$$

The materials needed may then be calculated by setting up an algebraic equation to solve for each ingredient, knowing that each is determined as a percentage of dry mass of wood furnish (Equation 7.3):

$$W_{panel} = W_o \times 0.06 W_o \times 0.01 W_o = 1.07 W_o = 2,925 \text{ g} \qquad (7.3)$$

Solve for W_o = 2,734 g, so $W_{rs} = 0.06 W_o$ = 164 g and $W_{sw} = 0.01 W_o$ = 27 g.

The amount of liquid resin applied, will of course, be greater than 6% by weight, as there will be significant water content of the resin. The amount of liquid resin is therefore given by the expression in Equation 7.4:

$$W_{rsol} = \frac{W_{rs}}{\% \text{ resin solids in resin solution}/100} \qquad (7.4)$$

where

W_{rsol} is the mass of the "resin solution" or the combined mass of resin solids plus the carrier liquid (typically water). W_{rsol} represents the "mass of resin to spray" or blend with the furnish. Typical solids contents for liquid commercial resins are 65% for urea formaldehyde (UF) and 45–55% (or more) for phenol formaldehyde (PF). Let's say we are making our lab panel with UF resin at 65% solids, thus the amount of liquid resin we need to blend with our wood furnish is $W_{rsol} = \frac{164\,g}{65\%/100} = 252$ g.

A final point to note is that although the furnish input may be at very low moisture content, it would be unusual to be completely oven dry. Thus, to find the actual amount of wood fiber furnish to use, W_M, the wood fiber needs to be adjusted (after calculation of other material needs on a dry basis) for moisture content, that is, apply the moisture content equation to solve for $W_M = W_0 ((1 + M)/100)$. In so doing, it may be necessary to include the amount of water in the resin solution. We may also need to overcalculate our total requirements somewhat to account for material loss or wastage in the blending, handling and mat layup process. Depending on lab conditions and skill of the operator, a material requirement adjustment factor could be on the order of 2–4%. This is determined by experience.

Ordinarily, resin and other component solids blended with the wood furnish are calculated as a percentage of the dry mass of wood, as illustrated above. One exception is in the calculation of hardener for example, as required to cure UF resin. Given the fact that hardener addition is small and since its addition is required for the express purpose of interacting with the resin solids, it is calculated as a percentage of the resin solids. Extending our example, assume we need to add 1% ammonium chloride solids to our resin, and that our ammonium chloride is in an aqueous solution of 20%. The mass of hardener solids is therefore $0.01 \times W_{rs} = 0.01 \times 164$ g $= 16.4$ g and the amount of liquid hardener solution required is 16.4 g/20% = 32.8 g.

7.6.2 Clamping or Consolidation Pressure

The bonding of solid materials such as solid wood to be joined for furniture, lumber for glulam or veneer for plywood requires application of pressure to effect intimate contact of the adhesive and substrate. Likewise, the consolidation of substrates such as strands, particles, or fibers, requires that pressure be applied. Pressure is force per unit area ($P = F/A$) measured in units such as lb/in.^{2} or N/m^2, wherein 1 N/m^2 = 1 Pascal (1 Pa) and 1 N = 100 g force. The pressures required for wood- and fiber-composite consolidation are usually expressed in mega Pascals, MPa (1 MPa =1 $\times$ 10^6 Pa = 1 N/mm^2) or pounds per square inch (psi). Consolidation of solid wood, lumber, or veneer typically requires pressures of less than 1.5 MPa (approximately 215 psi), whereas composites made of smaller furnish elements generally require 2–8 MPa (290–1,160 psi). MDF and hardboard are consolidated in the 5–8 MPa range, particleboard 1–4, and OSB and PSL are consolidated nearer 5 MPa.

As an example, consider the amount of force required to consolidate the laboratory particleboard panel considered given as the foregoing example of material requirement calculations. If we specify that the panel is to be consolidated at 2.25 MPa (326.25 psi), then the force required of the laboratory press is $F = P \times A = 2.25$ MPa $\times$ 360,000 mm^2 = 810,000 N = 81,000 kg or 81 metric tons (about 89 English tons of 2,000 lb per ton).

7.6.3 Glue Application Rate for Lumber and Veneer Substrates

Although glue application to solid substrates such as lumber or veneer could be calculated or specified similarly to the manner used for strand, particle, or fiber composites, the typical method is different. Ordinarily, the glue application is specified as mass of resin solids per unit area, for example, g/m^2, or in the English system lb/Msf where Msf is "1,000 square feet." The application rate may be given either in terms of amount of glue per unit area of single glue line or in amount per double glue line, that is, rate/sgl, rate/dgl or rate/Msf sgl or rate/Msf dgl. For example, PF resin application for plywood manufacturing may be specified with a spread rate of 100–244 g/m^2, depending on PF solids content of the glue mixture (resin plus extenders, fillers, and other additives), manufacturing conditions, species to be bonded, and so on. To convert g/m^2 spread rate to lb/Msf, divide by 4.882 [10].

7.7   Measurement Conventions for Production Capacity and Output

Quantification of manufacturing output provides one of the metrics required for production management and business functions. In this context, production is the actual amount of product made in a given period of time. Capacity is the amount of product that a factory (mill) could produce if running full time at full design production rate. Capacity utilization is the ratio of production to capacity.

Although there is a trend toward the use of cubic volume measurements regardless of product type, other quantities are in use. To the uninitiated, these measures may be confusing and even misleading. Therefore, it is important to be aware of the specialized units of measure that are often employed in the wood-using and related industries. The actual measures used are dependent upon the type of product under consideration, as well as the intended purpose for the data. In broad terms, most wood-based composite products designed for use in the construction industry may be classified as either lumber-like or panel products. The measurement conventions for these two classes of products differ somewhat, and are illustrative, but not comprehensive examples of conventions used to quantify production capacity and output.

7.7.1   Measures for Lumber- and Timber-Like Products

Engineered products that replace or substitute for solid-sawn lumber and timbers are scaled in similar fashion to their solid-sawn counterparts. In North America, the word lumber is generally understood as boards and framing dimension stock, with timbers being larger sawn beams and columns. Engineered products may exceed by a large margin the size of solid-sawn members, due to the ability to fabricate large glued-up members.

7.7.1.1   *Log Rules: A Starting Point for Understanding Lumber Volume Estimation*

The estimation of tree or log volume is of obvious and practical importance to the wood-using industries. However, the accurate accounting of the volume of logs has proved a significant challenge, because logs are not generally perfect cylinders, but may be eccentric, curved, tapered, or characterized by other attributes that complicate their measurement. Furthermore, the need to estimate the volume of square, rectangular, or prismatic shapes (e.g., sawn timbers and lumber) extracted from such logs presents an interesting case study in the nature of the challenges and anomalies often associated with estimation of volumetric quantities in the wood-using industries. For example, Freese [17] listed well over 200 log rules developed for the estimation of log volume in a variety of historic and geographic circumstances. Not all of these historic rules have survived the test of time, but the number of historic and newly developed rules in use today remains high. For example, Briggs [18] described at least 24 log rules in contemporary application in the US Pacific Northwest, including rules used for export of logs to various parts of the Pacific rim. Although log rules are outside of our specific consideration here, awareness of some of the challenges of log volume estimation (log scaling) serves to illustrate the potential scope of the subject of wood product volume estimation.

7.7.1.2   Conventions for Estimating Solid-Sawn Lumber Volume

As a basis for the estimation of product volume of lumber-like composites, it is important to consider some of the conventions applicable to solid sawn lumber. In particular, it is necessary to be cognizant of the difference between nominal and actual dimensions, and their usage in volume estimation procedures, particularly with respect to board-foot scaling.

Within North America, lumber volumes are traditionally and contemporarily estimated by board-foot scaling. A board-foot is defined as a volume of wood equivalent to one inch thick by one foot wide by one foot long. Since one foot is 12 in. (30.5 cm), this amounts to 1 in. × 12 in. × 12 in. = 144 in.^{3} (2.54 cm × 30.5 cm × 30.5 cm = 2,360 cm^3). The concept is based on the historic idea of trying to estimate how many one-inch-thick pieces of lumber could be sawn from an ideally circular log. This notion is representative of a diagrammatic approach to volume estimation. Now, consider the fact that if one cuts from a fresh log, a piece of lumber that is exactly one inch thick, the final product will not be one inch thick, due to the fact that wood shrinks as it dries. Thus, the concept of the nominal dimension came into being. Nominal means, "in name only." Nominal lumber dimensions differ from actual dimensions mainly due to wood shrinkage and other losses, such as those attributable to sawing variation and planning or surfacing the lumber.

In the case of softwood construction lumber, the North American practice is to specify nominal thickness and width dimensions as multiples of 2 in., with the actual dimensions of the finished lumber something less than nominal. Lumber in this category is called dimension lumber or simply dimension. For example, a nominal "2 × 4" used for wall framing is actually 1.5 in. thick by 3.5 in. wide (3.8 × 8.9 cm) when dry and planed. Dimension lumber is to 4.5 in. nominal thickness and 16 in. nominal width. Timbers are defined as solid-sawn products that have a nominal thickness of 5 in. or greater [19].

The scaling of softwood construction lumber is calculated according to nominal thickness, nominal width, and actual length rounded down to either nearest foot or nearest multiple of 2 ft. As an example, consider the board-foot (bf) volume of a nominal softwood 2 × 4, 8 feet long. The volume is calculated with nominal dimensions as follows: 2 in. × 4 in. × 96 in. = 768 in.^{3}/144 in.^{3} per board-foot = 5.33 bf. This would be the volume estimation used for production reporting and purchasing specifications. The actual volume of this piece of dimension lumber is 1.5 in. × 3.5 in. × 96 in. = 504 in.^{3}, which is 504/768 = 66% of the nominally scaled board-foot volume. This approach to lumber scaling illustrates but one of the oddities of volume estimation or reporting as practiced in the wood-using industries.

Throughout most of the world, metric measurements are used, thus a nominal 2 × 4 in. the US is elsewhere a nominal 50 × 100 (mm × mm), with actual finished cross-sectional dimensions specified as 38 mm × 89 mm. Though the units of measure differ, the concepts of nominal and actual dimensions of lumber still apply.

7.7.1.3   *Volume Estimation of Engineered Lumber-Like Products*

The foregoing discussion of volume estimation of solid-sawn lumber is germane to the topic of adhesive-bonded wood materials, in that board-foot volume estimation and reporting is used for certain engineered wood products, specifically, those classified as glulam. It is common practice within North America to report aggregate production figures for glulam in terms of million board feet. This arises from the fact that glulam is fabricated from dimension lumber. Since softwood construction dimension lumber is scaled on a nominal board-foot volume basis, it is logical to also scale the product, glulam, on the same basis as the raw material input. Thus, the final measure is a reflection of the volume of raw material input into the fabrication process, not a precise accounting of the final manufactured product volume.

LVL, manufactured by the adhesive bonding of wood veneer, is volumetrically tallied in more direct, cubic volume measures. That is, although LVL may be used in some of the same applications as glulam, its volume is not reported in terms of board feet. In fact, glulam is the only engineered wood product for which board-foot volume tally is customary. Rather, LVL production is commonly quantified in either cubic feet or cubic meters, with industry aggregate production reported in million cubic feet, thousand cubic meters, or multiples thereof. In similar fashion, SCL and CLT volumes are recorded on the basis of cubic feet or cubic meters. I-joist production is commonly reported in million linear feet.

7.7.2   Measures for Panel Products

7.7.2.1   *Volumetric Measures*

Volumetric measures are direct and simple to understand and use. Multiplication of thickness, width, and length in appropriate units yields volume, typically in m^3. Most of the world's countries account for production of panel products with cubic volume measurement, with some multiple thereof, for example, thousand cubic meters, $m^3 \times 10^3$ or mm^3 or million cubic meters, $m^3 \times 10^6$ to report aggregate production figures. Production efficiency may be measured in terms of cubic meters of log input per cubic meter of final product output. Although the use of the SI (metric) system is becoming more common in the United States, particularly with respect to government-reported industry statistics, the older square-footage basis is still used routinely. It is therefore important to be familiar with both systems and to be able to convert units between the two.

7.7.2.2   *Modified Surface Measures*

In the United States, wood-based panel production has been historically reported on a square-foot (sf) basis, standardized to a given thickness of panel, that is, sf/unit thickness. One of the rationales for this type of measurement is to provide a ready estimate of the surface area, for example, of walls to be sheathed, for a given quantity of panel output. US reports of manufacturing facility output of structural panels (construction plywood and OSB) is conventionally given in square feet of panel (sf), or some multiple, for example, Msf (thousand square feet) or MMsf (million square feet) on a 3/8 inch thickness basis. Aggregate industry production, for example that provided in APA-Engineered Wood Association (APA) quarterly reports, is shown in billion (10^9) square feet, 3/8 in. basis. Nonstructural panels are usually reported on differing thickness bases. Particleboard and MDF are conventionally reported as

Table 7.2   *Conversion factors for wood-based panel manufacturing capacity and output metrics.*

From/To	Sf, 3/8	Msf, 3/8	Sf, 3/4	Msf, 3/4	ft.³	m³
Sf, 3/8	1	0.001	0.5	0.0005	0.03125	0.000885
Msf, 3/8	1,000	1	500	0.5	31.25	0.885
Sf, 3/4	2	0.002	1	0.001	0.0625	0.00177
Msf, 3/4	2,000	2	1,000	1	62.5	0.177
ft.³	32	0.032	16	0.016	1	0.0283
m³	1,130	1.130	565	0.565	35.315	1

Sf, square feet; Msf, 1,000 square feet.

sf of panel (or multiple thereof) on a 3/4 inch thickness basis. Insulation board is measured as sf, 1/2 inch thickness basis, and hardboard is reported as sf, 1/8 inch thickness basis. Conversion factors for typical modified surface measures to cubic feet or cubic meters are given in Table 7.2.

Contemporary US Government reports and United Nations Food and Agriculture Organization (FAO) data tend to be in cubic meters. To overcome the limitation of volumetric measures only, selected panel production data may also be reported as modified metric surface measures.

7.8   Technology of Inorganic-Bonded Materials

The primary focus of this book is on fundamental concepts related to wood and natural fiber materials bonded with organic adhesives. The concepts of adhesion applicable to organic polymer resins combined with organic polymer substrates is germane both to adhesive bonding systems and to the interaction of organic fiber with polymer matrices in fiber/plastic composites, the latter a topic of Chapter 8. We now turn our attention to materials that have very different properties, that is, those which combine wood and natural fiber with inorganic matrices. The hydration reactions of the inorganic matrix govern the bonding of such materials. In particular, systems utilizing Portland cement and gypsum as the binding matrix are of commercial and industrial significance, and merit discussion in the overall context of wood and natural fiber composites. We will begin with a brief overview of the historic development of this interesting and useful class of materials.

7.8.1   A Brief History of Inorganic-Bonded Materials

Inorganic-bonded materials have a long and varied history, first manufactured commercially in the early 1900s. Table 7.3 provides a historic overview of the development in inorganic-bonded wood composites [20]. The first low-density inorganic-bonded wood composite panels, utilizing a magnesite binder, were developed in Austria in 1914. Currently, these products are manufactured by Heraklith (now Knauf Insulation) in Europe and by Tectum in the United States. They are mainly for interior applications. These low-density panels are often used as insulation boards because of their excellent thermal and acoustic properties. The early magnesite-bonded boards were generally of low quality because the magnesite matrix was sensitive to moisture [21, 22]. Soon after, cement-bonded wood composites became popular with the development of wood wool cement board (WWCB), which had better water resistance. The production of WWCB had already spread over Europe in the 1930s, several years after

Table 7.3   *Historic overview of inorganic-bonded products, with common abbreviations for major product categories.*

Approximate decade of development	Product	Approximate density, kg/m^3	Country
1900	Gypsum-bonded wood shavings board	400	Austria
1910	Magnetite-bonded wood wool board (WWCB)	400	Austria
1920	Wood wool cement board (WWCB)	400	Austria
1930	Wood chips cement board	600	The Netherlands
1960	Coarse (crushed) wood particle cement board	500–700	Austria
1970	Cement-bonded particle board (CBPB)	1,250–1,400	Switzerland
1990	High density wood wool cement board	900	Philippines
2000	Wood strand cement board (WSCB)	1,000–1,100	The Netherlands

wood wool gypsum boards and wood wool magnesite boards were successfully produced and used in Austria. After 1950 WWCB has prevailed worldwide [23].

WWCBs are generally used for core applications and for exposed decorative ceilings and roof decks. Since Portland cement-bonded WWCBs are fully moisture-resistant, they are also used successfully as substitutes for conventional concrete floors and walls, and most recently for thick and very large wall elements. The worldwide demand for these large wall elements is expected to increase because of increasing heating and cooling costs and greater demand for living comforts. Now, there are plants available that are capable of producing several types of cement-bonded products such as WWCB, wood strand cement board (WSCB), and large wall elements on one combined manufacturing line.

There was no strong demand to manufacture wood-cement panels for industrial application until the mid-1930s. In the 1940s, resin-bonded particleboard technology was gradually developed; much of its technological advancement is also applicable to cement-bonded wood particleboards. Cement-bonded particleboard (CBPB) is a high-density product that was developed in the 1970s to replace asbestos-cement board for structural applications. Common uses of CBPB in Europe include facades, electrically heated and raised floors, permanent shuttering of concrete floors and walls, and fire- and moisture-resistant furniture.

The density of CBPB (approximately 1,250–1,400 kg/m^3) is much higher than that of resin-bonded particleboard (usually less than 750 kg/m^3). This has constrained the use of CBPB in many applications. Since CBPBs are made from short and thick particles, it has relatively low flexibility and strength. It has relatively high expansion when exposed to moisture and significant shrinkage upon drying. It is necessary to pre-drill pilot holes for mechanical fasteners due to high density of the panels. Therefore, manufacturers, builders, and architects were interested in developing a product that could better meet structural requirements and overcome these hurdles. This was attained with the development of WSCB, which accepts screws without pre-drilling, and exhibits lighter weight, greater bending strength, more flexibility, and less expansion and shrinkage due to moisture than CBPB [20]. Gypsum fiberboard and particleboard were developed after cement-bonded wood composites with the increased demand for fire-proof interior building materials. Gypsum-bonded panels are frequently used to finish interior wall and ceiling surfaces. In the United States, these products are generically called "dry wall" because they replace wet plaster systems [24].

7.8.2 Cement-Bonded Materials

Wood and other plant fiber in combination with cement has been practiced for well over one hundred years. Cement imparts durability, moisture resistance, and prevention of biological degradation to the plant fiber, whereas the fracture resistance of cement is improved by addition of lignocellulosic strands, particles, or fibers. Successful combination of these disparate materials depends upon an understanding of their chemical interactions and resultant performance. Readers interested in details beyond those provided in this text are referred to the excellent review of cement-bonded wood composites by Frybort et al. [25].

7.8.2.1 Portland Cement Production Primer

Cement is an important building material worldwide. Portland cement, or ordinary Portland cement (OPC), is the most common type, produced by grinding together Portland cement clinker with calcium sulfate and other minor constituents. Clinker is produced by calcining the necessary raw materials at 1,450°C. OPC is mixed with coarse gravel or crushed stone with sand to make concrete. OPC is also used for mortar and stucco.

7.8.2.2 Cement Cure Chemistry

The most important compounds present in cement are tricalcium silicate ($3CaO \cdot SiO_2$), dicalcium silicate ($2CaO \cdot SiO_2$), tricalcium aluminate ($3CaO \cdot Al_2O_3$), and calcium oxide (CaO). The $2CaO \cdot SiO_2$ and $3CaO \cdot SiO_2$ react slowly with water to yield $Ca(OH)_2$ and calcium silicate hydrate (CSH) gel. This reaction not only helps in holding the material together but also makes the concrete less pervious to water. The hardening process is mainly due to the formation of $CaCO_3$ when $Ca(OH)_2$ reacts with CO_2 from the atmosphere. Compounds formed as cement undergoes hydration and crystallization are insoluble in water, which is why cement can harden underwater [26, 27].

7.8.2.3 Shorthand Notation

Since the compounds in cement science are complex, a simplified representation is often used in cement nomenclature:

$C = CaO$ = calcium oxide (lime)
$S = SiO_2$ = silicon dioxide (silica)
$A = Al_2O_3$ = aluminum oxide (alumina)
$F = Fe_2O_3$ = ferric oxide
$H = H_2O$ = water
$\bar{S} = SO_3$
$\bar{C} = CO_2$
$M - MgO$
$N = Na_2O$
$K = K_2O$

7.8.2.4 Cement Compounds

The major phases of Portland cement are tricalcium silicate ($3CaO \bullet SiO_2 = C_3S$), known as alite; dicalcium silicate ($2CaO \bullet SiO_2 = C_2S$) or belite; tricalcium aluminate

Table 7.4   *Chemical composition of different types of cement, percent by weight.*

Composition	I	II	III	IV	V
C_3S	50	42	60	25	40
C_2S	25	32	15	50	40
C_3A	12	6	10	5	4
C_4AF	8	12	8	12	10
Fineness (cm^2/g)	3,500	3,500	4,500	3,500	3,500

($3CaO \bullet Al_2O_3 = C_3A$), aluminate; tetracalcium alumino ferrite ($4CaO \bullet Al_2O_3 \bullet Fe_2O_3 = C_4AF$), or ferrite; and calcium sulfate dehydrate ($CaSO_4 \cdot 2H_2O = C\bar{S}H_2$).

In a commercial clinker these phases do not exist in a pure form. The $3CaO \bullet SiO_2$ phase is a solid solution containing Mg and Al and is called *alite*. In the clinker, it consists of monoclinic or trigonal form, whereas synthesized $3CaO \bullet SiO_2$ is triclinic. Alite is the most important constituent of normal Portland cement, constituting 50–60% and promoting strength development. The $2CaO \bullet SiO_2$ phase occurs in the β (belite) forms and contains, in addition to Al and Mg, some K_2O. The aluminate phase C_3A constitutes 4–12% in most Portland cements and is substantially modified by ionic substitution. The ferrite phase, designated C_4AF, is a solid solution of variable composition from C_2F to C_6A_2F.

7.8.2.5   *Cement Types and Composition*

ASTM C-150 [28] describes five major types of Portland cement. They are

Type I   Regular or ordinary Portland cement
Type II   Moderate sulfate resistant cement
Type III   High early strength cement
Type IV   Low heat of hydration cement
Type V   Sulfate resistant cement

The general chemical composition, fineness, and compressive strength characteristics of different types of cement are shown in Tables 7.4 and 7.5.

7.8.2.6   *Hydration Reactions*

Portland cement, when combined with water, immediately reacts exothermically in a process called hydration which leads to solidification. The hydration reactions are shown in

Table 7.5   *Compressive strength, MPa (psi), of different types of cement.*

Time	I	II	III	IV	V
1 day	6.89 (1,000)	5.52 (800)	13.79 (2,000)	3.45 (500)	5.52 (800)
28 days	34.47 (5,000)	34.47 (5,000)	48.26 (7,000)	34.47 (5,000)	31.03 (4,500)
90 days	44.82 (6,500)	44.82 (6,500)	51.71 (7,500)	44.82 (6,500)	44.82 (6,500)

Equations 7.5–7.8 [29]. In these equations, observe that the shorthand notations presented in Section 7.8.2.3 are used.

$$2C_3S + 6H \rightarrow C_3S_2H_3 + 3CH \tag{7.5}$$

$$2C_2S + 4H \rightarrow C_3S_2H_3 + CH \tag{7.6}$$

$$C_3A + 3C\bar{S}H_2 + 26H \rightarrow C_3A \cdot 3C\bar{S} \cdot 32H \tag{7.7}$$

$$C_3A \cdot 3C\bar{S} \cdot 32H + 2C_3A + 4H \rightarrow 3(C_3A \cdot C\bar{S} \cdot 12H) \tag{7.8}$$

7.8.2.7 Heat of Hydration: Inhibitory Effect of Wood Species

Heat of hydration is a major characteristic relative to the efficiency of the hardening process, and in inorganic-bonded wood and fiber composites, it is indicative of the compatibility of cement and lignocellulosic material. Accordingly, the maximum hydration temperature and the time required to reach that maximum varies when cement is mixed with fiber from different plant species. An example showing the heat of hydration of unmodified cement as compared to a cement-wood mixture is given in Figure 7.4.

Inspection of Figure 7.4 reveals that addition of wood to cement suppresses both the rate of hydration reaction (as evidenced by lower slope of the heating curve) and the maximum hydration temperature ($T_2 < T_2'$), with a concurrent increase in time to maximum hydration temperature ($t_2 > t_2'$). Neat cement achieves a hydration temperature typically in excess of 80°C. "Good suitability" of a wood-cement mixture is generally indicated by a hydration temperature of at least 60°C (Sandermann and Kohler, as cited by Frybort et al. [25, 31]). The

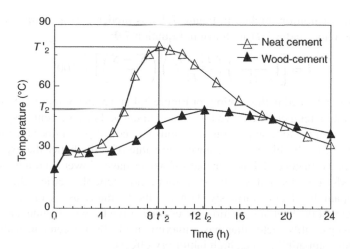

Figure 7.4 *Heat of hydration as a function of time for neat (unmodified) Type I Portland cement and a Portland cement-cork oak (Quercus suber) wood mixture. T = maximum hydration temperature of wood-cement mixture; T' = maximum hydration of neat cement. The time to maximum hydration temperature for wood-cement and neat cement are represented by t and t', respectively. Adapted from [30] with permission of Springer Science + Business Media © 1990.*

Table 7.6   Inhibitory index, I, for some North American wood species.

Species name	Inhibitory index
Softwoods	
Lodgepole (*Pinus contorta*)	2.6
Western white pine (*Pinus monticola*)	3.9
Grand fir (*Abies grandis*)	6.7
Engelmann spruce (*Picea engelmannii*)	15.0
Western redcedar (*Thuja plicata*)	24.2
Ponderosa pine (*Pinus ponderosa*)	37.9
Douglas-fir (*Pseudotsuga menziesii*)	53.9
Western hemlock (*Tsuga heterophylla*)	63.9
Western larch (*Larix occidentalis*)	118.3
Hardwoods	
Chestnut oak (*Quercus prinus*)	14.0
Yellow-poplar (*Liriodendron tulipifera*)	40.0
White oak (*Quercus alba*)	73.7
Water oak (*Quercus nigra*)	76.5
Post oak (*Quercus stellata*)	85.5
Black tupelo (*Nyssa sylvatica*)	92.0
Black oak (*Quercus velutina*)	106.7
Sweetgum (*Liquidambar styraciflua*)	125.7
Scarlet oak (*Quercus coccinea*)	136.0
Southern red oak (*Quercus falcata*)	138.5
Red maple (*Acer rubrum*)	212.4

Source: Softwood data drawn from Hofstrand et al. [32]; hardwood data from Moslemi and Lim [35], and tabulated by Defo et al. [34]. Reproduced from [34] with permisson of the Forest Products Society.

hydration reaction inhibition implicit in Figure 7.4 was modeled by Hofstrand et al. [32] by means of an "inhibitory index" (I) as shown in Equation 7.9:

$$I = \left[\left(\frac{t_2 - t_2'}{t_2'} \right) \left(\frac{T_2' - T_2}{T_2'} \right) \left(\frac{S' - S}{S'} \right) \right] \times 100 \qquad (7.9)$$

where t and T are as defined in the caption for Figure 7.4 and S and S' represent the maximum slope of the hydration temperature curve for wood-cement and neat cement, respectively. The inhibitory index is useful for comparison of various sources of wood fiber for use in cement composites (Table 7.6). In this case, the lower the value of I, the better the expected wood-cement compatibility. In general, it is seen that softwoods have a lower overall inhibitory effect than do hardwoods. A related and possibly superior or more reliable metric is the compatibility factor. The compatibility factor, variously shown in the literature as C_A or F_c, is the ratio of the surface integral of the heat of hydration versus time curve for wood-cement to neat cement [33, 34]. The compatibility factor thus has a maximum of 100 for cement, with low values indicative of poor compatibility (i.e., high inhibitory effect).

The inhibition (or lack of compatibility) of wood with cement has been attributed to sugars, hemicelluloses, and extractives, or the degradation products of these wood components. It has been shown that treatments such as cold- or hot-water extraction of wood generally reduce inhibition/increases compatibility. Similar beneficial effects have also been seen for other natural fiber-cement mixtures. For example, cold-water extraction improved hydration of rattan-cement systems [36].

7.8.2.8   *Carbon Dioxide Treatment of Portland Cement-Bonded Composites*

During the manufacture of a cement-bonded wood composite, the cement hydration process normally requires from 8 to 24 hours to develop sufficient board strength and cohesiveness to permit the release of consolidation pressure. By exposing the cement to CO_2 the initial hardening stage can be reduced to less than 5 minutes. The phenomenon results from the chemical reaction of CO_2 with calcium hydroxide to form calcium carbonate, thus accelerating the process of hardening. This also expedites the formation of CSH gel. Reduction of initial cure time of cement-bonded wood composite is not the only advantage of using CO_2 injection. The inhibiting effects of sugars, tannins, and others on cement hydration are also greatly reduced. This is especially important for a variety of lignocellulosic materials [21, 37]. In addition, research has demonstrated that CO_2-treated composites can be twice as stiff and strong as untreated composites. Finally, CO_2-treated composites do not experience efflorescence, a condition whereby calcium hydroxide migrates to the surface of the material. Thus, the desired appearance of the final product is maintained [38, 39].

7.8.2.9   *Cement Process Technology and Products*

Cement-bonded composites are made by blending proportionate amounts of the fibrous materials with cement in the presence of water and allowing cement to cure or "set " to make a rigid composite. Asbestos fibers were formerly used to reinforce cement products. However, asbestos was first supplemented by and has now been totally replaced with cellulose fibers. It is also not uncommon to use various synthetic fibers capable of providing specific properties to the final products for different end applications. Portland cement is the main binder of fiber-cement composites. Additives are also incorporated to enhance certain properties in the final products. Depending on curing temperatures, fiber cement composites can be either low temperature, air-cured or high temperature, autoclave cured. Air-cured formulations of nonfibrous materials contain large amounts of Portland cement combined with no or little amounts of very fine clays, silica fume ground limestone or fly ash. For example, a representative air-cured formulation may contain 10% cellulose fibers and 90% Portland cement. Another formulation may contain 8% cellulose fibers, 2% PVA fibers, 10% silica fume, and 80% Portland cement. Autoclaved formulations normally contain fine silica with other fine minerals and lesser quantities of cement. For example, an autoclaved formulation contains 5–10% cellulose fibers, 35–50% Portland cement and 40–60% silica.

In general, the following cement-bonded wood composites are considered to be the main products on the market today:

1. Wood wool cement board (WWCB)
2. Cement bonded particle board (CBPB)
3. Wood strand cement board (WSCB)
4. Cement bonded fiber board (CBFB).

In view of space limitations, we will highlight in the following sections, only two of the products listed above, namely WWCB and WSCB.

Figure 7.5   *Wood wool cement board (WWCB), used for its thermal insulation or acoustic properties. Inorganic-bonded materials in general are also resistant to degradation by moisture, decay, and wood-destroying insects.*

7.8.2.10   *Production Technology of Wood Wool Cement Board (WWCB)*

Wood wool cement board (Figure 7.5) is made with excelsior and Portland cement. The density of WWCB could be as low as approximately $400\ kg/m^3$. Wood wool is shredded wood excelsior that is in the form of long, narrow, and curved wood strands [40]. It has been used as reinforcement for inorganic binders dating to the beginning of the nineteenth century. Gypsum- and magnesite-bonded excelsior boards were manufactured in 1905 and 1915, respectively [22]. However, the first commercial production of WWCBs was initiated and patented in Austria in 1927. More plants were established in the early 1990s due to the large demand for the product in meeting housing requirements. In recent years the excellent insulating and acoustic properties of WWCB have attracted an increased interest from architects, builders and board producers.

For the production of WWCB, small logs are cut into blocks of 50 cm (20 in.) length. These blocks are shredded to long wood wool strands. The strands are dipped in a salt solution to improve the setting of the cement. The wet wood wool, together with cement powder, is fed into a continuous mixer. From the mixer the mixture is transported to the distribution machine, which spreads a continuous mat of material onto molds. After passing a pre-press roll, the molds are separated by a circular saw and stacked. The stacks (kept under pressure) are stored for 24 hours for setting of the cement. After this initial setting stage, the boards are taken from the molds for further curing, while the molds can be cleaned and oiled for re-use. The raw materials for WWCB are round wood (logs), Portland cement, water and a small amount of salt solutions. In general, hardwood is less suitable. The salt solution depends on the wood species used. Commonly used solutions include sodium silicate, magnesium chloride, and calcium formate [41].

7.8.2.11   *Production Technology of Wood Strand Cement Board (WSCB)*

Wood strand cement board (WSCB), a class of recently developed panels composed of long and thin wood strands bonded with Portland cement, has structural strength, medium density, and adequate elasticity [42]. Logs with diameters as small as 70–270 mm (2.75–10.63 in.) are harvested, debarked, and seasoned outside for at least one month up to one year, depending on species, to reduce the effect of sugars and extractives on the setting of the cement. Water-soaking of cut strands is another option for removal of sugars and extractives from certain tropical hardwoods. In this method, strands cut from green logs can be soaked in warm water for up to a few hours to remove the water-soluble cement-setting inhibitors [42]. This water soaking method might not be practical for a large-scale production because a substantial amount of strands have to be treated and it may present a water treatment issue.

For the raw materials, the selection of cement-compatible wood species is critical for successful production of WSCB. Some species, such as alder (*Alnus glutinosa*) and larch (*Larix decidua*), have demonstrated limited compatibility with cement because they contain excessive amounts of tannins and/or sugars that retard the setting of cement. Therefore, it is critical to select proper species or to remove the tannins and sugars before proceeding with board manufacture. Generally, pine, spruce, and several hardwoods work very well for WSCB production. The use of low-density wood is advantageous for production of low-density WSCB. Further, it is known from practice that low-density wood provides improved machinability, with sufficient mechanical and physical properties at an acceptable product density [42].

7.8.2.12   *Applications/Markets of Cement-Bonded Materials*

Successfully marketed Portland cement-bonded composites consist of both low-density products made with excelsior and high-density products made with particles and fibers. The low-density products may be used as interior ceiling and partition wall panels in commercial buildings. Along with the previously mentioned advantages, they offer sound control and can be quite decorative. In some parts of the world, these panels can be installed and function as complete wall and roof decks. In general, the exterior of the panels is finished with stucco, while the interior panel is with plaster. High-density panels can be used as flooring, roof sheathing, fire doors, load bearing walls, and cement forms. Fairly complex molded shapes can be molded or extruded. Also, decorative roofing tiles or nonpressure pipes can be made from high-density cement bonded materials. As an example of structural use of inorganic-bonded material, WWCBs are used successfully for thick and very large wall elements. Panels in 60 cm widths are in common use in Western Europe. Earthquake-resistant structures have been built with the incorporation of WWCB panel systems.

7.8.3   Gypsum-Bonded Materials

Gypsum panels are frequently used to finish interior wall and ceiling surfaces. In the United States, these products are generically called "drywall" because they replace wet plaster systems. Other names for gypsum panels include plasterboard, wallboard, gypsum board, or sheetrock. To increase the bending strength and stiffness of the gypsum panel, it is frequently wrapped in paper, which provides a tension surface. An alternative to wrapping the gypsum with fiber to increase strength and stiffness is to put the fiber within the panel, in which case, the product is called gypsum fiberboard. Several firms in the United States and Europe are doing this with recycled paper fiber [38]. There is no reason that other lignocelluloses cannot be used. Gypsum is widely available and does not have the highly alkaline character of cement.

Experimentally, sisal and coir have been successfully used in gypsum panels [43]. Wood in particle, as opposed to fibrous form, is used in the manufacture of gypsum particleboard.

7.8.3.1   Gypsum Cure Chemistry

Gypsum rock is converted into gypsum plaster by driving off some of the chemically combined water. Heating gypsum at the neighborhood of 120°C for one hour results in a hemi-hydrate ($CaSO_4 \cdot \frac{1}{2}H_2O$) – with three quarters of the water removed. This hemi-hydrate is generally called beta hemi-hydrate. Gypsum hemi-hydrate is also known as Plaster of Paris. Prolonged heating over several hours results in the formation of anhydrite with practically none of the chemically combined water left. Anhydrite sets more slowly and is a slightly stronger plaster than hemi-hydrate, but with the drawback of added production cost. In practice, a simple production system would most likely give a mixture of the hemi-hydrate and anhydrite phases. Much of the commercial plaster produced industrially today is Plaster of Paris [24]. Gypsum plaster sets by chemically combining with water to form solid calcium sulfate dehydrate and release heat (Q) in the curing reaction, as shown in Equation 7.10 [44].

$$CaSO_4 \cdot \frac{1}{2}H_2O + 1.5H_2O \rightarrow CaSO_4 \cdot 2H_2O + Q \tag{7.10}$$

7.8.3.2   Gypsum Process Technology and Products

Gypsum panels are normally made from a slurry of gypsum, water, and lignocellulosic materials as reinforcement. In large-scale production, the slurry is extruded onto a belt. The belt carries the slurry through a drying oven to drive off the water and facilitate the cure of the gypsum. The panel is then cut to length, and trimmed if necessary.

Gypsum particleboard is made of gypsum and reinforced with wood particles. It has two smooth surfaces and a core of gypsum mixed with wood particles. The technology was developed in Germany and Finland. The process echnology of gypsum particleboard drew upon experience and machinery mainly from gypsum wallboard and wood particleboard.

Gypsum fiberboard is a product that is relatively new to the North American markets, being introduced in about 1990. Gypsum fiberboard products were developed over the past 20 years in Germany, where the product has been quite successful, capturing about 20–25% of the total gypsum board market. There are a number of competing processing technologies. What all of these have in common is the fact that the finished board is "paperless," that is, it does not have any paper facings as does the conventional gypsum board. Instead, gypsum fiberboard consists of 18% ground waste newsprint/magazine fibers uniformly dispersed throughout the gypsum matrix. It is this recycled paper fiber that provides the reinforcement for the matrix instead of the paper skins [24]. The production process of gypsum fiberboard is similar to gypsum particleboard.

References

1. Stark NM, Cai Z, Carll C. Wood-based composite materials, panel products, glued-laminated timber, structural composite lumber, and wood–nonwood composite materials, In: *Wood Handbook*. Madison, WI: Centennial Edition. USDA Forest Products Laboratory; 2010. pp 11-1–11-28.
2. Chapman KM. Wood-based panels: particleboard, fibreboards and oriented strand board. In: Walker JCF, editor. *Primary Wood Processing Principles and Practice*. Dordrecht, The Netherlands: Springer; 2006. pp 427–475.

3. Moody RC, Hernandez R. Glued-laminated timber. In: Smulski S, editor. *Engineered Wood Products A Guide for Specifiers, Designers and Users*. Madison, WI: PFS Research Foundation; 1997. pp 1–39.

4. Glulam product guide. Form No. X440D. APA–The Engineered Wood Association, Tacoma, WA. 2008.

5. ANSI/AITC *American National Standard: Structural Glued Laminated Timber*, A190.1-07. Centennial, CO: American Institute of Timber Construction; 2007.

6. Yeh B, Gagnon S, Williamson T, Privu C, Lum C, Kretschmann D. The North American product standard for cross-laminated timber. *Wood Design Focus*, 2012;22(2):13–21.

7. ANSI/APA. *American National Standard: Standard for Performance-Rated Cross-Laminated Timber*, PRG-320-2012. Tacoma, WA: APA–The Engineered Wood Association; 2012.

8. Baldwin RF. *Plywood and Veneer-Based Products. Manufacturing Practices*. San Francisco: Miller Freeman Inc.; 1995.

9. Shi S, Walker J. Wood-based composites: plywood and veneer-based products. In: Walker JCF, editor. *Primary Wood Processing Principles and Practice*. Dordrecht, The Netherlands: Springer; 2006. pp 391–426.

10. Sellers T Jr. *Plywood and Adhesive Technology*. New York: Marcel Dekker, Inc.; 1985.

11. Nelson S. Structural composite lumbers. In: Smulski S, editor. *Engineered Wood Products A Guide for Specifiers, Designers and Users*. Madison, WI: PFS Research Foundation; 1997. pp 147–172.

12. Irle MA, Barbu MC, Reh R, Bergland L, Rowell RM. Wood composites. In: Rowell RM, editor. *Handbook of Wood Chemistry and Wood Composites*. Boca Raton, FL: CRC Press, Taylor & Francis Group; 2013. pp 321–411.

13. Spelter H, McKeever D, Alderman M. *Status and Trends: Profile of Structural Panels in the United States and Canada*. Research Note FPL-RP-636. Madison, WI: U.S. Department of Agriculture, Forest Service, Forest Products Laboratory; 2006.

14. Suchsland O, Woodson GE. *Fiberboard Manufacturing Practices in the United States*. Agriculture Handbook No. 640. U.S. Department of Agriculture, Forest Service; 1986. pp 261.

15. Shmulsky R. Jones PD. Structural composites. In: *Forest Products and Wood Science: An Introduction*, 6th ed. Chichester, UK: Wiley-Blackwell; 2011. pp 321–363.

16. Shmulsky R, Jones PD. Nonstructural composites. In: *Forest Products and Wood Science: An Introduction*. Chichester, UK: Wiley-Blackwell; 2011. pp 365–396.

17. Freese F, *A Collection of Log Rules*. Gen. Tech. Rep. FPL-01. Madison, WI: U.S. Department of Agriculture, Forest Service, Forest Products Laboratory; 1973.

18. Briggs DG. *Forest Products Measurements and Conversion Factors: With Special Emphasis on the U.S. Pacific Northwest*. Institute of Forest Resources Contribution No. 75. Seattle, WA: College of Forest Resources, University of Washington; 1994.

19. Technology NIoS. *American Softwood Lumber Standard*, Voluntary Product Standard PS 20-10. Washington, DC: U.S. Government Printing Office; 2010.

20. van Elten GJ. History, present and future of wood cement products, in International Inorganic Bonded Composite Materials Conference and Exhibits, Vancouver, Canada; 2004.

21. Hermawan D. Manufacture of Cement-Bonded Particleboard using Carbon Dioxide Curing Technology [PhD Dissertation], Kyoto University; 2001.

22. Kossatz G, Lempfer K, Sattler H. Wood based panel with inorganic binder Holzforschung Austria, Annual Report. 1983.

23. van Elten GJ. Innovation in the production of cement-bonded particleboard and wood-wool cement board. in: 5th International Inorganic Bonded Wood and Fiber Composite Materials Conference, Spokane, Washington; 1996.

24. Green G. Gypsum fiber board. *Construction Dimensions*, 1991;(May):22–29.

25. Frybort S, Mauritz R, Teischinger A, Müller U. Cement-bonded composites: a mechanical review. *BioResources*, 2008;3(2):602–626.

26. MacLaren DC, White MA. Cement: its chemistry and properties. *Journal of Chemical Education*, 2003;80(6):623–635.

27. Taylor HFW. *Cement Chemistry*. London: Thomas Telford; 1997.
28. ASTM. Standard test methods for evaluating properties of wood-base fiber and particle panel materials, D1037-06a. West Conshohocken, PA: ASTM International; 2010.
29. Wang T, editor. *Encyclopedia of Wood Industry*. Beijing: China Forestry Press; 2002.
30. Hachmi M, Moslemi AA, Campbell AG. A new technique to classify the compatibility of wood with cement. *Wood Science and Technology*, 1990;24(4):345–353.
31. Sandermann W, Kohler R. Über eine kurze Eignungsprüfung von Holzern fur zementgebundene Werkstoffe. *Holzforschung*, 1964;18:53–59.
32. Hofstrand AD, Moslemi AA, Garcia JF. Curing characteristics of wood particles from nine northern Rocky Mountain species mixed with Portland cement. *Forest Products Journal*, 1984;34(2):57–61.
33. Hachmi M, Moslemi AA. Correlation between wood-cement compatibility and wood extractives. *Forest Products Journal*, 1989;39(6):55–58.
34. Defo M, Cloutier A, Riedl B. Wood-cement compatibility of some Eastern Canadian woods by isothermal calorimetry. *Forest Products Journal*, 2004;54(10):49–56.
35. Moslemi AA, Lim YT. Compatability of southern hardwoods with Portland cement. *Forest Products Journal*, 1984;34(7/8):22–26.
36. Adefisan OO, Fabiyi JS, McDonald AG. Hydration behaviour and infrared spectroscopy of pre-treatments effect on Portland cement—*Eremospatha macrocarpa* and *Laccosperma secundiflorum* systems. *Journal of Applied Sciences*, 2012;12(3):254–262.
37. Geimer RL, Souza MR, Moslemi AA, Simatupang MH. Carbon dioxide application for rapid production of cement particleboard. in Inorganic Bonded Wood and Fiber Composite Material Conference, Spokane, Washington; 1992.
38. English B, Chow P, Bajwa DS. Processing into composites. In: Rowell RM, Young RA, Rowell JK, editors. *Paper and Composites from Agro-Based Resources*. New York: Lewis Publishers, CRC Press; 1997. pp 287–291.
39. Simatupang MH, Seddig N, Habighorst C, Geimer RL. Technologies for rapid production of mineral-bonded wood composite boards. In: Moslemi AA, editor. *2nd International Inorganic Bonded Wood and Fiber Composite Materials Conference*, October 14–17, 1990, Moscow, Idaho Madison, WI: Forest Products Research Society; 1991. pp 18–27.
40. van Elten GJ. Production of wood wool cement board and wood strand cement board (Eltoboard) in one plant and applications of the products. in *10th International Inorganic-Bonded Fiber Composites Conference*, São Paulo, Brazil; 2006.
41. van Elten GJ. Newly developed Eltomation plant capable to produce all types of wood wool cement board (WWCB), wood strand cement board (WSCB-EltoBoard) and large WWCB wall elements, in *11th International Inorganic-Bonded Fiber Composites Conference*, Madrid, Spain; 2008.
42. Aro M. Wood strand cement board. in *11th International Inorganic-Bonded Fiber Composites Conference*, Madrid, Spain, 2008. pp 169–179.
43. Mattone R. Comparison between gypsum panels reinforced with vegetable fibers: their behavior in bending under impact. In: *2nd International Rilem Symposium: Vegetable Plants and their Fibres as Building Material*. Salvador: Bahia, Brazil; 1990.
44. Cunningham WA, Dunham RM, Antes LL. Hydration of gypsum plaster. *Industrial and Engineering Chemistry*, 1955;44(10):2402–2408.

8

Natural Fiber and Plastic Composites

8.1 Introduction

Although fiber–plastic composites are not strictly adhesive-bonded materials, concepts of adhesion are crucial to the material properties and performance of this increasingly important class of a material. Industries that traditionally rely on processes described in Chapter 7 may either employ fiber-plastic manufacturing directly in their own operations, or these materials may be used as components of assembled components. Materials covered in this chapter include introduction to plastics, natural fibers and their temperature-related performance, plastic composite processing technology, overcoming incompatibility of synthetic polymers and natural fibers, melt compounding natural fibers and thermoplastics, and performance of natural fiber and plastic composites.

8.1.1 Synthetic Petrochemical Polymers

Polymers are substances that are made of molecules with high molar masses and a large number of repeating monomer units [1–3]. There are both naturally occurring and synthetic petrochemical polymers. Naturally occurring polymers include proteins, starches, cellulose, and latex. Synthetic petrochemical polymers, produced commercially, have a wide range of properties and uses. Depending upon the intermolecular force, synthetic polymers are classified into two main types—plastics and rubbers (Figure 8.1). Plastics are rigid materials at service temperatures, and rubbers are flexible, low-modulus materials, which exhibit long-range elasticity. The mixing of plastics and rubbers leads to thermoplastic elastomers, which normally have low modulus and are used as impact modifiers for other polymers [3]. Plastics are further subdivided into thermoplastics and thermosets. A thermoplastic is a polymer that becomes moldable above a specific temperature, and returns to a solid state upon cooling. Most thermoplastics have a high molecular weight, whose chains associate through intermolecular

Introduction to Wood and Natural Fiber Composites, First Edition. Douglas D. Stokke, Qinglin Wu and Guangping Han.
© 2014 John Wiley & Sons, Ltd. Published 2014 by John Wiley & Sons, Ltd.

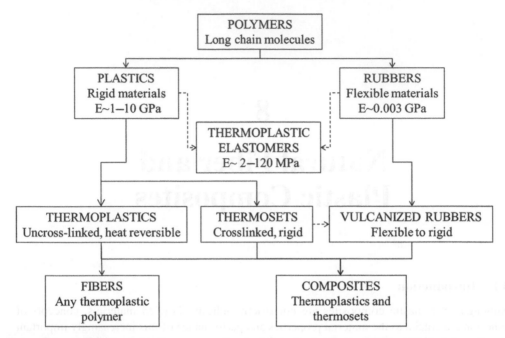

Figure 8.1 *Classification of polymers according to intermolecular force property. Adapted from figure 3 http://www.open.edu/openlearn/science-maths-technology/science/chemistry/introduction-polymers/content-section-1.2 © The Open University.*

forces. This property allows thermoplastics to be remolded because the intermolecular interactions spontaneously reform upon cooling. Thermosets are defined as polymers that become irreversibly hard on heating or by the addition of special chemicals. This hardening involves a chemical change (curing) and hence scrap thermoset cannot be recycled except as a filler material. The curing process invariably involves a chemical reaction, which connects the linear molecules together to form a single macromolecule (i.e., cross-linking). Thermoset polymers are harder, stronger, and more brittle than thermoplastics and have better dimensional stability. Most of the cross-linked and network polymers, which include vulcanized rubbers, epoxies, phenolic, and polyester resins are thermosets.

Depending on the arrangement of monomer units, polymers are classified as linear, branched, or cross-linked polymers [4]. A linear polymer is one in which the monomer units are linked in a chain-like manner. A branched polymer occurs when there are extensions of linked monomer units that protrude from various points on the chain. The resulting polymer tends to have lower strength and lower viscosity compared with a linear polymer of the same molecular weight. This type of polymer is characterized by a large degree of branching, forcing the molecules to be packed rather loosely forming a low-density material. Polymer molecules can also be interconnected, or cross-linked. Polymers can be both crystalline and amorphous in nature. Coiling or folding in a linear or branched polymer causes certain sections with the proper molecular arrangement to be crystalline, while the more randomly arranged areas remain amorphous.

Glass transition temperature, T_g, and melting point temperature, T_m, are two important material properties for thermoplastics [2]. Physical properties of a thermoplastic change drastically

Table 8.1   *Selected properties of common thermoplastics.*

Polymer name	Glass transition temp, T_g (°C)	Melting temp, T_m (°C)	TGA decomp temp[a] (°C)	Linear CTE[b] (μm/m°C)	Flexural modulus (GPa)
Acrylonitrile butadiene styrene (ABS)	110–125		375	65–95	2.07–4.14
Polymethyl-methacrylate (PMMA)	85–110	160	313	50–90	2.24–3.17
Acrylonitrile (AN)	95	135		66	3.45–4.07
Polytetra-fluoroethylene (PTFE)	126	327	525	70–120	0.525
Polyvinylidene fluoride (PVDF)	−60 to −20	170–180	470	70–142	1.72–2.89
Nylon 6	40–87	210–220	400	80–83	2.69
Nylon 6,6	50	255–265	426	80	2.83–3.24
Polycarbonate (PC)	140–150		473	68	2.35
Polybutylene terephthalate (PBT)		220–287	386	60–95	2.28–2.76
Polyethylene terephthalate (PET)	73–80	245–265	414	65	2.41–3.10
Polyether-etherketone (PPEK)	150	334	575	40–108	3.86
Polyetherimide (PEI)	215–217			47–56	3.31
Low-density polyethylene (LDPE)	−125	98–115	459	100–200	1–1.55
High-density polyethylene (HDPE)	−120	130–137	469	59–110	2.40–3.30
Polyimide (PI)	210–433	310–365		45–56	3.10–3.45
Polyphenylene oxide (PPO)	100–142		400	38–70	2.25–2.76
Polyphenylene sulfide (PPS)	88	285–290	508	49	3.79
Polypropylene (PP)	−20	160–175	417	81–100	1.17–1.72
Polystyrene (PS)	74–109		351	50–83	2.62–3.38
Polysulfone (PSO)	190		510	56	2.69
Polyethersulfone (PES)	220–230			55	2.40–2.62
Polyvinyl chloride (PVC)	75–105	100–260	265	50–100	2.07–3.45

Source: From References 5 and 6.
[a] TGA decomp temp—decomposition temperature from thermal gravimetric analysis at a heating rate of 20°C/min [7]
[b] Linear CTE—linear coefficient of thermal expansion.

without an associated phase change above T_g and below T_m. Within this temperature range, most thermoplastics are rubbery due to alternating rigid crystalline and elastic amorphous regions, approximating random coils. Table 8.1 summarizes some important properties for common plastics.

Among plastics, polyethylene is composed of chains of repeating CH_2 units (Figure 8.2a). When ethylene is polymerized at moderate pressures (15–30 atm), the end product is high-density polyethylene (HDPE). Under these conditions, the polymer chains grow to very great length in a linear fashion (i.e., linear polymer). HDPE is hard, tough, and resilient. Most HDPE is used in the manufacture of containers, such as milk bottles and laundry detergent jugs. When ethylene is polymerized at high pressures (1000–2000 atm) and elevated temperatures

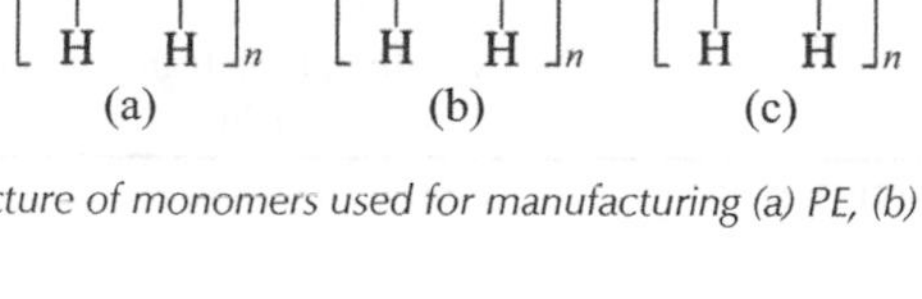

Figure 8.2   *Chemical structure of monomers used for manufacturing (a) PE, (b) PP, and (c) PVC polymers.*

(190–210°C), the product becomes branched low-density polyethylene (LDPE). This form of ethylene is relatively soft, and most of it is used in the production of plastic films, such as those used in shopping and sandwich bags.

Polypropylene's (PP) molecular structure has a methyl group, $-CH_3$, on alternate carbon atoms of the chain (Figure 8.2b). PP is slightly more brittle than polyethylene, but softens at a temperature about 40°C higher. PP is used extensively with and without reinforcing fillers in the automotive industry for interior trim such as instrument panels, in food packaging such as yogurt containers, and in natural fiber polymer composites such as siding and docking material. It can be formed into fibers of very low absorbance and high stain resistance, used in clothing and home furnishings, especially carpeting.

Poly vinyl chloride (PVC) is a polymer similar to polyethylene, but having chlorine atoms at alternate carbon atoms on the chain (Figure 8.2c). PVC is produced by polymerization of the monomer vinyl chloride [8]. PVC is rigid and somewhat brittle. It can be made softer and more flexible by the addition of plasticizers (e.g., phthalates). Most PVC produced annually is used in the manufacture of industrial pipes. It is also used in the production of "vinyl" siding for houses and clear plastic bottles. When it is blended with a plasticizer such as a phthalate ester, PVC becomes pliable and is used to form flexible articles such as raincoats and shower curtains.

8.1.2   Bio-Based Polymers

Polymers from renewable resources have attracted an increasing amount of attention over the last two decades, predominantly due to environmental concerns, and finite petroleum resources [9,10]. Generally, polymers from renewable resources can be classified into three main groups (Table 8.2):

- Natural polymers, such as starch, protein, and cellulose;
- Synthetic polymers from natural monomers, such as polylactic acid (PLA); and
- Polymers from microbial fermentation, such as polyhydroxybutyrate (PHB).

Similar to numerous petroleum-based polymers, many properties of bio-based polymers can be improved through blending and composite formation.

Table 8.2   *Main types, properties, and applications of bio-based polymers.*

Polymer types		Properties	Applications
Natural polymers	Starch-based polymers	Short shelf-life, biodegradable and compostable	Films, molding, extrusion
	Cellulose derivatives	Biodegradable, limited compostable, good mechanical properties	Films, injection molding
Polyhydroxyalkanoates		Long shelf-life, varying biodegradability, limited compostable	Molding, films
Polylactic acid (PLA)		Weatherproof, high strength, high modulus	Films, molding, extrusion, fibers

Source: Adapted from [10] © 2010 ETC/SCP.

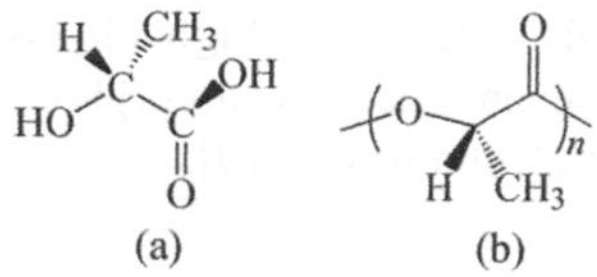

Figure 8.3   *Two forms of lactic acid: L-lactic acid (a) and D-lactic acid (b).*

One of the most promising bio-based polymers is PLA, which is an aliphatic semi-crystalline polyester derived from agricultural products. The building monomer of PLA, lactic acid (2-hydroxy propionic acid) is obtained by fermentation from plants such as corns or sugar beets [11]. Lactic acid exists in two different optically active L- and D-enantiomers (Figure 8.3). PLA exhibits different properties with varying proportion of each of the two stereoisomers in the final polymer.

Direct poly-condensation of lactic acid and the ring-opening polymerization of the cyclic lactide dimer (Figure 8.4) are two possible routes for polymerization of lactic acid. Poly-condensation is simple, but the equilibrium is reached at a fairly low molecular weight due to the difficulty of removing water [12, 13]. Thus, the ring opening polymerization is preferred to produce high molecular weight PLA.

Poly(lactic acid) homopolymers, generally available, containing only a small percentage of D-enantiomers, have a glass-transition and melt temperature of about 55°C and 175°C, respectively, and require processing temperatures in excess of 185–190°C [12, 14]. However, PLA homopolymers have a very narrow processing temperature window, which makes it difficult for them to be used as an alternative in many commercial polymers applications [15]. To improve their thermal properties, a stereocomplex is formed from enantiomeric PLAs, poly(L-lactic acid) (PLLA), and poly(L-lactide) (i.e., poly(D-lactic acid) (PDLA)) due to the strong interaction between PLLA and PDLA chains [16]. The stereocomplexed PLLA/PDLA blend has a melting temperature of 220–230°C (i.e., about 50°C higher than those of pure PLLA and PDLA), and can retain a nonzero strength in the temperature range up to the melting

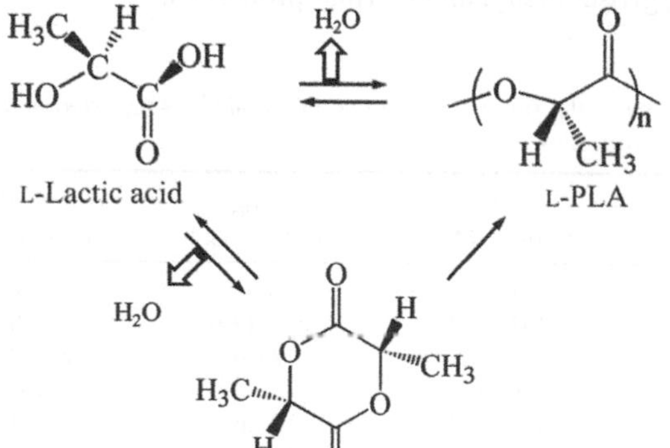

Figure 8.4   *Ring-opening polymerization routes to PLA [13]. Reproduced from [13] with permission of John Wiley & Sons, Inc © 2000.*

point [17]. Moreover, the PLLA/PDLA blend has a higher hydrolysis resistance compared with that of the pure PLLA and PDLA, even when it is amorphous-made, due again to the strong interaction between PLLA and PDLA chains [18, 19].

8.2 Natural Fibers and Their Temperature-Related Performance

Natural plant fibers suitable for polymer composite processing can be generally classified as [20] seed fibers (e.g., cotton and kapok), bast fibers (e.g., flax, hemp, jute, kenaf, and ramie), leaf fibers (e.g., sisal, pineapple, banana, palm, and Manila hemp), fruit fibers (e.g., coconut), wood fibers (softwood and hardwood), and grass and reed fibers (e.g., bamboo, wheat, rice, rye, oat, and corn) (recall Table 1.1). The raw material from various sources are processed through different technologies (steam explosion, grinding, etc.) to produce material in either fiber or particulate form for polymer composite applications. The physical, chemical, mechanical, and thermal properties of the materials dictate greatly their performance in the composites.

8.2.1 Physical, Mechanical, and Chemical Properties

The common fiber form materials include cotton, jute, flax, hemp, ramie, and sisal. Table 8.3 lists some published data on selected physical and mechanical properties of common natural fibers. The properties vary significantly with fiber sources and methods at which the fibers are extracted. Comprehensive reviews of cellulose-based fibers and their composites were made [20, 21]. The long natural fibers have been used to reinforce polymer composites through production techniques including hand lay-up, resin spray and press-molding, resin transfer molding, pultrusion, filament winding, bulk and sheet molding, and nonwoven processing [20].

The material used in wood-plastic composites (WPCs) is most often in particulate or flour form. The raw material from various sources (e.g., wood sawdust, planer shavings, sanding dust and scraps, and rice husks) are collected and hammer-milled to form fine powders, which are classified through the standard size screens. Clemons [22] reviewed primary sources, physical and chemical properties, costs and availability, and applications of wood flour in relation to WPCs. Some of the important properties include fiber/particle morphology, chemical composition, density, hygroscopicity, and thermal properties.

Table 8.3 *Physical and mechanical properties of common natural fibers reported in the literature used as reinforcing fillers in polymer composites.*

Fibers	Density (g/cm^3)	Tensile strength (MPa)	Young's modulus (GPa)	Elongation at break (%)	Moisture absorption (%)
Coir	1.2	175	4–6	30	10
Cotton	1.5	287–597	5.5–12.6	7–8	8–25
Flax	1.5	345–1035	27.6	2.7–3.2	7
Hemp	1.5	690		1.6	8
Jute	1.3	393–773	27.6	1.5–1.8	12
Ramie	1.5	400–938	61.4–128	3.6–3.8	12–17
Sisal	1.5	511–635	9.4–22	2–2.5	11
Softwood kraft	1.5	1000	40		
Viscose cord		593	11	11.4	

Source: Adapted from [20, 21] with permission of Elsevier © 1999.

Table 8.4 *Chemical composition reported in the literature for typical fibers used as fillers in wood-plastic composites.*

Species	Cellulose (wt%)	Hemi-cellulose (wt%)	Lignin (wt%)	Ash (wt%)	References
Pine (softwood)	40–45	25–30	26–34	0.2–0.5	[21, 22]
Maple (hardwood)	45–50	21–36	22–30	0.4	[21, 22]
Bamboo	42.3–49.1	24.1–27.7	23.8–26.1	1.3–2	[24]
Rice straw	41–57	33	8–19	8–38	[24–26]
Rice husk	35–45	19–25	20	14–17	[24–26]
Bagasse	40–46	24.5–29	12.5–20	1.5–2.4	[24, 27]
Cotton stalk	43.1	26.9	27.3	1.3	[28]
Jute	61–71.5	12–20.4	11.8–13	2	[21]
Hemp	70.2–74.4	17.9–22.4	3.7–5.7	1.5	[29]
Kenaf	51–59	21.5	15–19	5	[29, 30]

Source: Adapted from [23] with permission of Elsevier © 2008.

The natural fibers mainly consist of natural lignocellulosic polymers. Some information on chemical composition reported in the literature for various fibers is summarized in Table 8.4. Understanding the chemical composition of the fibers has an important implication on the thermal and mechanical performance of the composites.

Among the major chemical components, cellulose is the essential structural element of all natural fibers. The cellulose molecules consist of glucose units linked together by β-1,4-glycosidic bonds in long chains, and these chains are linked together in bundles called microfibrils through hydrogen bonding and van der Waals forces. Solid cellulose forms a microcrystalline structure with both crystalline and amorphous regions. The tensile strength (TS) of the cellulose microfibrils is estimated as high as 7.5 GPa. Hemicelluloses are polysaccharides bonded together in relatively short, branching chains. They are intimately associated with the cellulose microfibrils, embedding the cellulose in a matrix. Hemicelluloses are highly hydrophilic. Lignin is a complex hydrocarbon polymer with both aliphatic and aromatic components, which gives rigidity to the natural fiber. Lignin is a three-dimensional polymer with an amorphous structure and a high molecular weight. Lignin is less affinity for water, and is thermoplastic in nature (e.g., softening at temperatures around 90°C and flowing at temperatures around 170°C).

8.2.2 Thermal Degradation

Natural fibers are sensitive to temperature change and show thermal degradation at about 200°C [31, 32]. During composite processing, natural fiber particulates are compounded into melt polymer matrix by means of an extruder or a kneader at elevated temperatures. It is often of practical significance to understand and predict the thermal decomposition process of natural fibers. This knowledge will help better design composite process. For example, understanding the degradation process of natural fibers through kinetic analysis can have an important bearing on the determination of curing temperatures in the case of thermosets and extrusion temperatures in thermoplastic composites [31].

Typical thermal degradation of natural fibers through thermo-gravimetric analysis (TGA) and derivative TG (DTG) analysis is shown in Figure 8.5. The process consists of a low-temperature degradation peak associated with degradation of hemicellulose (at about 270°C),

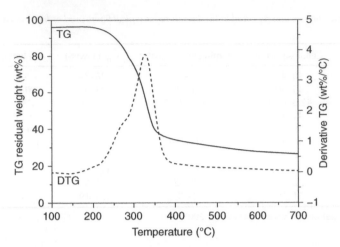

Figure 8.5   *Typical thermo-gravimetric decomposition process of natural fibers (i.e., heat-treated bamboo fibers with a heating rate of 5°C/min).*

and a high-temperature main peak observed at about 330°C (thermal decomposition of cellulose). The main degradation of lignin, which has a good thermal stability, occurs at temperatures above about 350°C.

The fundamental rate equation used in all TGA kinetic studies is generally described as:

$$\frac{da}{dt} = k\ f(\alpha) \tag{8.1}$$

where k is the rate constant and $f(\alpha)$ is the reaction model, a function depending on the actual reaction mechanism. Equation 8.1 expresses the rate of conversion, da/dt, at a constant temperature as a function of the reactant concentration loss and rate constant. In this study, the conversion rate α is defined as:

$$\alpha = (W_o - W_t)/(W_o - W_f) \tag{8.2}$$

where W_t, W_o, and W_f are time t, initial, and final weights of the sample, respectively. The rate constant k is generally given by the Arrhenius equation:

$$k = A\ \exp(-E_a/RT) \tag{8.3}$$

where E_a is the apparent activation energy (kJ/mol), R is the gas constant (8.3145 J/K mol), A is the pre-exponential factor (min^{-1}), T is the absolute temperature (K). A combination of Equations 8.1 and 8.3 gives the following relationship:

$$\frac{da}{dt} = A\ \exp\left(-\frac{E_a}{RT}\right) f(\alpha) \tag{8.4}$$

For a dynamic TGA process, introducing the heating rate, $\beta = dT/dt$, into Equations 8.4 and 8.5 is obtained as:

$$\frac{da}{dt} = \left(\frac{A}{\beta}\right) \exp\left(-\frac{E_a}{RT}\right) f(\alpha) \tag{8.5}$$

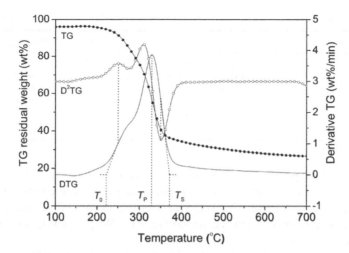

Figure 8.6 *Determination of decomposition characteristic parameters of heat-treated fibers using 150°C heat-treated bamboo fibers as an example at a heating rate of 5°C/min [36].*

Equations 8.4 and 8.5 are the fundamental expressions of analytical methods to calculate kinetic parameters on the basis of TGA data. Two of the most common "model-free" methods for determining apparent activity energy include the Kissinger model [33]:

$$\ln\left(\frac{\beta}{T_p^2}\right) = \ln\left(\frac{AR}{E_a}\right) + \left(\frac{1}{T_p}\right)\left(\frac{-E_a}{R}\right) \tag{8.6}$$

and the Flynn–Wall–Ozawa (FWO) model [34, 35]:

$$\log(\beta) = \log\left(\frac{AE_a}{Rg(\alpha)}\right) - 2.315 - 0.4567\frac{E_a}{RT} \tag{8.7}$$

In the Kissinger method, $\ln(\beta/T_p^2)$ is plotted against $1/T_p$ for a series of experiments at different heating rates with the peak temperature, T_p, obtained from the DTG curve. The iso-conversional FWO method is the integral method, which leads to $-E_a/R$ from the slope of the line determined by plotting $\log \beta$ against $1/T$ at any certain conversion rate.

Thermal decomposition characteristic parameters can be calculated from TG, DTG, and D^2TG (second time derivatives) curves. Figure 8.6 shows the calculation method of the parameters by using 150°C treated bamboo fiber at a heating rate of 5°C/min as an example. The onset temperature of decomposition, T_o, and the shift temperature of decomposition, T_s, is obtained by drawing the tangent line of the DTG with the point of the first and the last peak of the D^2TG curve and making the value of DTG zero. The peak temperature, T_p, is determined by the maximum of the DTG which is the fastest speed of the decomposition. The decomposition characteristic properties of heat-treated bamboo fibers at 150°C are shown in Table 8.5.

Table 8.5 shows that T_o, T_p, and T_s of the bamboo fibers increase with the increasing heating rate as a result of the true temperature of the samples lagging behind the surrounding environment temperature. The decomposition temperature ranges from 250°C to 292°C. The

Table 8.5 *Decomposition characteristics parameters of heat-treated bamboo fibers at 150°C for 2 hours [36].*

Heating rate	T_o	T_p	T_s	T_s-T_o	WL_o	WL_s	WL_s-WL_o	Residue
(°C/min)	(°C)	(°C)	(°C)	(°C)	(%)	(%)	(%)	(%)
5	249.8	329.2	351.3	101.5	8.2	58.1	49.9	27.6
10	261.3	339.7	359.8	98.5	7.5	60.3	52.8	24.2
20	271.9	353.4	374.6	102.7	7.1	61.7	54.6	23.3
30	283.4	362.2	383	99.6	7.9	60.4	52.5	23.5
40	291.9	365.9	385.9	94	7.6	57.4	49.9	23.6

values of differential temperatures between the onset and shift points (i.e., $T_s - T_o$) indicate that the main thermal decomposition fraction happens in a temperature range from 94°C to 103°C. The weight loss fraction ($WL_s - WL_o$) ranges from about 50–55% among the heating rates used. There is no notable difference among the WL_o, WL_p, and WL_s values as a result of the same chemical content.

The decomposition activation energies of various heat-treated bamboo fibers can be evaluated with the above data and degradation models including the Kissinger and FWO methods. Figure 8.7 shows a typical iso-conversional plot from the FWO method using 150°C treated bamboo fibers as an example. The conversion rates ranging from 0.2 to 0.8 levels led to plots with relatively constant slopes, and the range can be used to calculate the E_a value (about 164.8 kJ/mol in this case).

Yao et al. [23] studied thermal decomposition process of ten selected natural fibers. It was shown that all fibers had the similar TG and DTG curves as a result of being similar cellulosic material. The result shows that a stable apparent activation energy range of 160–170 kJ/mol is

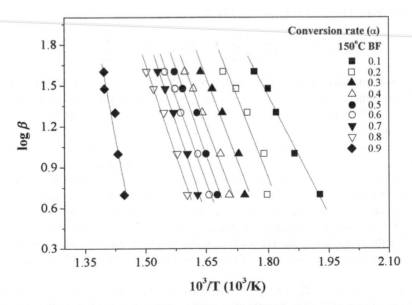

Figure 8.7 *Typical iso-conversional plots of Flynn–Wall–Ozawa method using 150°C heat-treated bamboo fibers as an example [36].*

Table 8.6 *Apparent activation energy of select fibers calculated by Kissinger method.*

Natural fiber	E_a (kJ/mol)	R^2
Bagasse	161.1	0.9986
Bamboo	161.6	0.9764
Cotton stalk	146	0.9828
Hemp	171.1	0.9994
Jute	165.6	0.9996
Kenaf	157.7	0.9971
Rice husk	167.4	0.9953
Rice straw	176.2	0.9659
Wood-Maple	153.7	0.9905
Wood-Pine	159.3	0.9970
All fiber average[a]	161.8 (8.2)	0.9910 (0.0112)

Source: Adapted from [23] with permission of Elsevier © 2008.
[a]Values from ten types of fibers with mean value and standard deviation.

suggested for the most of selected fiber throughout the polymer processing temperature range (Table 8.6). The activation energy values provide a way of ranking the thermal decomposition behavior of various fibers; and allow developing a simplified approach to predict the thermal decomposition behavior of natural fibers in relation to polymer composite processing.

The fractional conversion can be written in the following form for the isothermal TGA [37]:

$$\alpha(t) = 1 - [1 - t(1 - n)A \, e^{-x}]^{1/(1-n)} \qquad (8.8)$$

where the value of reduced activation energy ($x = E_a/RT$) is constant in the isothermal condition. Use of average values of $E_a = 116$ kJ/mol, $n = 3.01$, and $\ln A = 18.65 \ln s^{-1}$ for the selected natural fibers, the isothermal conversion curves are predicted at different temperatures (Figure 8.8 in comparison with measured sample weight loss data).

There is a quite good agreement between experimental and predicted sample weight loss data of the fibers. Similar analysis can be applied to nonisothermal degradation of natural fibers as well [38]. The models obtained can be used to provide a guideline for incorporating natural fibers into engineering thermoplastics (such as nylon, PET) for polymer composite processing.

8.3 Plastic Composite Processing Technology

The processing of fiber-reinforced plastic composites can be divided into two main steps, compounding and forming, which can be done on a continuous or a separate flow. In this process, temperature has to be sufficiently high all along to have the thermoplastic material in the molten state. This allows adequate mixing in the compounding step and shaping in the forming step. Natural fibers as fillers for polymer composites are known to degrade starting at relatively low temperature, and increasingly with higher temperatures. Thus the processing temperature has to be kept as low as possible, and the choice of thermoplastics for such composite systems is limited to these with low melting point, commonly below 180°C [39].

The purpose of compounding is to mix polymers, fillers, fibers, and additive for achieving necessary dispersion and distribution, to avoid extensive polymer filler or fiber degradation, and to produce granules with regular shape and good processability. The compounding is

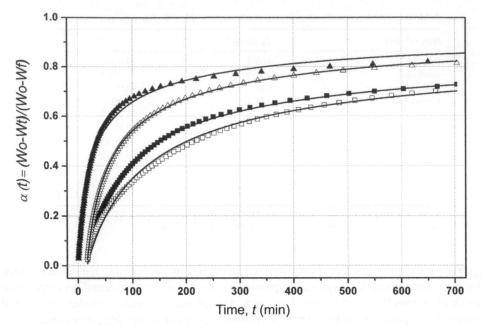

Figure 8.8 *A comparison of experimental (symbols) and simulated (solid line) isothermal α−t curves at different temperatures: (□) 235°C; (■) 245°C; (△) 255°C; (▲) 265°C. Adapted from [37] with permission of Elsevier © 2013.*

generally done in a continuous screw extruder, but is also possible in a batch mechanical blender [26, 39, 40]. Thermoforming of the final product can be performed through a die (extrusion), in a cold mold (injection molding), between calenders (calendering), or between two mold halves (compression molding). In-line processing is also possible, referring to a one-step production, in which the compounding extruder feeds directly the thermoforming machine.

8.3.1   Extrusion: A Fundamental Processing Platform

A typical screw extruder consists of feeders, modular barrels, screws, a gear box, a heating and cooling unit, and a centralized control unit for adjusting extrusion speed, feeding rate, temperature, and other process parameters [41]. The extruding screw system, including screws and barrels, is the core component of the extruder, which helps mix, devolatilize, and perform reactions for wide-range plastic applications. Barrels provide an access for feeding, venting, side-stuffing, and liquid addition. Each barrel section is heated with an individual temperature control unit, and is internally cored for high-density cooling as well. Rotating screws impart enough shear force and energy to mix components and produce a continuous blending fluid in a barrel [41].

An extruding screw is usually made up of three zones along its axial direction: (1) feeding zone, (2) melting zone, and (3) melt pumping zone [42]. In the feed zone, polymers and fillers are usually in a solid state. In the melting zone, most polymer is melted, while the fillers and

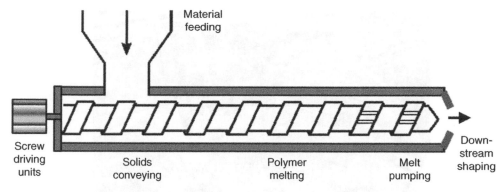

Figure 8.9 *Schematic of a single-screw extrusion system.*

part of the polymer retain in a solid state. The fillers are softened at high temperature, thus forming a plastic flow with melted polymer in the mixing chamber. Under the pumping action of the screw, the blends are compressed in each screw cell. Finally, the softened fillers are individually surrounded by the totally molten polymer, thus forming a continuous fiber–matrix blending fluid. This fluid is pumped toward the injection-molding die or the extrusion die, or connected to a pelletizer after air- or water-cooling [41].

The screw modes are normally divided into single screws and twin screws. Both systems are extensively used for processing fiber and polymer composites. In a single-screw extruder (Figure 8.9), only one screw is used inside extruder barrels. The goal of a single-screw extrusion process is to build pressure in a polymer melt so that it can be extruded through a die or injected into a mold. For single screw extruders (SSEs), the barrel-length-to-diameter ratios (L/D) range from 20 to 30 [43]. The advantages of single-screw extrusion include high pressure built-up and low cost, while its disadvantages are limited self-cleaning and mixing ability, low flexibility, and selective material intake.

Twin screws are classified into co-rotating screws and counter-rotating screws according to the rotating directions of the screws (Figure 8.10). According to the contacting situations of the two screws, twin screws are divided into separate, tangential, and intermeshing types. In

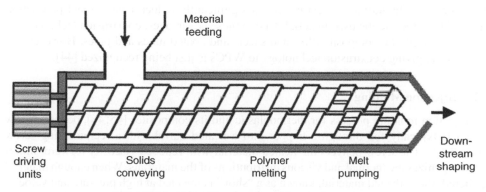

Figure 8.10 *Schematic of a twin-screw extrusion system.*

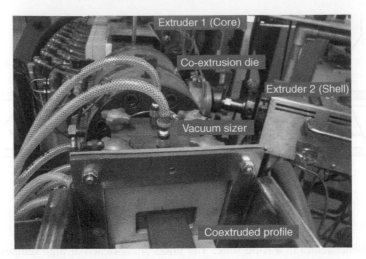

Figure 8.11   *A laboratory-scale co-extrusion system for processing fiber–polymer composites.*

addition, the configurations of the screws themselves may be varied using forward conveying elements, reverse conveying elements, kneading blocks, mixing elements, and other designs in order to achieve particular mixing characteristics. Twin screw extrusion is used extensively for mixing, compounding, or reacting polymeric materials. The L/D may range from 39 to 48 according to the modular construction [43]. The advantages of twin-screw extrusion include high mixing ability, high flexibility, high throughput, and self-cleaning, while its disadvantages are low pressure built-up, and high costs.

Coextrusion is a process in which two or more polymer materials are extruded and converged upon a single feedblock or die to form a single multi-layer structure including flat, annular, or core–shell profiles (Figure 8.11). Single-screw extruders, twin-screw extruders or their combination can be used in a co-extrusion system. It has become one of the most advanced plastic processing technologies in creating multi-layer composites with different complementary layer characteristics, and in making properties of final products highly "tunable". For example, target composite properties such as oxygen and moisture barrier, shading and insulation, and mechanical properties can be adhibited by incorporating one or more layers with target properties. In addition, coextrusion can significantly reduce material and production costs, and help recycle the used material. Due to these advantages, coextrusion technology is widely used in plastic composite films and sheets and coated tubes and pipes. However, the potential of applying coextrusion technology to WPCs is just being recognized [44].

8.3.2   Injection Molding

In an injection molding process for thermoplastic and its composites [45, 46], pelletized raw material is typically fed through a hopper into a heated barrel with a reciprocating screw (Figure 8.12). The screw delivers the heated material with reduced viscosity forward, mixes and homogenizes the thermal and viscous distributions of the materials. When enough material is gathered, the collected material, known as a "shot," is injected at high pressure and velocity into the part-forming cavity (i.e., mold). The material within the mold cools under the clamping

Injection Molding Machine

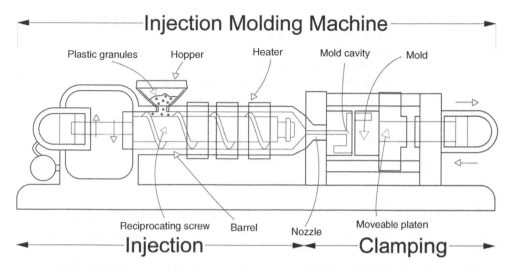

Figure 8.12 *A laboratory-scale injection molding machine (Left: mold, Middle: clamping, and Right: injection and control). Created by Brendan Rockey, University of Alberta Industrial Design, for Injection Molding Wikipedia article, with permission.*

pressure by the use of cooling lines circulating water or oil from a thermo-circulator. The screw reciprocates and acquires material for the next cycle. Once the required temperature is achieved, the mold opens and the part is de-molded. Then, the mold closes and the process is repeated.

8.3.3   Compression Molding

Compression molding is a process in which a heated polymer is compressed into a preheated mold taking the shape of the mold cavity and cured due to heat and pressure applied to the material (http://www.substech.com, [46]). During this process (Figure 8.13), a pre-weighed amount of a polymer mixed with additives and fillers (known as charge) is placed into the lower half of the mold. The charge may be in form of powders, pellets, putty-like masses or pre-formed blanks. The charge is usually preheated prior to placement into the mold. Preheated

Figure 8.13 *A laboratory compression molding machine (Left: press and right: mold)*

polymer becomes softer resulting in shortening the molding cycle time. The upper half of the mold moves downward, pressing on the polymer charge and forcing it to fill the mold cavity. The mold, equipped with a heating system, provides curing (cross-linking) of the polymer (if a thermoset is processed). The mold is opened and the part is removed from it by means of the ejector pin. If thermosetting resin is molded, the mold may be opened in the hot state–cured thermosets maintain their shape and dimensions under these conditions.

8.3.4   Thermal Forming

Thermal forming is a process used to shape thermoplastic sheet into discrete parts. The basic steps involve [47]:

• heating a thermoplastic sheet until it softens;
• forcing the hot and pliable material against the contours of a mold by using mechanical (mechanical thermoforming), air (pressure thermoforming) or vacuum (vacuum thermo-forming) pressure; and
• holding against the mold and allowing to cool, thus retaining the shape of the plastic.

Thin-gage thermoforming is primarily used for the manufacture of disposable cups, containers, lids, trays, blisters, clamshells, and other products for the food, medical, and general retail industries. Thick-gage thermoforming includes parts as diverse as vehicle door and dash panels, refrigerator liners, utility vehicle beds, and plastic pallets. The advantages of the process include low machine cost, low temperature requirement, low mold cost, low pressure requirement, large parts easily formed, and fast mold cycles. The disadvantages include high cost of raw material (sheet form), high scrap, limited part shapes, inherent wall thickness variation and high internal stress of final parts.

8.4   Overcoming Incompatibility of Synthetic Polymers and Natural Fibers

8.4.1   Introduction

Most polymers, especially thermoplastics, are nonpolar (hydrophobic) substances, which are not compatible with polar (hydrophilic) natural fibers. Poor adhesion between polymer and the fiber can result, leading to cohesion fracture of the composites as shown in Figure 8.14 [48]. In order to improve the affinity and adhesion between natural fiber and thermoplastic matrices in production, chemical coupling agents (bonding agents or compatibilizers) have been used

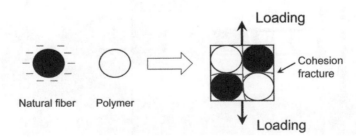

Figure 8.14   *Schematic of fiber and polymer bonding without coupling treatment.*

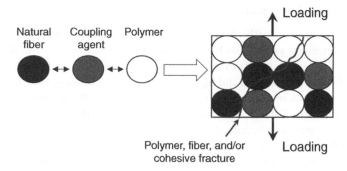

Figure 8.15   *Schematic of fiber and polymer bonding with coupling treatment.*

[49, 50]. Coupling agents act as bridges (Figure 8.15) that link the fiber and thermoplastic polymers by covalent bonding and secondary bonding, including hydrogen bonding and van der Waals forces [51, 52].

Much work has been done on coupling agents in natural fiber polymer composites [53, 54]. Several reviews have been published on coupling agents and coupling treatment for WPCs [55, 56]. Sections 8.4.2–8.4.4 provide a summary of some published work.

8.4.2   Coupling Agents: Definition

Coupling agents comprise bonding agents, compatibilizers, and dispersing agents [55]. Bonding agents act as bridges that link the fibers and thermoplastic through covalent bonding, polymer chain entanglement, and/or strong secondary interactions as in the case of hydrogen bonding [55, 56]. Compatibilizers are used to provide compatibility between immiscible fiber and polymer phases through reduction of the interfacial tension. They help lower the surface energy of the fiber, and make it nonpolar, more similar to the plastic matrix. Some bonding agents, such as maleated polypropylene (MAPP), and maleated styrene–ethylene/butylenes–styrene (SEBS-MA) also act as compatibilizers in the composites [57]. Dispersing agents reduce the interfacial energy at the fiber–matrix interface to help uniform dispersion of the fiber in a polymer matrix without aggregation and thereby facilitate the formation of new interfaces [58]. Thus, a functional distinction between bonding agents, compatibilizers, and dispersing agents should be noticed. However, all bonding agents, compatibilizers, and dispersing agents are normally lumped together as coupling agents to reduce confusions.

8.4.3   Coupling Agents: Classification and Function

Coupling agents are classified into organic, inorganic, and organic–inorganic types [55]. Organic agents include isocyanates, anhydrides, amides, imides, acrylates, chlorotriazines, organic acids, monomers, polymers, and copolymers. Inorganic coupling agents include silicates (e.g., layered montmorillonite clay). Organic–inorganic agents include silanes and titanates.

8.4.3.1   *Organic Coupling Agents*

Organic coupling agents normally have bi- or multifunctional groups in their molecular structure. These functional groups, such as (–N=C=O) of isocyanates, (–COO–) of maleic anhydrides, and (–Cl–) of diclorotriazine derivatives, interact with the polar groups (mainly hydroxyl groups (–OH)) of cellulose and lignin to form covalent bonding or hydrogen bonding [51, 59–61]. Alternatively, organic coupling agents can modify the polymer matrix by graft copolymerization, thus resulting in strong adhesion, even cross-linking, at the interface.

Anhydrides such as maleic anhydride (MA), acetic anhydride (AA), succinic anhydride (SA) and phthalic anhydride are popular coupling agents. AA, SA, and phthalic anhydride have two functional groups, that is, carboxylate groups (–COO–), which can link the fiber through esterification or hydrogen bonding. But MA is an α, β-unsaturated carbonyl compound, containing one carbon–carbon double bond (C=C) and two carboxylate groups (–COO–). This conjugated structure greatly increases the graft reactivity of the carbon–carbon double bond on the heterocyclic ring with the polymer matrix through the conjugate addition under a radical initiator, resulting in cross-linking or strong adhesion at the interface. Therefore, MA is the best coupling agent in these anhydrides. MA is also used to modify the polymer matrix by graft copolymerization. The formed copolymers, for example, MAPE, MAPP, SEBS-MA and styrene/maleic anhydride (SMA), are used as coupling agents [57, 62, 63]. A hypothetical coupling structure of MAPE is shown in Figure 8.16 [64, 65].

Similar to MA, acrylic acids and methacrylates such as methacrylic acid (MAA), methyl methacrylate (MMA), epoxypropyl methacrylate (EPMA), and glycidyl methacrylate (GMA) also contain the α, β-unsaturated carbonyl structure, which may lead to cross-linking or strong interfacial adhesion. Organic acids such as abietic acid (ABAC) and linoleic acid (LAC) contain dienes and carboxylate groups in their molecular structure, which are helpful to form strong adhesion in the interfacial region. In addition, the reactive allylic group (–CH$_2$–) in LAC might graft to the polymer matrix [66]. Lacking chemical bonding at the interface,

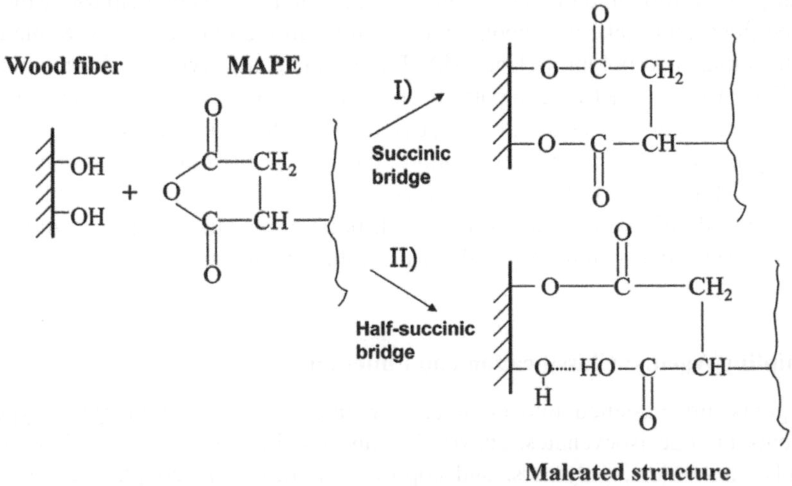

Figure 8.16   *A hypothetical coupling structure of MAPE between wood and polymer. Reprinted from [65] with permission of Taylor & Francis Ltd © 2005.*

titanium di(dioctylpyrophosphates)oxyacetate (KR 138S) and Na_2SiO_3 perform poorly [50, 67]. Na_2SiO_3 is usually required to mix with organic coupling agents.

Isocyanates as organic agents link wood fiber through the urethane structure, which is more stable than esterification [60, 68]. Due to the difference in molecular structure, the reactivity of isocyanates decreases in the following order: poly[methylene(polyphenyl isocyanate)] (PMPPIC), toluene 2,4-diisocyanate (TDIC), hexamethylene diisocyanate (HMDIC), ethyl isocyanate (EIC) [66, 69]. The delocalized π-electrons of the benzene rings in PMPPIC and TDIC lead to the stronger interaction with PS and other polymer matrices compared with HMDIC and EIC without π-electrons. Moreover, the cellulose phase and the polymer phase (PS or PVC) are continuously linked by PMPPIC at the interface, while the discrete nature of TDIC, HMDIC, and EIC make them inferior in this respect [52]. As a result, PMPPIC is the best coupling agent in these isocyanates, while TDIC has better coupling effectiveness than HMDIC and EIC.

8.4.3.2   *Inorganic Coupling Agents*

Inorganic coupling agents possibly act as a dispersing agent to counteract the surface polarity of wood fiber and improve the compatibility between wood fiber and polymer [55, 70]. Maldas and Kokta [70] demonstrated that inorganic sodium silicate in combination with PMPPIC as a coating material for wood fibers helped improve tensile and impact strengths, tensile elongation, and modulus of WPCs. The coating on the wood fiber surface helps reduce agglomeration of the hydrophilic cellulosic fibers, and leads to better dispersion of the fillers into the thermoplastic matrix. The use of nanosized, layered, silicate particles (i.e., clay) which have a larger surface area (~750 m^2/g) and a higher aspect ratio (>100) than conventional, macro-sized fillers to reinforce polymers and their composites, has drawn a great deal of attention in recent years. Adding a small amount of nanoclay can improve a number of composite properties such as stiffness and strength, thermal and dimensional stability, flame retardance, and barrier properties [71].

8.4.3.3   *Organic–Inorganic Agents*

Organic–inorganic agents are hybrid organic–inorganic compounds in structure [72]. The functionality of the organic part in these agents determines their coupling effectiveness. Organic–inorganic coupling agents are intermediate in function compared to organic and inorganic agents. Silanes, represented as $R–Si(OR')_3$, have better performance in organic–inorganic coupling agents recently used, because the attachment of silanes to hydroxyl groups of cellulose or lignin is accomplished either directly to the alkoxy group ($–OR'$) attached to silicon or via its hydrolyzed products (i.e., silanol) by the hydrogen bonds or ether linkage [73]. The functional group (R-) in silanes also influences the coupling action. Silane A-172 and A-174 both contain a vinyl group in the molecular structure; silane A-186 and A-187, an epoxy group; while silane A-1100, an amino group. When in contact with PVC, polar methacryloxy groups in silane A-174 form a polar chain that is more hydrophilic than that of A-172, resulting in poor adhesion. But for other matrices the α, β-unsaturated carbonyl structure of acrylic groups in A-174 may help form strong adhesion, even cross-linking, at the interface. Silane A-186 and A-187 with an epoxy group link cellulose and lignin by ether linkage whereas NH_2 groups of A-1100 offer mostly hydrogen bonding which is a weaker force [73].

Dichlorotriazines and derivatives have multifunctional groups in their molecular structure. These groups have different functions in the coupling reaction [59]. On the heterocyclic ring, the reactive chlorines react with the hydroxyl group (–OH) of wood fiber and give rise to the ether linkage between the cellulose phase and the coupling agent. The electronegative nitrogen may link the hydroxyl group through hydrogen bonding. On the alkyl chain, the carbon–carbon double bond (C=C) form covalent bonds with the polymer matrix by grafting. At the same time, the electronegative nitrogen in the amino groups and oxygen in the carboxylate groups also link the cellulose phase through hydrogen bonding.

8.4.4   Coupling Agents: Coupling Mechanism

Chemical coupling plays a very important role to improve interfacial adhesion for natural fiber polymer composites. Coupling efficiency varies with coupling agents and processes. Coupling agent performance is mainly influenced by acid number, molecular weight, backbone structure, and concentration of the coupling agent [64]. Coupling agents with larger molecular weight and moderate acid number and concentration have better performance at the interface. The primary bonding forms include covalent bonding (e.g., esterification), secondary bonding, and mechanical connection (Figure 8.17).

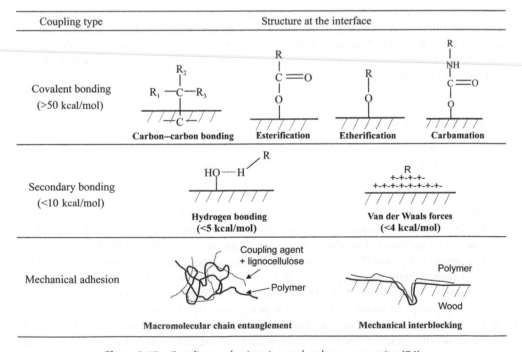

Figure 8.17   *Coupling mechanisms in wood–polymer composites [74].*

8.5   Melt Compounding Natural Fibers and Thermoplastics

8.5.1   Challenges for Melt Blending of Natural Fibers

A proper mixing of fillers with a polymer matrix is an important issue in the plastic industry. In almost all polymer processes, the quality of finished products is largely determined by the mixing quality of the filled plastics [75]. Natural fibers as a filler in polymer composites have advantages compared to inorganic fillers, including lower density, greater deformability, less abrasiveness to equipment, are renewable and biodegradable, and lower in cost [76]. However, natural fibers are significantly different from other fillers (e.g., glass and Kevlar fibers) for the reinforcement of plastic products. Most polymers, especially thermoplastics, are nonpolar (hydrophobic) substances, while natural fibers are polar (hydrophilic). These two phases are not compatible, thus resulting in poor fiber distribution during compounding. Moreover, these incompatible fiber–matrix blends also produce poor adhesion at the interface. Therefore, coupling agents, compatibilizers and/or dispersing agents are usually required for a better distribution of the fibers in the matrix and for enhanced mechanical properties of the composites [39, 55].

Compounding natural fiber or flour with polymers is a challenge. Natural material is light compared to thermoplastics and tends to bundle and bridge, making continuous and uniform metering very difficult [40]. Natural fiber is a hygroscopic material. During compounding at high temperatures, the fibers, usually at higher moisture content relative to other fillers such as glass and carbon fibers, generate water vapor. Thus, it is necessary to remove the water vapor to achieve proper polymer wetting of the fibers, good physical properties, absence of voids, and good pelletizing performance [39, 40]. Natural fibers are subject to degradation at high temperatures. The melting temperature of most thermoplastics is higher than the degradation temperature of the fibers. Thus, longer mixing time and high mixing temperatures will usually result in poor mechanical properties of the finished products. Additionally, the physical performance and cosmetic appearance of the compounded material will depend upon the extent of the fiber degradation [40]. The shape, size, species, thermal behavior, and viscoelasticity of the fibers also influence the mixing quality of fiber–matrix blends.

Blending natural fiber and plastic matrix is usually accomplished through a mixer (e.g., a screw extruder, a batch kinetic mixer, or a roll mill). The structure of mixing shafts (e.g., screws and rotors), flow profiles in the mixing chamber, the distribution of shear force, and energy of these systems will determine the mixing quality and influence the mechanical properties of the finished composite products. The introduction of natural fiber also alters the rheological properties of filled plastics as influenced by blending temperature, torque, and length, and by moisture content, surface characteristic, and weight fraction of the fibers.

8.5.2   Compounding Processes

Based on the coating and grafting methods used in processing natural fiber and polymer composites, the compounding is generally divided into three basic processes [55]:

- Coupling agents directly coated on the fiber and polymer during mixing (one-step method);
- Coupling agents coated or grafted on the surface of the fiber, polymer or both before mixing in the second process (two-step method); and
- Part of the polymer and the fiber furnish treated with a coupling agent, and then mixed with untreated fiber and polymer (one-step method).

Generally, all three processes are suitable for melt-blended composites. The one-step method is relatively simple and inexpensive. In the two-step processes, the resulting mixtures are usually ground to a certain mesh size (e.g., 20-mesh) for subsequent melt-blending formation [77]. Less coupling agent and less mixing time are required to obtain good adhesion between the fibers and polymers in a two-step process [78]. In addition, the two-step process helps increase the interface area [50, 60], thus resulting in improved mechanical properties of the composites.

The volume fraction of natural fibers in a composite may be up to 80% for melt-blending processes such as extrusion and transfer molding. Higher volume fraction of the fibers causes production problems (e.g., aggregation and blocking of the extrusion barrel). Therefore, the volume fraction of the fibers is usually controlled to approximately 50%. Coupling agents usually account for 2–8% by weight of the fibers for melt-blending formulations [60, 76]. Accordingly, coupling agents typically account for 1–3% of the total weight of a composite.

Usually, mixing temperature is controlled at less than 200°C for most coupling treatments to avoid decomposition and degradation of the fibers and some thermoplastic matrices [77, 79]. Maldas and Kokta [77] reported that the maximum improvement in mechanical properties of chemi-thermomechanical pulp (CTMP) and PS composites was achieved when the mixing time was 15 min at 175°C. For melt-blended composites, the blends are required to re-mix 5–10 times (about 6–8 min) during compounding to achieve a better distribution of coupling agents at the interface, when directly mixing coupling agents with polymer and wood fiber [60]. Rotation speed has similar influence on the coupling effectiveness as does mixing time. It was reported that moderate mixing speeds were preferred for better fiber length distribution and coupling effectiveness [79].

8.5.3   Compounding Principle

Compounding natural/wood fibers and thermoplastics is a process of "solid–solid mixing" [80]. Thermoplastics are usually in powder, granular, pelletized, or diced form, whereas wood filler is in the form of short fibers, particulate or flour. During compounding process, the fiber and polymer are "pumped" into a mixing space or unit (such as screws and the gap of rolls). Under the temperature close to the melt point of the matrix, a laminar flow of the fiber and polymer blends is formed in this mixing space. In general, this flow is nonsteady and non-Newtonian. The melted polymers randomly rotate and move along the wall of this mixing unit. Lumps of the fibers are uncoiled and elongated by laminar convective flow. The interfacial distance between solid fibers and melted polymers is decreased to some extent by shear stress. Finally, the fibers are separated and re-distributed in the matrix. The compounding process is accomplished when the dispersion of the fibers is uniform in the matrix.

The compounding process can significantly change the morphology of natural fibers from the intense mixing and shearing process. Figures 8.18 and 8.19 show a morphological comparison and actual fiber dimension distributions of sugarcane bagasse fibers before and after compounding with HDPE [27].

According to the cumulative distribution function (CDF), the sugarcane fiber distribution in the length and diameter before and after compounding fitted a lognormal distribution as follows (Equation 8.19):

$$f(X, \mu, \sigma) = \frac{1}{X\sqrt{2\pi\sigma^2}} \exp\left[-\frac{(\ln X - \mu)^2}{2\sigma^2}\right] \tag{8.9}$$

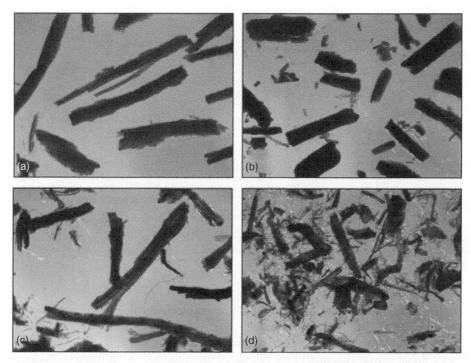

Figure 8.18   *A comparison of fiber morphology before and after compounding of the sugarcane bagasse fibers: (a) and (b) + 20-mesh raw bagasse fibers before and after compounding (20X); (c) and (d) 0 + 20-mesh extracted before bagasse fibers before and after compounding (20X). Reproduced from [27] with permission of John Wiley & Sons, Inc © 2006.*

where $f(X)$ is the probability density function (PDF) of X. The independent variable, X, is the fiber length or diameter with the unit of millimeter ($X > 0$). In Equation 8.1, μ and σ are the location and shape parameters, respectively. For both length and diameter, the sample distributions after compounding had a left shift with respect to those before compounding (Figure 8.19). For the length, the broad distribution range before compounding was narrowed to a small range after compounding (Figure 8.19a), thus the concentration of the short fibers significantly increased. For the diameter, the distribution range after compounding also shifted to the left, but was close to that before compounding (Figure 8.19b). The shift in the fiber dimension distribution was probably ascribed to the shear stresses and friction force between the fibers and HDPE resins during compounding. Hence, the amount of short fibers with a small diameter significantly increased after compounding.

8.5.4   Melt Rheological Properties

Melted fiber–matrix blends in a mixing chamber are a non-Newtonian fluid. The rheological properties of this fluid, including shear rate (γ), shear stress (τ), and viscosity (η), are usually determined by a thermodynamic rheometer [81–83]. For simulation of practical fluid behavior of natural fiber and polymer blends in a mixer, a capillary rheometer, configured in either shear

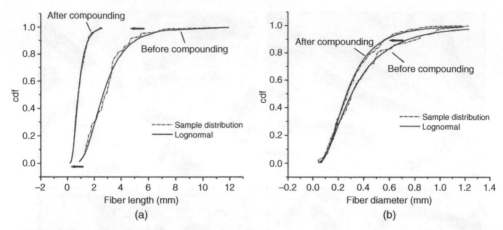

Figure 8.19   *Dimension distributions of raw bagasse fibers before and after compounding. (a) fiber length and (b) fiber diameter. Vertical axes represent the cumulative distribution function (CDF). Reproduced from [27] with permission of John Wiley & Sons, Inc © 2006.*

or elongational rheometry mode, is usually used to evaluate the rheological properties of these melts (Figure 8.20).

The shear stress at the wall (τ_w) and the apparent shear rate at the wall (γ_a) of the capillary die (Figure 8.20a—shear rheometry) are calculated using the following formulas (Equations 8.10 and 8.11), respectively:

$$\tau_w = \frac{F}{4A_p(L/D)} \tag{8.10}$$

$$\gamma_a = \frac{2}{15} \frac{V d_p^2}{D^a} \tag{8.11}$$

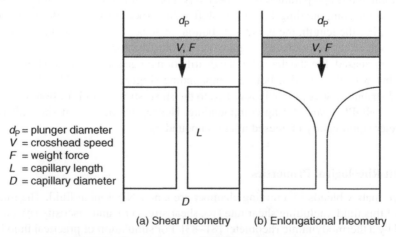

d_P = plunger diameter
V = crosshead speed
F = weight force
L = capillary length
D = capillary diameter

(a) Shear rheometry   (b) Enlongational rheometry

Figure 8.20   *A schematic of capillary rheometer heads.*

where A_p = cross-section area of the plunger (m²), D = diameter of the capillary (m), F = weight force (kg), L = length of the capillary (m), and V = crosshead speed of the plunger (m/s). The shear rate at the wall (γ_w) of the capillary is calculated from the apparent shear rate using the Rabinowitsch correction [84]:

$$\gamma_w = \frac{(3n' + 1)}{4n'}\gamma_a \qquad (8.12)$$

where n' is the flow behavior index. It can be determined by the slope of the linear plot of $\log[\tau_w]$ versus $\log[\gamma_a]$. The melt viscosity is then calculated as follows:

$$\eta = \frac{\tau_w}{\gamma_w} \qquad (8.13)$$

The melt viscosity, η, of isotactic-polypropylene (i-PP) and wood fiber composites decreases with increasing shear rate, γ_w, and at any given shear rate the viscosity increases with increase in the weight fraction of wood fibers (Figure 8.21a). However, the $\eta–\gamma_w$ plot tends to converge at very high shear rate values and wood has little effect on shear viscosity when γ_w is larger than 10,000 1/s. The dependence of η versus γ_a for wood fiber–PP melts at high weight fraction of wood fibers (45 wt%) seems also linear, when γ_a is between 1 and 1000 1/s [83].

Temperature is an important factor influencing the rheological properties of the filled polymer melt. The increase in temperature leads to decrease the viscosity of the fiber–polymer melt (Figure 8.21b) because molecular motion is facilitated at higher temperatures due to

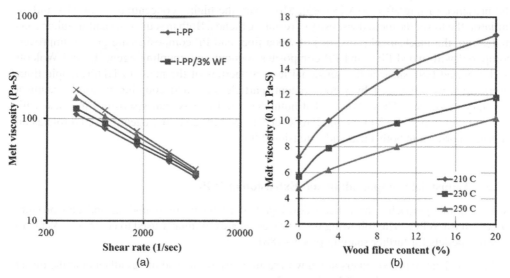

Figure 8.21 *Variation of melt viscosity with shear rate (a) and with wood fiber content (b) using a capillary die of L/D = 33.6 and applied wall shear stress range of 40–250 kPa. Reproduced from [82] with permission of John Wiley & Sons, Inc © 1989.*

availability of greater free volume. Maiti and Hassan [82] used the Arrhenius expression to relate melt viscosity with melt temperature for i-PP/WF composites (Equation 8.14):

$$\eta = A \, \exp\left(\frac{E_a}{RT}\right) \tag{8.14}$$

where A = melt material constant, E_a = activation energy, R = universal constant, and T = absolute melt temperature. The calculated activation energy values for viscous flow of i-PP/WF composites are 5.02, 5.76, 6.05, and 6.38 kcal/mol for the PP, PP/WF (3%), PP/WF (10%), and PP/WF (20%), respectively. The increased activation energy for filled blends shows that fiber–polymer melts are less temperature sensitive than unfilled polymer melts because only the polymer fraction in the blends contributes toward free volume change [82].

Collier and coworkers [85] investigated the influence of coupling agents, MA and MAPP, on the fluid characteristics for wood fiber and PP blends. The melt viscosity of wood fiber and PP blends without a coupling agent increases with the concentration of wood fiber, and the higher the concentration of wood fiber, the higher the melted viscosity of blends. While the viscosity difference between the blends with a coupling age and virgin PP significantly decreases. It implies that the compatibility and adhesion between wood fiber and PP is greatly improved by the introduction of MA and MAPP as a coupling agent. However, there is a different behavior for simultaneous and sequential maleation [85]. For sequential maleation with MAPP, the melted viscosity of blends increases with the concentration of wood fiber at a given concentration of MAPP. For simultaneous maleation, the melted viscosity of blends treated with MA decrease with the concentration of wood fiber at low shear rate (less than $0.1 \, \text{s}^{-1}$). When the shear rate is larger than 0.1, the viscosity of the melted blends decreases considerably. Therefore, simultaneous maleation has more influence on the melted viscosity of blends than sequential maleation. It is due to the degradation of PP by an initiator and the presence of ungrafted MA [85, 86]. Moreover, the higher concentrations of MA and an initiator, the lower the melted viscosity of resulting blends. Both in sequential and simultaneous maleation, the mechanical properties of wood fiber and PP composites are greatly improved compared with wood fiber and PP composites without any coupling agent. Li and Wolcott [87] reported both shear and extensional flow properties of the melts of HDPE-maple flour composites as influenced by wood content, particle size, and coupling treatment using a capillary rheometry. They showed that both shear and extensional viscosities increase with wood content up to the 70 wt% loading level. The MAPE coupling agent played a role as an internal lubricant in both shear and extensional flows.

8.5.5   Industrial Compounding and Extrusion of WPC

A variety of approaches have been developed to perform compounding and end product extrusion for WPCs. The following three processes are commonly used to compound WPC to produce pellets for further shaping process [88].

1. Pre-dried natural/wood fibers along with resin, regrinds, and additives all enter at the throat of a twin-screw extruder (TSE) (Figure 8.22A).
2. Pre-dried natural/wood fibers enter at the throat of a TSE, while resin, regrind, and additives are introduced along the TSE barrel (Figure 8.22B).

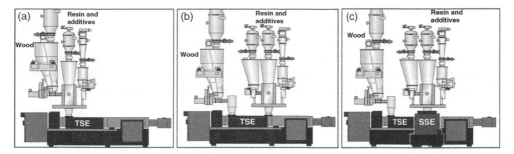

Figure 8.22   *Three approaches commonly used for pelletizing wood-plastic composites. See text for descriptions. Adapted from [88] © 2011 K-Tron.*

3. Natural/wood fibers enter at the throat of a TSE and are dried within the extruder. Resin, regrind, and additives are introduced at the throat of an SSE, which injects polymer melt along the barrel of the TSE (Figure 8.22C).

The following three processes are commonly used to compound WPC and subsequently shaping the profiles [88].

1. A continuous mixer discharges dry natural/wood fibers at the throat of a TSE, while resin, regrind, and additives are introduced along the TSE barrel. A profile die is used to produce end product profiles (Figure 8.23A).
2. A high-speed compounding extruder supplies compounded blends to multiple single or twin-screw profiling extruders to produce end product profiles. Alternatively, the compounding extruder can be used to feed a gear pump to produce the end product profile (Figure 8.23B).
3. Multi-extrusion application with wood and resin, regrind, and additives on main TSE with flanking SSEs discharging to die head (Figure 8.23C).

8.6   Performance of Natural Fiber and Plastic Composites

The use of wood-filled plastic composites as a building material or structural member requires detailed information on their engineering properties. Extensive research has been done to

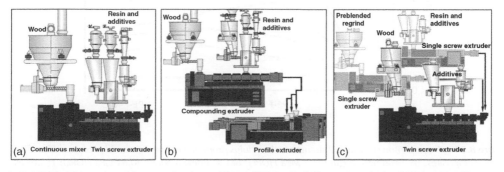

Figure 8.23   *Three approaches commonly used for pelletizing and shaping wood-plastic composites. See text for descriptions. Adapted from [88] © 2011 K-Tron.*

evaluate the effect of formulation and processing parameters on the composite performance. However, engineering performance data on wood fiber–plastic composites is still limited, inconclusive, or proprietary in nature. How different quantities of wood flour and other additives affect the structural performance of the products often remains an uncertainty associated with the composite material. This section presents a summary of mechanical, thermal expansion, biological resistance, and UV performance properties of the composites.

8.6.1   Mechanical Properties

Mechanical properties of wood/natural fiber plastic composites are highly formulation dependent. The quality and amount of base resin, wood/natural fibers, coupling agents, and other processing additives play a significant role in determining the final composite properties [89–91]. The following examples briefly illustrate the influence of composite constituents and composite structure on its mechanical properties.

A comparison of fractional strength properties of injection-molded bagasse fiber-filled HDPE composites as influenced by coupling treatments is shown in Figure 8.24. The base composite formulation (PE-F) consisted of 70 wt% HDPE and 30 wt% bagasse fiber. The formulation PE-F-M contained 70 wt% HDPE, 30 wt% bagasse fiber, and 2 wt% maleic anhydride-grafted polyethylene (MAPE). The formulation of PE-F-M-S contained 70 wt% HDPE, 30 wt% bagasse fiber, 2 wt% MAPE, and 5 wt% SEBS-MA. The loading rate for MAPE and SEBS-MA was based on the base composite (HDPE + bagasse fiber) weight. With the addition of 30 wt% bagasse fiber, bending strength (BS), bending moduluss (BM), TS, and tensile moduluss (TM) increased, while the impact strength and tensile extension

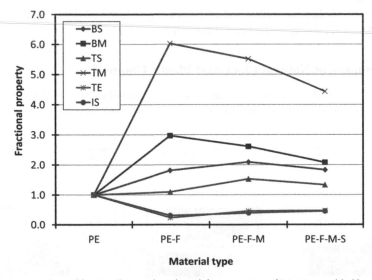

Figure 8.24   *A comparison of fractional strength and modulus properties of injection-molded bagasse fiber-filled HDPE composites. BS, bending strength (21.7 MPa HDPE); BM, bending modulus (0.64 GPa HDPE); TS, tensile strength (17 MPa HDPE), TM, tensile modulus (0.37 GPa HDPE); TE, tensile extension at break (11.7% HDPE); and IS, impact strength (12.7 kJ/m² HDPE).*

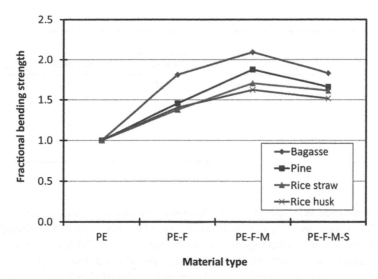

Figure 8.25 *A comparison of fractional bending strength properties of injection-molded natural fiber-filled HDPE composites as influenced by coupling treatments. BS, bending strength (21.7 MPa HDPE); BM, bending modulus (0.64 GPa HDPE).*

at break decreased. The TM and BM had the largest increase (6 and 3 times, respectively), showing a stiffening effect of the added fiber. The reduction of IS and TE indicates that the use of natural fiber increased the brittleness of the composites. The use of MAPE as coupling agent led to some reduction of TM and BM due to the low modulus value of the MAPE itself. However, the addition of MAPE led to some increase of BS, TS, TE, and IS, indicating improved compatibility between HDPE matrix and the fiber. Further addition of the impact modifier SEBS-MA led to reduction of BS, TS, BM, and TM, but some slight improvement of TE and IS.

A comparison of fractional BS of injection-molded natural fiber-filled HDPE composites as influenced by fiber type (and coupling treatment) is shown in Figure 8.25. The type of natural fibers used had a distinct effect on the BS of the composites. In comparison with pine, rice straw, and rice husk, more fiber-like bagasse fiber material performed the best in both BS and modulus.

The quality of base resin used to manufacture the composite had a large effect of the final composite properties. Table 8.7 shows a comparison of mechanical (and thermal expansion) properties of two extruded WPC formulations. The WPC-I system contained more recycled-LDPE and less virgin HDPE compared with WPC-II system, while other processing conditions remained the same. As a result of low performance properties of the recycled LDPE, both bending modulus of elasticity (MOE) and modulus of rupture (MOR) of the WPC-I system were lower than those from WPC-II system.

Co-extruded WPC with a core–shell structure has been recently developed and used to enhance performance characteristics of WPC [44, 92]. The shell layer, made of thermoplastics unfilled or filled with minerals and other additives, plays a critical role in enhancing overall composite properties. Investigations have been done to develop a stabilized shell layer by blending HDPE and additives including a compatibilizer, a photostabilizer, and a nanosized

Table 8.7 *Mechanical and thermal expansion properties of two WPC formulations.*

Composite type	Material make-up	Property		
		MOE (GPa)	MOR (MPa)	LCTE[a] (10^{-5}/°C)
WPC-I	Recycled-LDPE: HDPE: wood: lubricant: MAPE = 30: 10: 50: 6: 4 wt%	2.16 (0.05)	22.32 (0.54)	8 (0.42)
WPC-II	Recycled-LDPE: HDPE: wood: lubricant: MAPE = 10: 30: 50: 6: 4 wt%	3.23 (0.07)	31.04 (0.41)	5.50 (0.26)

[a]LCTE, linear coefficient of thermal expansion.

TiO2 on the co-extruded WPC [93] by using combined wood and mineral fillers [44]. Carbon nanotubes (CNTs) were used in the shell layer; and glass fibers and precipitated calcium carbonate (PCC) modified by silane agents have also been used [94]. With a proper combination of constituting layers, one can achieve a balance of such properties as low weight, high strength, high stiffness, wear resistance, biological resistance, unusual thermal expansion characteristics, and appearance.

Table 8.8 shows a property comparison of co-extruded WPC materials with varying glass fiber contents and thicknesses in the shell/cap layer [94]. For each of the two core composite

Table 8.8 *Mechanical properties of coextruded composites with two different core quality and shell thickness including various glass fiber content in shell layer[a].*

Shell No: GF content (%): modulus (GPa)	Shell thickness (mm)	Core: WPC-I			Core: WPC-II		
		MOE (GPa)	MOR (MPa)	LCTE (10^{-5}/°C)	MOE (GPa)	MOR (MPa)	LCTE (10^{-5}/°C)
Shell 1	0.8	1.88	21.42	8.1	2.85	31.57	5.8
0:	1	1.77	21.52	8.5	2.76	31.40	6.1
0.85	1.2	1.66	21.25	8.8	2.56	32.31	6.4
	1.6	1.56	20.99	9.2	2.27	28.51	6.5
Shell 2:	0.8	2.16	25.68	8.2	2.99	32.76	5.6
10:	1	2.07	25.49	8.3	2.94	33.76	5.8
1.30	1.2	1.96	25.36	8	2.60	32.53	6
	1.6	2.08	26.89	8.3	2.57	32.53	6.1
Shell 3:	0.8	2.38	24.68	8	3.22	34.50	5.5
20:	1	2.55	27.62	7.8	3.20	35.69	5.6
2.30	1.2	2.66	30.55	7.4	3.06	36.49	5.7
	1.6	2.70	30.49	7.7	3.24	38.24	5.8
Shell 4:	0.8	2.94	31.21	6.8	3.55	36.32	5.5
30:	1	3.13	34.48	7	3.66	38.90	5.4
3.60	1.2	3.34	36.62	6.2	3.80	43.14	5.2
	1.6	3.83	39.34	6	4.11	46.85	4.8
Shell 5:	0.8	3.23	32.65	6.3	3.97	41.86	4.6
40:	1	3.44	35.65	5.5	4.06	43	4.7
5.80	1.2	3.70	36.49	5.7	4.13	44.87	4.5
	1.6	4.66	42.49	4.5	4.95	48.59	3.8

[a]Core Formulations—WPC I- Recycled-LDPE: HDPE: wood: lubricant: MAPE = 30: 10: 50: 6: 4 wt%; WPC II- Recycled-LDPE: HDPE: wood: lubricant: MAPE = 10: 30: 50: 6: 4 wt%. LCTE—Linear coefficient of thermal expansion.

systems, composite modulus increased with the increase of shell modulus at a given shell thickness. When the shell modulus was less than the core modulus, the increase of shell thickness led to reduced composite modulus. For example, when the shell thickness was doubled from 0.8 to 1.6 mm, modulus for composites with pure HDPE-shell was reduced from 1.88 to 1.56 GPa for weaker core (WPC-I), and from 2.85 to 2.27 GPa for moderate strength core (WPC-II), respectively. Thus, a thinner shell should be preferred for this system. When the shell modulus was higher than the core modulus, the opposite is true. The increase of shell thickness led to increased composite modulus. For example, when the shell thickness was doubled from 0.8 to 1.6 mm, modulus for composites with 30% glass fiber-filled shell was increased from 2.94 to 3.83 GPa for weak core, and from 3.55 to 4.11 GPa for moderate core, respectively. Thus, thicker shells should be preferred for this system.

Engineering plastics such as polyamide 6 and poly(ethylene terephthalate) (PET), offer excellent modulus, toughness, and thermal stability. Using engineering plastics as matrix to enhance WPCs was tried as early as 20 years ago [48]. However, severe thermal degradation happens to reinforcing cellulosic fibers at temperatures needed to process engineering thermoplastics with a melting temperature over 220°C [95]. Techniques of using purified pulp fibers with a cellulose content greater than 80%, and adding the fibers into pre-melted polymer (nylon) at downstream of a twin-screw extruder has also been attempted [95] with moderate success. Unfortunately, the cellulose content of unprocessed wood/natural fibers is generally lower than 50% and the purification process adds significant costs to the manufacturing process. A new kind of *in situ* polymer blends, designated as microfibrillar blends (MFBs), provides another promising route to enhance properties of general thermoplastics using engineering thermoplastics [96, 97]. Using this concept, wood flour was successfully mixed into the pre-prepared MFBs of HDPE/PET through a two-step extrusion process as shown in Figure 8.26 [98]. During the first extrusion, the HDPE-PET MFB pellets were made at the melting temperature of PET (extrusion temperature profile = 130, 150, 190, 240, 270,

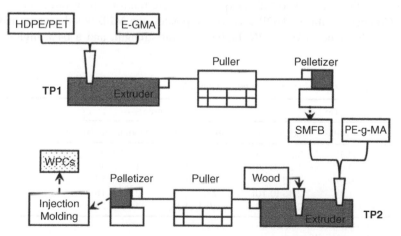

Figure 8.26 *Schematic illustration of process for preparing wood-plastic composites based on HDPE/PET MFBs. Temperature profile 1 (TP1): 130, 150, 190, 240, 270, 270, 270, 250, 250, and 250°C; temperature profile 2 (TP2): 130, 150, 160, 170, 190, 180, 170, 170, 160, and 160°C. Adapted from [98] with permission of Elsevier © 2011.*

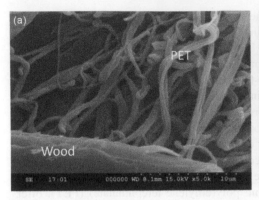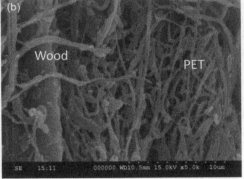

Figure 8.27 *Morphology of the dispersed PET phase in HDPE/PET/wood (45/15/40 w/w) composites. (a) No E-GMA and (b) 2.5% E-GMA. The fracture surfaces were etched in hot xylene for 15 min. Adapted from [98] with permission of Elsevier © 2011.*

270, 270, 250, 250, and 250°C from the hopper to a strand die). The extrudates were continuously hot stretched using a puller and a strand pelletizer and fine PET fibers were formed in the strand (Figure 8.27). During the second extrusion, the formed MFB pellets were melt blended with dried wood flour using the same extruder and a temperature profile of 130, 150, 160, 170, 190, 180, 170, 170, 160, and 160°C. Since the extrusion temperature was lower than the melting temperature of PET, the formed PET fibers were maintained in the matrix (Figure 8.27a). The use of a coupling agent like ethylene–glycidyl methacrylate (E-GMA) copolymer helps enhance their bonding to the HDPE matrix (Figure 8.27b).

The inclusion of PET fibrils (less than 500 nm) in the blend helped increase mechanical properties of the blend, especially the moduli (Table 8.9). The addition of 40 wt% wood flour during the second extrusion process did not influence the size and morphology of PET fibrils. The fibrils and wood fibers had a synergic reinforcement effect on composite properties. Compared with the HDPE/wood composites, the MFB/wood system had a 65% higher TS, a 95% higher TM, a 42% higher flexural strength, and a 64% higher flexural

Table 8.9 *Comparison of mechanical properties of HDPE, MFB, and their composites.*

System[a,b,c]	Flexural[d]		Tensile[d]		Impact strength[d] (kJ/m^2)
	Strength (MPa)	Modulus (GPa)	Strength (MPa)	Modulus (GPa)	
H	22.4 (0.3)	0.72 (0.03)	21.9 (0.2)	1.02 (0.09)	27.2 (1.1)
M	36.2 (0.6)	1.28 (0.06)	26.5 (0.5)	1.50 (0.39)	4 (0.3)
M/E	34.6 (0.9)	1.15 (0.02)	27.3 (0.6)	1.20 (0.14)	5.8 (0.2)
H/W/P	43 (1.4)	2.15 (0.10)	23.7 (1.9)	2.27 (0.16)	3.8 (0.2)
M/E/W/P	60.9 (0.6)	2.94 (0.08)	39.1 (0.2)	4.43 (0.27)	2.9 (0.2)

Source: Adapted from [98] with permission of Elsevier © 2011.
[a] H, HDPE; M, MFB; P, PE-g-MA; W, wood
[b] MFB, HDPE/PET (75/25 w/w); MFB/E-GMA, HDPE/PET/E-GMA (75/25/2.5 w/w)
[c] PE-g-MA, 1 wt% based on the total weight of plastics and wood flour added; and 40 wt% based on the total weight of plastic and wood flour.
[d] The values in parentheses are standard deviation.

modulus, respectively. The technology offers a way to use engineering plastics such as PET for high-performance WPC manufacturing.

8.6.2 Thermal Expansion Properties

A growing demand for structural WPC elements characterized by low thermal expansion has recently appeared in applications such as curved structure design, where minimum dimension movement is required. Thus, thermal expansion and contraction caused by external temperature variations are considered some of the most important performance properties for WPC [99, 100].

Thermal expansion and contraction of materials are usually measured by the linear coefficient of thermal expansion (LCTE). The LCTE (α_L, 1/°C) is calculated as shown in Equation 8.15,

$$\alpha_L = \frac{1}{L}\frac{dL}{dT} \tag{8.15}$$

where L is the linear dimension of the test sample and dL/dT is the rate of change in the linear dimension per unit temperature. The LCTE is a vital engineering property for the composites' structural applications. A low LCTE value is a desirable property in order to achieve dimensional stability. LCTE of crystalline polymers depends on various parameters, including processing conditions, thermal history, and degree of crystallinity and crystallite size. Filler-aspect ratio, volume fraction, orientation, and distribution in the matrix also play a deciding role in determining the LCTE of composites.

Table 8.10 shows typical LCTE values of common fillers, plastics, and filled plastic composites. LCTE values for wood (along the grain) and mineral (e.g., calcium carbonate) fillers are about 20 times lower than those of plastics and about 10 times lower than those of WPCs [100]. LCTE measured across the grain (radial and tangential) is proportional to wood specific gravity and is about 5 to 10 times higher than those along the grain. Thus, the use of filler can help reduce LCTE of WPCs. However, the effect of fillers on the LCTE is far from being

Table 8.10 *Typical linear coefficient of thermal expansion (LCTE) values of common fillers, plastics, filled plastic composites.*

Materials	LCTE (10^{-6} m/m°C)
Wood, oak parallel to grain	4.6–5.9
Wood, oak across to grain	30.9–39.1
E-glass	5
Mica	3
Polyvinyl chloride (PVC)	50.4
Polyethylene (PE)	
High-density PE	109–198
Low-density PE	180–400
Ultrahigh molecular weight PE	230–360
Polypropylene (PP)	100–200
Nylon, type 6/6, molding compound	80
Polypropylene—glass-fiber reinforced	32
Polyethylene—glass-fiber reinforced	35.8

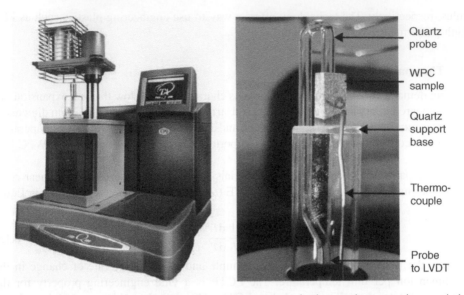

Quartz
probe

WPC
sample

Quartz
support
base

Thermo-
couple

Probe
to LVDT

Figure 8.28   *LCTE measurement with a TMA (left: TA Q400 TMA, and right: sample stage with a wood-plastic composite sample).*

understood. It appears that the largest effect is caused by the degree of anisotropy of the filler and the filler orientation in the flow (and in the final product). The characteristics of polymer (e.g., density and degree of crystallinity) and extrusion conditions (e.g., speed, pressure, and temperature profile) are also important for controlling the LCTE values [100].

Tests for determining LCTE can be done with small samples using a thermomechanical analyzer (TMA). Figure 8.28 shows a TA Q400 TMA system (left). The sample is placed on a quartz base (Figure 8.28, right) and an extension quartz probe is then placed on the top surface of the sample. A loading of 5 g force is normally applied to the probe to ensure a proper contact of the probe and the sample. The change in the length of the sample with temperature is measured using a linear variable differential transformer (LVDT) with a sensitivity of ± 0.02 μm. The length and temperature data are recorded and analyzed with TAs Universal Analysis software.

Typical dimensional changes from TMA measurements in relation with temperature change for neat HDPE and bamboo fiber-filled HDPE composites are shown in Figure 8.29. The sample dimension increased as the temperature increased and decreased as the temperature decreased. The LCTE is represented by the slope of the linear portion of each curve. The neat plastic had an obvious larger dimensional change than that of filled composites for a given temperature change. A significant residual deformation is seen for the neat plastic at the end of the final heating cycle. The dimensional change of the neat resin seemed to be more significant at a lower temperature range ($<20°C$) than that at a higher temperature range ($>20°C$), compared with filled composites. This could be due to complex stress development in the mixed plastics during the cooling process. Composites with bamboo fillers had much smaller dimensional changes, showing the restricting effect of the fillers in response to temperature changes. And residual deformation of filled composites was also significantly lowered.

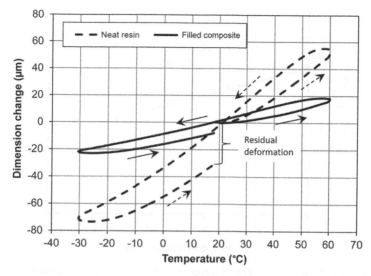

Figure 8.29 *Typical dimensional changes from TMA measurements in relation with temperature change for neat recycled PP/HDPE resin and bamboo fiber-filled composites. Composite formulation: 60 wt% resin and 40 wt% fiber.*

Figure 8.30 shows LCTE as a function of wood flour loading level in compression-molded WPCs (Vos 1998). The LCTE was measured using larger samples. The mean LCTE for specimens decreased by 70% as wood fiber content increased from 0% to 60% [101].

The use of mineral filler can also help lower the thermal expansion coefficient of WPCs. Figure 8.31 shows LCTE plots for the composite systems with hybrid bamboo (40%) and PCC

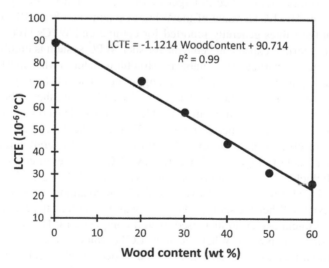

Figure 8.30 *Effect of wood contents on thermal expansion (17–60°C) of wood flour-filled recycled-PP&PE composite systems. Line shows linear regression fit [101].*

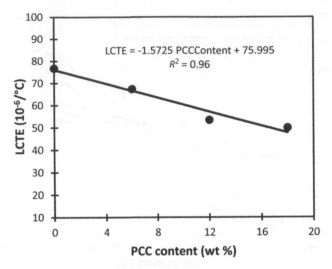

Figure 8.31   *Effect of PCC contents on thermal expansion (20–60°C) of bamboo flour-filled recycled-PP&PE composite systems (40% bamboo filler); Line shows regression fit.*

fillers. The measured LCTE values decreased linearly with increased PCC content. At the 18% PCC loading levels, the LCTE reduction for the temperature ranges of 20–60°C is 34.9%. This is attributed to a small LCTE value of PCC itself and reduced overall plastic volume in the composite.

The quality of base resin can also influence the LCTE value of WPC. Table 8.7 shows a comparison of LCTE values for two extruded WPC formulations. The use of more recycled-LDPE in the WPC-I system, which led to poor bonding among recycled-LDPE, HDPE, and wood fibers in the composite, resulted in larger thermal expansion for the system compared with WPC II system. The LCTE value of the WPC-I system (i.e., $8 \times 10^{-5}/°C$) is significantly higher than the values generally reported for commercial WPCs [100]. Thus, further improvement of the thermal expansion behavior for such WPC system is highly needed.

Layered composites are particularly suitable for structural material when low thermal expansion properties are required [102]. Some high-strength fiber materials (e.g., glass fibers, GFs) show a very low or even negative LCTE. Therefore, by embedding these fibers in the composite matrix, which has a high and positive LCTE, it is possible to obtain a material with satisfactory mechanical characteristics and low LCTE [103]. The developed materials can be then used as shell or core layers in co-extruded WPC to control overall composite thermal expansion (Figure 8.32, left). Figure 8.32 (right) shows LCTE values of co-extruded WPC as a function of shell thickness and modulus for a core system with $E = 2.26$ GPa and LCTE = $8 \times 10^{-5}/°C$. The LCTE values of the co-extruded WPC varied almost linearly with changes in the shell layer thickness. When the shell LCTE value was greater than the core LCTE (S1/S2), an increase in shell thickness led to increased overall composite LCTE. As such, thin shells are preferred for such composites. When the shell LCTE values were smaller than the core LCTE (S3/S4/S5), an increase in shell thickness led to decreased overall composite LCTE. Thick shells are preferred for minimizing the overall composite LCTE for these composites. The composite LCTE values converged to the core LCTE value when the shell thickness

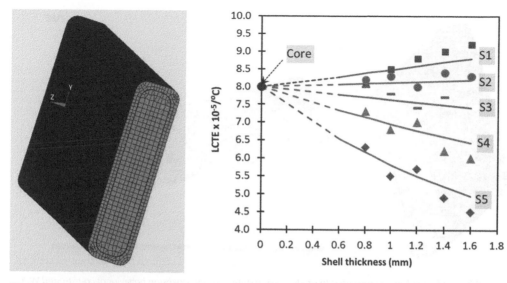

Figure 8.32 *LCTE values of co-extruded WPC (a) as a function of shell thickness and modulus for a core system with E = 2.26 GPa and LCTE = 8 × 10⁻⁵/°C. (b) Lines show predicted values by a finite element model. Shells S1, S2, S3, and S4 contain 0, 10, 20, 30, and 40% short glass fibers with 0.85, 1.30, 2.30, 3.60, and 5.80 GPa, respectively [104].*

approached zero (i.e., core-only composites). The use of strong shells with low LCTE values (e.g., shell 5 with 40% GF) helped lower overall composite LCTE, especially for the system with a large LCTE value.

The thermal expansion behavior of laminated composites can be analyzed following an approach based on classical lamination theory, CLT [102, 105, 106]. However, the CLT cannot be directly applied to co-extruded composite material with a fully capped shell layer, due to the restraining effect of the composite shell. Figure 8.33 shows a finite element model used to predict the steady state thermal expansion behavior of co-extruded WPC [104]. The predicted lines (Figure 8.32b) compare well with the experimental values for various shell systems. The model provides a way to optimize raw material composition to minimize the thermal expansion behavior of co-extruded WPC.

8.6.3 Biological Resistance Properties

Since WPC formulations contain wood and/or natural fibers (a nutrient source), this material is susceptible to decay fungi, molds, algae, termites, and marine borers. Extensive tests have been done with decay resistance of WPC products based on weight loss and mechanical property loss from the laboratory (AWPA E10) and field (AWPA E7) tests. It has been shown that high fiber content and large fiber size led to more weight and strength loss from the decay fungi [107,108]. Schirp et al. [108] compared decay resistance of two WPC formulations with yellow poplar and redwood using *Tinea versicolor* and *Gloeophyllum trabeum* fungi (Table 8.11).

Tinea versicolor caused a weight loss of approximately 56% in yellow poplar, 3% in redwood, 1% in WPC A (49% wood loading), and 6% in WPC B (70% wood). *Gloeophyllum*

(a)

(b)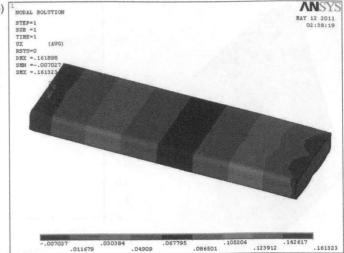

Figure 8.33   *A finite element mesh used to predict steady state thermal expansion behavior of co-extruded WPC (a) and predicted thermal strain field (b) [104].*

trabeum caused a weight loss of approximately 25% in yellow poplar, 3% in redwood and yellow poplar, but no noticeable weight losses in either WPC formulation. Maple filler was used in the WPC formulations, which did not represent an ideal substrate for *G. trabeum*. Thus, the decay performance of WPC material is strongly formulation dependent [108].

Termite test data is more limited for WPC [109]. The USDA FPL field test of extruded WPC showed visible termite nibbling on in-ground stakes after 3-year exposure in Saucier, Mississippi [107]. The Michigan Tech field test of extruded WPC in Hilo, Hawaii showed little termite activity on WPC test blocks over a 27-month period [108]. To date, no standards have been written for the testing of termite resistance for WPC. However, AWPA E1 tests were designed to test termite resistance of solid wood. Although it was not developed for testing WPC material, the standard has been used by the industry to evaluate these materials.

Table 8.11   *Weight loss (based on wood fraction) of two WPC formulations, yellow poplar, and redwood, following 3 months of incubation with Tinea versicolor and Gloeophyllum trabeum in agar (Each value represents the average of 16 replicates).*

Materials	Fungi	Density (kg/cm^3)	Weight loss (%)
A: Wood/HDPE/additive: 49/45/6	*T. versicolor*	1.05 (0.01)	1.17 (1.25)
	G. trabeum	1.07 (0.01)	−1.41 (0.13)
B: Wood/HDPE/additive: 70/24/6	*T. versicolor*	0.99 (0.02)	6.32 (0.47)
	G. trabeum	0.98 (0.01)	0.38 (0.74)
Yellow poplar	*T. versicolor*	0.21 (0.03)	55.94 (5.59)
	G. trabeum	0.40 (0.05)	25.21 (5.83)
Redwood	*T. versicolor*	0.44 (0.03)	3.03 (0.51)
	G. trabeum	0.45 (0.01)	2.76 (1.56)

Source: Adapted with permission from Reference 108. Copyright 2008 American Chemical Society.

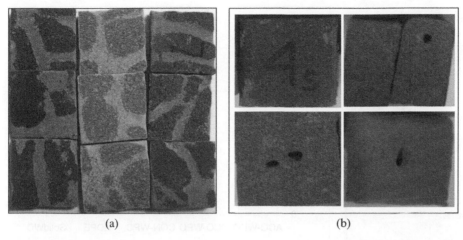

(a)   (b)

Figure 8.34 *Typical termite damage modes of WPC samples from AWPA E1 test. (a) Laboratory-made WPC; (b) commercial WPC.*

Figure 8.34 shows typical damage mode for various samples after the AWPA E1 test ((a) laboratory-made panel; (b) commercial panels) [109]. Damage on unexposed, as-extruded surfaces is mainly nibbling. The nibbling helps expose more fibers and leads to more attack, especially for materials made with larger particle sizes. Thus, the extruded surface provides some protection for the material from termite attack. A breaking of the surface layer due to surface nibbling, weathering, machining, and other damage can lead to more termite attack on the material. Damage on machined, interior surface is by both nibbling and through-holes as shown. AWPA E1 calls for a sample thickness of 6 mm, while most commercial WPC is made with greater thickness. Thus, a decision is needed on the sample thickness, as machining of WPC has a large influence on the test result. Termites attacked machined sides for samples with as-extruded thickness (e.g., 17 mm). Termite mortality rate, sample weight loss, and sample damage rating from the AWPA E1 test for WPC are strongly formulation dependent. WPC material treated with an effective preservative system should be able to resist termite attack on both as-extruded exterior surface and machined interior surface.

Pressure-treated wood is widely used for durable outdoor applications. However, there are some environmental concerns when wood is pressure-treated. The proper disposal of the treated wood after its service life poses a significant industrial problem. Recycling treated wood fiber into WPC offers advantages in recovering valuable wood resources and in helping to create WPC products that are less bio- and photodegradable. Previous work in the field has been limited to chromated copper arsenate (CCA)-treated wood under compression molding [110]. A large quantity of alkaline copper quaternary (ACQ)- and micronized copper quaternary (MCQ)-treated wood is available. Successful development of WPCs from pressure-treated wood materials requires detailed information about the manufacturing variables and understanding of the coupling system and coupling efficiency between treated wood fiber and plastics in the composite.

HDPE-based composites with ACQ- and MCQ-treated wood fibers were manufactured through injection molding [111]. Decay tests with the brown rot fungus *G. trabeum* (ATCC

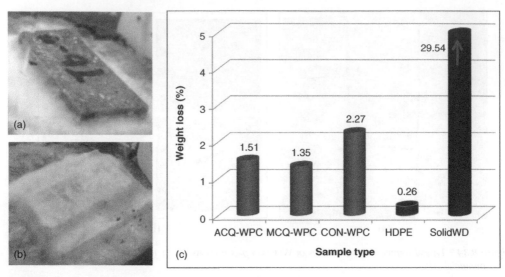

Figure 8.35 *Photographs of decay samples. (a) MCQ-WPC; (b) CON-WPC; and (c) WPC sample weight losses of different composite samples after decay test with Gloeophyllum trabeum.*

11539) were performed in accordance with the AWPA Standard Method of Testing Wood Preservatives by Laboratory Soil-Block Cultures (E10-06). The samples were exposed to the fungus for 16 weeks. For each type of the sample, three replications were conducted, and the mass loss data was collected after the close of the test. The average mass losses for ACQ-treated wood-HDPE (ACQ-WPC), MCQ-treated wood-HDPE (MCQ-WPC), untreated wood-HDPE (CON-WPC), pure HDPE (HDPE), and the solid wood control (SolidWD) are presented in Figure 8.35. The ACQ- and MCQ-treated wood-HDPE composites had lower mass losses than the untreated wood-HDPE composite, indicating the enhanced decay-resistant performances of polymer composites from pressure-treated wood materials.

Chemical leaching from WPC made of treated wood fibers is a concern for practical use of the material [110]. Co-extruded materials with a fully capped structure (Figure 8.36, left) can help lower the leaching rate. The solid shell/cap layer with no or less wood loading prevents the core layer from leaching chemical from treated wood fibers (Figure 8.36, right), leading to long-term protection of the product. By using a shell layer, water absorption and thickness swelling of the co-extruded samples can also be significantly reduced in comparison with pure core layer [44].

8.6.4   UV Resistance Properties

Wood/natural fiber plastic composites are susceptible to color change and weathering resulting from ultraviolet (UV) radiation [112]. The radiation of the sun, particularly the UV portion, is mainly responsible for limiting the lifetime of WPCs exposed to the environment. UV radiation is extremely destructive to polymeric materials and the energy in UV radiation is strong enough to break molecular bonds of the materials. This behavior in the polymeric materials brings

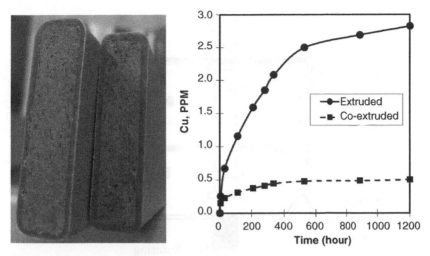

Figure 8.36   *Co-extruded WPC with MCQ-treated wood fibers (a) and Cu leaching data for the extruded and co-extruded composites (b).*

about thermal oxidative degradation, which results in embrittlement, discoloration, or overall reduction in mechanical properties.

For polyethylene, research has shown that the carbonyl groups are the main UV light-absorbing species responsible for its photo initiation reactions [113]. In one proposed reaction, formation of free radicals occurs in the presence of UV light, and the free radicals attack the polymer. The attack may lead to termination through cross-linking or chain scission [114]. In another possible reaction mode, both carbonyl and vinyl groups under UV lights are produced and chain scission occurs [114]. The yellowing of wood is mainly due to the breakdown of lignin to water-soluble products, leading to the formation of carboxylic acids, quinones, and hydropcroxy [115]. In understanding the reaction pathways for wood photo-degradation, it was suggested that the hydroquinones and paraquinones are a redox couple [116]. Under irradiation, the process begins with oxidation of the hydroquinones to form paraquinones, which are further reduced to hydroquinones as the degradation cycle repeats [113].

Test devices with artificial light sources are generally used to accelerate the photodegradation process [117]. The light sources in the test devices include filtered carbon arcs, filtered xenon arcs, and metal halide or fluorescent UV lamps. The xenon arc is strongly preferred as a light source to approximate natural sunlight, because the spectral output of the xenon lamp closely matches the full spectrum of sunlight radiation in the UV, visible, and infrared (IR) wavelengths. Xenon has been extensively adopted by the automotive, polymer additive, and textile industries in particular, and is generally replacing carbon arc technology. Figure 8.37 shows a schematic of a rotating xenon arc weathering and light stability tester and Figure 8.38 shows mounted WPC samples for testing. This device features a central light source, positioned vertically, with a filter system surrounding it, and a controlled water spray. The test samples are mounted facing the light on the rotating drum. A water cooling system for its lamp is used for the most popular xenon arc system.

The color change of the samples is normally measured using the CIELAB color system. Lightness (L) and chromaticity coordinates (a and b) are measured for replicate samples [112].

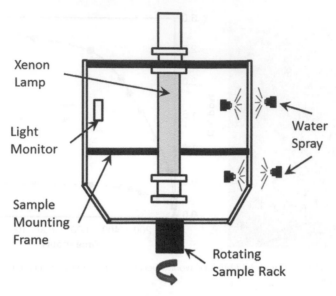

Figure 8.37   *A schematic of a rotating xenon arc weathering and light stability tester.*

The color change (ΔE_{ab}) is determined by Equation 8.16 with the procedure outlined in ASTM D 2244 (ASTM 2011):

$$\Delta E_{ab} = \sqrt{\Delta L^2 + \Delta a^2 + \Delta b^2} \tag{8.16}$$

where ΔL, Δa, and Δb represent the difference between the initial and final values of L, a, and b, respectively. In the CIELAB color system, the value L is a lightness factor. An increase

Figure 8.38   *WPC samples mounded on a rotating drum in a xenon arc weathering and light stability tester.*

in L means the sample is lightening (i.e., a positive ΔL for lightening and a negative ΔL for darkening). A positive Δa represents a color shift toward red, and a negative Δa represents a color shift toward green. A positive Δb represents a color shift toward yellow, and a negative Δb represents a color shift toward blue.

UV stabilizers protect WPCs from UV radiation present in sunlight. UV stabilizers are organic compounds which are radical scavengers or hydroxyperoxide decomposers, the most prominent of which are hindered amines, called hindered amine light stabilizers (HALSs). These additives protect WPCs both during processing and for the service life of the product and their use has improved aging process and appearance of most WPCs. The HALSs have been used for protecting polyolefin as free radical scavengers. It was reported that they photostabilize unfilled HDPE and PP. UV absorbers (UVA) are also used as photostabilizer in polyolefins such as HDPE and PP and their protection mechanism is based on the absorption of harmful

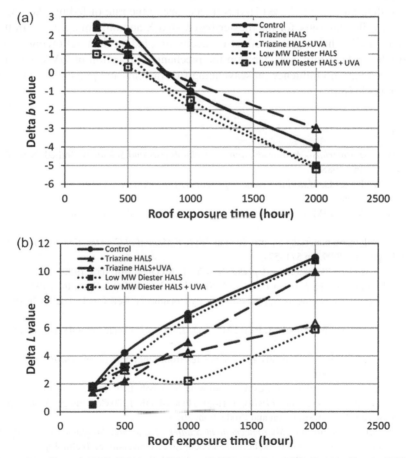

Figure 8.39 The effect of HALS (Triazine—Chimassorb 119 and low MW diester—Tinuvin 770) and UVA (benzotriazole—Tinuvin 360) on the Δb value (a) and ΔL value (b) of injection-molded wood-plastic composites [113]. Base WPC formulation = 40 wt% wood flour, 57.5 wt% HDPE, and 2.5 wt% coupling agent. HALS and UVA = 0.15 and 0.2 wt% of base composite weight. Adapted from [113] with permission of Elsevier © 2006.

UV radiation and its dissipation as heat. However, the photodegradation mechanism of WPCs is complicated because each component, namely wood flour and polymer may be means of a different mechanism. The mechanical properties of WPCs depend on the interaction and adhesion between the polymer matrix and the wood filler. Degradation of the interface caused by weathering may reduce the ability of the composite to effectively transfer stress between these components, resulting in lowered mechanical properties.

Figure 8.39 shows the effect of HALS (Triazine—Chimassorb 119 and low MW diester— Tinuvin 770) and UVA (benzotriazole–Tinuvin 360) on the Δb value (a) and ΔL value (b) of injection-molded WPCs [113]. Base WPC formulation included 40 wt% wood flour, 57.5 wt% HDPE, and 2.5 wt% coupling agent. HALS and/or UVA were added at 0.15 wt% and 0.2 wt% levels of base composite weight. It was shown that the addition of a benzotriazole UVA to both the triazine and low molecular weight diester-based HALS ultimately decreases the rate and degree of fading compared to the non-UVA stabilized triazine and diester samples. Also, the low molecular weight diester-based HALS seems to benefit from the addition of a UVA slightly more than the triazine-based HALS in decreasing the rate of fading. Adding a UVA also decreases yellowing only for the diester-based HALS and only marginally. In this way, the addition of a UVA has a greater effect on decreasing fading than yellowing. This result is expected when taking into consideration the mechanism of action of UVA which simply absorbs light and have no influence on scavenging free radicals and preventing the formation of chromophores [118].

References

1. Hall C. *Polymer Materials—An Introduction for Technologists and Scientists*, 2nd ed. New York: John Wiley & Sons, Inc; 1988.
2. Osswald TA, Menges G. *Materials Science of Polymers for Engineers*, 2nd ed. Cincinnati: OH: Hanser Gardner Publications; 2003.
3. Painter PC, Coleman MM. *Fundamentals of Polymer Science*, 2nd ed. Lancaster, PA: Technomic Pub Co; 1997.
4. Koutsos V. Polymeric materials: an introduction. In: *ICE Manual of Construction Materials*. Institution of Civil Engineers; 2009. pp 571–577.
5. Martino R.L. *Modern Plastics Encyclopedia*. New York: McGraw Hill, Inc.; 1989.
6. Brandrup J, Immergut EH. *Polymer Handbook*. New York: John Wiley & Sons, Inc.; 1975.
7. TA Instrument I. TA Instruments Application Library Documents. New Castle, DE TA Instrument Inc.; 2012.
8. Wilkes CE, Summers JW, Daniels A, Beard MT. *PVC Handbook*. Cincinnati, OH: Hanser Publications; 2005.
9. Yu L, Dean K, Li L. Polymer blends and composites from renewable resources. *Progress in Polymer Science*, 2006;31:576–602.
10. Almut J, Gunther J, Raschka A, Carus M, Piotrowski S, Scholz L. *Use of Renewable Raw Materials with Special Emphasis on Chemical Industry*. Hojbro Plads 4, DK-1200 Copenhagen K: European Topic Centre on Sustainable Consumption and Production; 2010.
11. Galland S. *Microstructure and Micromechanical Studies of Injection Moulded Chemically Modified Wood/Polylactic Acid Composites*. Skelleftea, Sweden: Lulea University of Technology; 2009.
12. Lim LT, Auras R, Rubino M. Processing technologies for poly(lactic Acid). *Progress in Polymer Science*, 2008;33(8):820–852.
13. Drumright RE, Gruber PR, Henton DE. Polylactic acid technology. *Advanced Materials*, 2000;12:1841–1846.

14. Garlotta D. A literature review of poly(lactic acid). *Journal of Polymers and the Environment*, 2002;9(2):63–84.
15. Tsuji H, Fukui I. Enhanced thermal stability of poly (lactide)s in the melt by enantiomeric polymer blending. *Polymer*, 2003;44:2891–2896.
16. Ikada Y, Jamshidi K, Tsuji H, Hyon SH. Stereocomplex formation between enantiomeric poly(lactides). *Macromolecules*, 1988;20:904–906.
17. Tsuji H, Ikada Y. Stereocomplex formation between enantiomeric poly(lactic acid)s: XI-mechanical properties and morphology of solution-cast film. *Polymers*, 1999;40:6699–6708.
18. Tsuji H. Autocatalytic hydrolysis of amoprhous-made polylactides: effects of L-lactide content, tacticity, and enantiomeric polymer blending. *Polymer*, 2002;43:1789–1796.
19. Tsuji H, Ikada Y. Stereocomplex formation between enantiomeric poly(lactic acid)s: 6-binary blends from copolymers. *Macromolecules*, 1992;25, 5719–5723.
20. Kozlowski R. Wladyka-Przybylak M. Uses of natural fiber reinforced plastics. In: Wallenberger FT, Weston N, editors. *Natural Fibers, Plastic and Composites*. Norwell, MA: Kluwer Academic Publishers; 2004. pp 249–271.
21. Bledzki AK, Gassan J. Composites reinforced with cellulose based fibers. *Progress in Polymer Science*, 1999;24:221–274.
22. Clemons CM. Wood flour. In: Xanthos M, editor. *Functional Fillers for Plastics*. Weinheim, Germany: Wiley-VCH Verlag GmbH & Co; 2010.
23. Yao F, Wu Q, Lei Y, Guo W, Xu Y, Thermal decomposition kinetics of natural fibers: activation energy with dynamic thermogravimetric analysis. *Polymer Degradation and Stability*, 2008;93(1):90–98.
24. Jackson MG. Review article: the alkali treatment of straws. *Animal Feed Science and Technology*, 1977;2:105–130.
25. Marti-Ferrer F, Vilaplana F, Ribes-Greus A, Benedito-Borras A, Sanz-Box C. Flour rice husk as filler in block copolymer polypropylene: effect of different coupling agents. *Journal of Applied Polymer Science*, 2006;99:1823–1831.
26. Sundstol F, Owen E. *Straw and Other Fibrous By-Products as Feed*. Amsterdam: Elsevier Science Publishers; 1984.
27. Lu JZ, Wu Q, Negulescu II, Chen JY. The influence of fiber feature and polymer melt index on mechanical properties of sugarcane fiber/polymer composites. *Journal of Applied Polymer Science*, 2006;102:5607–5619.
28. Nada AMA, El-Wakil NA, Hassan ML, Adel AM. Differential adsorption of heavy metal ions by cotton stalk cation-exchangers containing multiple functional groups. *Journal of Applied Polymer Science*, 2006;101:4124–4132.
29. Mohanty AK, Misra M, Hinrichsen G. Biofibres, biodegradable polymers and biocomposites: an overview. *Macromolecules*, 2000;276:1–24.
30. Shi J, Shi Q, Barnes HM, Pittman CU Jr. A chemical process for preparing cellulosic fibers hiearchically from kenaf bast fibers. *BioResources*, 2011;6(1):879–890.
31. Saheb DN, Jog JP. Natural fiber polymer composites: a review. *Advances in Polymer Technology*, 1999;18(4):351–363.
32. Bledzki AK, Sperber VE, Faruk O. *Natural and Wood Fiber Reinforcement in Polymers*, Rapra Review Reports, 2002;13(8):1–144.
33. Kissinger HE. Variation of peak temperature with heating rate in differential thermal analysis. *Journal of Research of the National Bureau of Standards*, 1956;57:217–221.
34. Flynn JH, Wall LA. General treatment of thermogravimetry of polymers. *Journal of Research of the National Bureau of Standards Section A—Physics and Chemistry*, 1966;70:487–493.
35. Ozawa T. A new method of analyzing thermogravimetric data. *Bulletin of Chemical Society Japan*, 1965;38:1881–1885.
36. Li Y, Du L, Kai C, Huang R, Wu Q. Bamboo and high density polyethylene composite with heat-treated bamboo fiber: thermal decomposition properties. *BioResources*, 2013;8(1):900–912.

37. Wu Q, Yao F, Xu X, Mei C, Zhou D. Thermal degradation of rice straw fibers: global kinetic modeling with isothermal thermogravimetric analysis. *Journal of Industrial and Engineering Chemistry*, 2013;19(2):670–676.
38. Yao F, Wu Q, Lei Y, Guo W, Xu Y. Thermal decomposition kinetics of natural fibers: global kinetic modeling with non-isothermal thermogravimetric analysis. *Journal of Applied Polymer Science*, 2009;114:834–842.
39. Caulfield DF, Clemons C, Jacobson RE, Rowell RM. Wood thermoplastic composites. In: Rowell RM, editor. *Handbook of Wood Chemistry and Wood Composites*. Boca Raton, FL: CRC Press; 2005. pp 365–378.
40. Stropoli TG. Compounding and pelletizing wood fiber and polymers with Buss kneaders. In: *Symposium of the Fourth International Conference on Woodfiber-Plastic Composites*. Madison, WI; 1997. pp 50–56.
41. Macado A, Martin C. Processing wood-polymer composites with twin-screw extruders. In: *Symposium of the Fourth International Conference in Woodfiber-Plastic Composites*. Madison, WI; 1997.
42. Agassant JF, Avenas P, Sergent JP, Carreau PJ. *Polymer Processing*. Berlin, Germany: Hanser Publishers; 1991.
43. Baird DG, Collias DI. *Polymer Processing: Principles and Design*. New York: John Wiley & Sons, Inc.; 1998.
44. Yao F, Wu Q. Coextruded polyethylene and wood-flour composite: effect of shell thickness, wood loading and core quality. *Journal of Applied Polymer Science*, 2010;118:3594–3601.
45. Rosato DV, Rosato MG. *Injection Molding Handbook*, 3rd ed. Norwell, MA: Kluwer Academic Publishers; 2000.
46. Todd RH, Allen DK, Alting L. *Manufacturing Processes Reference Guide*. New York: Industrial Press, Inc.; 1994.
47. Throne JL. *Understanding Transforming*. Cincinnati, OH: Hanser Gardner Publications, Inc.; 1999.
48. Klason C, Kubat J, Stromvall HE. The efficiency of cellulosic fillers in common thermoplastics. Part I: Filling without processing aids or coupling agents. *Journal of Polymeric Material*, 1984;10:159–187.
49. Chun I, Woodhams RT. Use of processing aids and coupling agents in mica-reinforced polypropylene. *Polymer Composites*, 1984;5(4):250–257.
50. Dalvag H, Klason C, Stromvall HE. The efficiency of cellulosic fillers in common thermoplastics. *Journal of Polymeric Material*, 1985;11:9–38.
51. Raj RG, Kokta BV, Maldas D, Daneault C. Use of wood fibers in thermoplastic composites. Part VI: Isocyanate as a bonding agent for polyethylene-wood fiber composites. *Polymer Composites*, 1988;9(6):404–411.
52. Maldas D, Kokta BV, Raj RG, Daneault C. Improvement of the mechanical properties of sawdust wood fiber-polystyrene composites by chemical treatment. *Polymer*, 1988;29:1255–1265.
53. Yang HS, Wolcott MP, Kim HS, Kim S, Kim HJ. Effect of different compatibilizing agents on the mechanical properties of lignocellulosic material filled polyethylene bio-composites. *Composite Structures*, 2007;79:369–375.
54. Li Q, Matuana LM. Surface of cellulosic materials modified with functionalized polyethylene coupling agents. *Journal of Applied Polymer Science*, 2003;88:278–286.
55. Lu JZ, Wu Q, McNabb HS Jr. Chemical coupling in wood fiber and polymer composites: a review of coupling agents and treatments. *Wood Fibre Science*, 2000;32(1):88–104.
56. Keener TJ, Stuart RK. Maleated coupling agents for natural fiber composites. *Composites: Part A*, 2004;35:357–362.
57. Oksman K, Lindberg H, Holmgren A. The nature and location of SEBS-MA compatibilizer in polyethylene-wood flour composites. *Journal of Applied Polymer Science*, 1998;69:201–209.
58. Rosen MJ. *Surfactants and Interfacial Phenomena*. New York: John Wiley & Sons, Inc.; 1978.
59. Zadorecki P, Flodin P. Surface modification of cellulose fibers Part II: The effect of cellulose fiber treatment on the performance of cellulose fiber treatment on the performance of cellulose-polyester composites. *Journal of Applied Polymer Science*, 1985;30:3971–3983.

60. Maldas D, Kokta BV. Improving adhesion of wood fiber with polystyrene by the chemical treatment of fiber with a coupling agent and the influence on the mechanical properties of composites. *Journal of Adhesion Science and Technology*, 1989;3(7):529–539.
61. Chtourou H, Riedl B, Ait-Kadi A. Reinforcement of recycled polyolefins with wood fibers. *Journal of Reinforced Plastics and Composites*, 1992;11:372–394.
62. Sanadi AR, Caulfield DF. Transcrystalline interphases in natural fiber-PP composites: effect of coupling agent. *Composite Interfaces*, 2000;7(1):31–43.
63. Sain MM, Kokta BV, Maldas D. Effect of reactive additives on the performance of cellulose fiber-filled polypropylene composites. *Journal of Adhesion Science and Technology*, 1993;7(1), 49–61.
64. Lu JZ, Wu Q, Negulescu II. Maleated wood fiber-high density polyethylene composites: coupling agent performance. *Journal of Applied Polymer Science*, 2005;96:93–102.
65. Lu JZ, Wu Q, Negulescu II. Maleated wood fiber-high density polyethylene composites: coupling mechanisms and interfacial characterization. *Composite Interfaces*, 2005;12(1–2):125–140.
66. Kokta BV, Maldas D, Daneault C, Beland P. Composites of polyvinyl chloride-wood fibers I. Effect of isocyanate as a bonding agent. *Polymer-Plastics Technology and Engineering*, 1990;29:87–118.
67. Maldas D, Kokta BV. Influence of phthalic anhydride as a coupling agent on the mechanical properties of wood fiber-polystyrene composites. *Journal of Applied Polymer Science*, 1990;41:185–194.
68. Maldas D, Kokta BV, Raj RG, Daneault C. Influence of coupling agents and treatments on the mechanical properties of cellulose fiber-polystyrene composites. *Journal of Applied Polymer Science*, 1989;37:751–775.
69. Kokta BV, Maldas D, Daneault C, Beland P. Composites of poly(vinyl chloride) and wood fibers Part III: Effect of silane as coupling agent. *Journal of Vinyl Technology*, 1990;12(3):146–153.
70. Maldas D, Kokta BV. Effects of coating treatments on the mechanical behavior of wood fiber-filled polystyrene composites. Part II: Use of inorganic salt/polyvinyl chloride and isocyanate as coating components. *Journal of Reinforced Plastics and Composites*, 1990;8:2–12.
71. Lei Y, Wu Q, Clemons C, Yao F, Xu Y, Influence of nanoclay on properties of HDPE/wood composites. *Journal of Applied Polymer Science*, 2007;106:3958–3966.
72. Bataille P, Ricard L, Sapieha S. Effects of cellulose fibers in polypropylene composites. *Polymer Composites*, 1989;10(2):103–108.
73. Kokta BV, Maldas D, Daneault C, Beland P. Composites of poly(vinyl chloride) and wood fibers. Part II: Effect of chemical treatment. *Polymer Composites*, 1990;11(2):84–89.
74. Lu JZ. *Chemical Coupling in Wood Plastic Composites*. PhD., Dissertation, Baton Rouge: Louisiana State University; 2003.
75. Grammann PL, Osswald TA, editors. Simulation of the melt mixing process of natural fiber-filled polymer composites: a boundary element approach. In: *First Wood Fiber-Plastic Composite Conference*. Madison, WI; 1993.
76. Myers GE, Chayadi IS, Coberly CA, Ermer S. Wood flour/polypropylene composites: influence of maleated polypropylene and process and composition variables on mechanical properties. *Journal of Polymeric Mater*, 1991;15:21–44.
77. Maldas D, Kokta BV. Influence of polar monomers on the performance of wood fiber reinforced polystyrene composites. Part I: evaluation of critical conditions. *Journal of Polymeric Mater*, 1990; 14(3–4): 165–189.
78. Stepek J, Daoust H. *Polymer/Properties and Applications*. New York: Springer-Verlag; 1983
79. Takase S, Shiriaishi N. Studies on composites form wood and polypropylene. *Journal of Applied Polymer Science*, 1989;37:645–659.
80. Matthews G. *Polymer Mixing Technology*. New York: Applied Science Publishers; 1982.
81. Czarnecki L, White JL. Shear flow rheological properties, fiber damage and mastication characteristics of aramid-, glass- and cellulose-fiber reinforced polystyrene melts. *Journal of Applied Polymer Science*, 1980;25:1217–1244.

82. Maiti SN, Hassan MR. Melt rheological properties of polypropylene-wood flour composites. *Journal of Applied Polymer Science*, 1989;37:2019–2032.

83. Rietveld JX, Simon MJ, editors. The influence of absorbed moisture on the processability and properties of a wood flour-filled polypropylene. In: *First Wood Fiber-Plastic Composite Conference*. Madison, WI; 1993.

84. Brydson JA. *Flow Properties of Polymer Melts*. Hoboken, NJ: John Wiley & Sons, Inc.; 1987.

85. Collier JR, Lu M, Fahrurrozi M, Collier BJ. Cellulosic reinforcement in reactive composite systems. *Journal of Applied Polymer Science*, 1996;61:1423–1430.

86. Maldas D, Kokta BV, Daneault C. Influence of coupling agents and treatments on the mechanical properties of cellulose fiber-polystyrene composites. *Journal of Applied Polymer Science*, 1989;37:751–775.

87. Li Q, Wolcott MP. Rheology of wood plastics melt. Part I: Capillary rheometry of HDPE filled with maple. *Polymer Engineering & Science*, 2005;45:549–559.

88. Anonymous. *Application Examples for Wood-Plastic Composites*. Pitman, NJ: K-Tron International, Inc.; 2011.

89. Wolcott MP, Englund K, editors. A technology review of wood-plastic composites. In: *Thirty-Third International Particleboard/Composite Materials Symposium*. Washington State University; 1999.

90. Lu JZ, Wu Q, Negulescu II. Maleated wood fiber-high density polyethylene. *Journal of Applied Polymer Science*, 2005;96:93–102.

91. Lei Y, Wu Q, Xu Y. Preparation and properties of recycled HDPE/natural fiber composites. *Composite Part A*, 2007;38:1664–1674.

92. Stark NM, Matuana LM. Coating WPCs using co-extrusion to improve durability. In: *Conference for Coating Wood and Wood Composites: Designing for Durability*. Seattle, WA; 2007. pp 1–12.

93. Jin S, Matuana LM. Wood/plastic composites co-extruded with multi-walled carbon nanotube-filled rigid poly(vinyl chloride) cap layer. *Polymer International*, 2010;59:648–657.

94. Kim BJ. *The Effect of Inorganic Fillers on the Properties of Wood Plastic Composites*. Baton Rouge: School of Renewable Natural Resources, Louisiana State University; 2012.

95. Caufield DF, Jacobson RE, Sears KD, Underwood J. Fiber reinforced engineering plastics. in *Second Annual International Conference on Advanced Engineered Wood Composites*. Bethel, Maine; 2001.

96. Evstatiev M, Fakirov S. Microfibrillar reinforcement of polymer blends. *Polymer*, 1992;3:877.

97. Lei Y, Wu Q, Zhang Q. Microfibrillar composites based on recycled high density polyethylene and poly(ethylene terephthalate): morphological and mechanical properties. *Composite Part A*, 2009;40:904–912.

98. Lei Y, Wu Q. High density polyethylene and poly(ethylene terephthalate) in situ sub-micro-fibril as matrix for wood plastic composites. *Composite Part A*, 2011;43:73–78.

99. Singh S, Mohanty A. Wood fiber reinforced bacterial bioplastic composites: fabrication and performance evaluation. *Composite Science and Technology*, 2007;67:1753–1763.

100. Klyosov AA. *Wood-Plastic Composites*. Hoboken, NJ: John Wiley & Sons, Inc.; 2007.

101. Vos DJ. *Engineering Properties of Wood-Plastic Composite Panels*. Madison: University of Wisconsin; 1998.

102. Jones RM. *Mechanics of Composite Materials*. New York: Mc-Graw Hill; 1975.

103. Zhu RP, Sun CT. Effects of fiber orientation/elastic constants on coefficients of thermal expansion in laminates. *Mechanics of Advanced Materials and Structures*, 2003;10:99–107.

104. Huang R, Xiong W, Xu X, Wu Q. Thermal expansion behavior of co-extruded wood plastic composites with glass-fiber reinforced shells. *BioResources*, 2012;7(4):5514–5526.

105. Hsueh CH, Ferber MK. Apparent coefficient of thermal expansion and residual stresses in multilayer capacitors. *Composites Part A*, 2002;33:1115–1121.

106. Halpin JC, Pagano NJ. The laminate approximation for randomly oriented fibrous composites. *Composite Materials*, 1969;3(4):720–724.

107. Clemons CM, Ibach RE. Effects of processing method and moisture history on laboratory fungal resistance of wood-HDPE composites. *Forest Products*, 2004;54(4):50–57.

108. Schirp A, Ibach RE, Pendleton DE, Wolcott MP. *Biological Degradation of Wood-plastic Composites (WPC) and Strategies for Improving the Resistance of WPC Against Biological Decay*. American Chemical Society; Washington D.C.: Development of Commercial Wood Preservatives Efficacy, Environmental and Health Issues; 2008.

109. Wu Q, Shupe T, Curole J, et al. Termite resistant properties of wood and natural fiber plastic composites— AWPA E1 Test Data. In: *IRG 40th Annual Meeting*, Beijing, China; 2009.

110. Kamdem DP, Jiang H, Cui W, Freed J, Matuana ML. Properties of wood plastic composites made of recycled HDPE and wood flour from CCA-treated wood removed from service. *Composite Part A*, 2004;35:347–355.

111. Shang L, Han G, Zhu F, et al. High-density polyethylene-based composites with pressure-treated wood fibers. *BioResources*, 2012;7(4):5181–5189.

112. Stark NM, Matuana LM. Ultraviolet weathering of photostabilized wood-flour-filled high-density polyethylene. *Journal of Applied Polymer Science*, 2003;90(10):2609–2617.

113. Muasher M, Sain M. The efficacy of photostabilizers on the color change of wood filled plastic composites. *Polymer Degradation and Stability*, 2006;91:1156–1165.

114. Wypych G. *Handbook of Material Weathering*. Toronto: Chemical Technology, Inc.; 1995.

115. Heitner C. Light induced yellowing of wood containing papers. In: Heitner C, Scaiano JC, editors. *Photochemistry of Lignocellulosic Materials*. American Chemical Society; 1993. pp 3–25.

116. Agarwal UP, editor. On the importance of hydroquinone-p-quinone redox system in the photoyellowing of mechanical pulps. In: *Tenth Annual International Symposium in Wood Pulping Chemistry*; 1999.

117. Brennan P, Fedor G, Roberts R. *Xenon arc Exposure Results: Rotating & Static Specimen Mounting Systems Compared*. Cleveland, OH: Q-lab Corporation, 2008.

118. Gugumus F. Possibilities and limits of synergism with light stabilizers in polyolefins. Part II: UV absorbers in polyolefins. *Polymer Degradation and Stability*, 2002;75(2):309–320.

Index

Introduction to Wood and Natural Fiber Composites, First Edition. Douglas D. Stokke, Qinglin Wu and Guangping Han.
© 2014 John Wiley & Sons, Ltd. Published 2014 by John Wiley & Sons, Ltd.

Printed and bound by CPI Group (UK) Ltd, Croydon, CR0 4YY

27/10/2024

14580169-0001